Asliddin Kalanov

Neurostomatology

Management of Orofacial Disorders

Springer

I dedicate this book in memory of my father, Bakhriddin Kalonov, an outstanding scientist and a unique person who inspired me with his devotion to science.

Preface

This book has been prepared to provide a comprehensive, clinically relevant, and evidence-based resource in the evolving field of neurostomatology. Disorders involving orofacial pain, cranial nerves, neurosensory disturbances, and related functional syndromes are frequently encountered in clinical practice; however, they often remain underrecognized, inconsistently classified, or addressed within fragmented disciplinary frameworks.

The purpose of this volume is to help bridge this gap by bringing together contemporary knowledge from neurology, dentistry, oral medicine, pain medicine, and related disciplines within a single, integrated reference. Particular emphasis has been placed on current classification systems, diagnostic reasoning, differential diagnosis, and contemporary therapeutic strategies, especially in relation to conditions that continue to present substantial diagnostic and management challenges in neurostomatological practice.

This book is intended for neurologists, dentists, oral medicine specialists, pain physicians, maxillofacial clinicians, researchers, and postgraduate trainees. It is designed to serve both as a practical guide for clinical application and as an academic resource for teaching, reference, and further study.

The chapters are organized to progress from foundational concepts to specific clinical syndromes, diagnostic frameworks, and management approaches. The text incorporates current literature, structured tables, and clinically relevant illustrations in order to enhance clarity and support practical application.

It is hoped that this volume will contribute to improved understanding, greater diagnostic accuracy, and more effective interdisciplinary care for patients with neurostomatological disorders.

Tashkent, Uzbekistan — Asliddin Kalanov

Acknowledgements

I would like to express my sincere gratitude to all those who supported the preparation of this book. I am especially thankful to my teachers, mentors, and colleagues for their guidance, encouragement, and valuable insights throughout this work.

I am particularly grateful to Prof. Ben A Sutter for his guidance, support, and encouragement. I also wish to express my sincere thanks to Grzegorz Zieliński, MD, PhD; Prof. John Radke; and Prof. Nick Yiannios for their support, encouragement, and valuable contributions to the development of this book.

I also wish to acknowledge the many researchers and clinicians whose scholarship and clinical dedication have advanced the field of neurostomatology and helped shape the scientific foundation of this volume.

My sincere appreciation also goes to my professional colleagues for their support during the preparation of the manuscript.

Finally, I am deeply grateful to my family for their patience, understanding, and constant encouragement throughout the writing process.

This book would not have been possible without their support.

Contents

1 Introduction

Neurostomatology is an interdisciplinary field of medicine that integrates the principles of neurology and dentistry in order to better understand, diagnose, and manage pain syndromes and other pathological conditions affecting the head and neck region. This discipline focuses on the complex functional and anatomical interactions between the central and peripheral nervous systems and the structures of the oral cavity, maxillofacial area, and craniofacial region. A comprehensive understanding of these interactions enables clinicians to develop more effective diagnostic strategies, targeted therapeutic approaches, and evidence-based preventive measures for a wide spectrum of neurostomatological disorders encountered in daily clinical practice.

The close relationship between neural structures and the tissues of the oral cavity plays a decisive role in the development and persistence of many clinical conditions. Disorders such as trigeminal neuralgia, various primary and secondary headache syndromes, temporomandibular joint dysfunction, facial pain syndromes, and other combined neurological and dental pathologies often occur without obvious structural changes detectable by routine clinical or radiological examination. In such situations, patients may undergo prolonged diagnostic procedures and repeated courses of ineffective treatment when the neurostomatological origin of symptoms is not adequately considered. Neurostomatology offers a unified clinical and pathophysiological framework that allows for the identification of neurogenic mechanisms underlying these conditions and supports more rational therapeutic decision-making.

Pain syndromes occupy a central position in neurostomatology, as pain remains one of the most frequent and disabling symptoms encountered in neurological and dental practice. Pain localized to the head and neck region can significantly impair essential physiological functions, including mastication, swallowing, speech articulation, and, in severe or advanced cases, even respiratory function. Persistent or recurrent pain not only affects somatic health but also has a substantial impact on the patient's emotional and psychological state. Chronic neurostomatological pain is often accompanied by anxiety, depressive symptoms, irritability, sleep disturbances, and

A. Kalanov, *Neurostomatology*, https://doi.org/10.1007/978-981-92-2552-1_1

emotional exhaustion, leading to decreased work productivity, reduced social activity, and difficulties in long-term social adaptation.

An important feature of neurostomatology is its relatively recent development as an independent scientific and clinical discipline. Advances in neurophysiology, functional diagnostics, and modern neuroimaging techniques have contributed significantly to its formation and ongoing development. The integration of knowledge from neurology, dentistry, maxillofacial surgery, otorhinolaryngology, and other related medical specialties allows for a truly multidisciplinary approach to complex diagnostic and therapeutic challenges. This collaborative model is particularly valuable in cases where pathological processes are primarily expressed through functional disturbances of the nervous system rather than through clearly defined morphological changes.

In addition to its diagnostic and therapeutic significance, neurostomatology plays an important role in disease prevention and early intervention. The identification of functional abnormalities, preclinical manifestations, and individual risk factors enables timely preventive measures and reduces the likelihood of disease progression and chronicity. Early diagnosis and targeted treatment not only improve clinical outcomes but also help minimize long-term physical, psychological, and social consequences for patients, thereby improving overall quality of life.

In this context, neurostomatology represents a rapidly evolving and highly promising field of modern medicine. Its integrative nature, focus on early detection, and emphasis on comprehensive patient-centered care make it an essential component in the management of complex head and neck disorders and pain syndromes across different clinical settings.

2 Brief Anatomy of Cranial Nerves and Craniofascial Ganglia

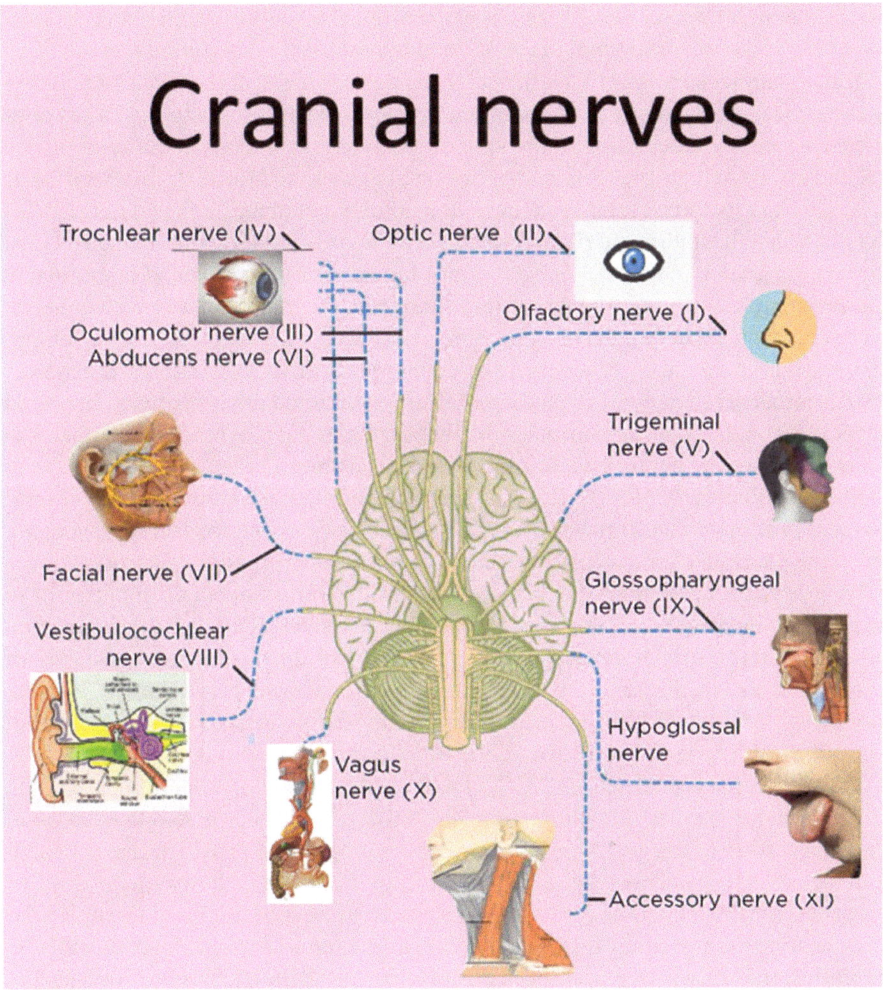

A. Kalanov, *Neurostomatology*, https://doi.org/10.1007/978-981-92-2552-1_2

In neurostomatology, all 12 cranial nerves are of clinical relevance, as they form an integrated functional network responsible for sensory perception, motor control, and autonomic regulation within the head and neck region. Each cranial nerve contributes in some way to the complex coordination of oral, facial, and pharyngeal functions. However, their degree of involvement in oral and maxillofacial activity is not uniform. Certain cranial nerves play a particularly dominant role in processes such as mastication, facial expression, speech articulation, swallowing, salivation, and sensory perception within the oral cavity. Because neurostomatological diseases most often manifest through disturbances in these functions, it is both clinically practical and scientifically justified to focus in greater detail on those cranial nerves whose impairment most frequently results in pain syndromes, motor dysfunction, and sensory abnormalities relevant to dental and neurological practice. These nerves include the trigeminal nerve (cranial nerve V), facial nerve (cranial nerve VII), glossopharyngeal nerve (cranial nerve IX), vagus nerve (cranial nerve X), and hypoglossal nerve (cranial nerve XII).

The trigeminal nerve (CN V) is the principal sensory nerve of the face and oral cavity and represents one of the most important structures in neurostomatology. It provides extensive sensory innervation to the skin of the face, the mucous membranes of the oral and nasal cavities, the teeth and periodontal tissues, and the temporomandibular joint. Through its three major branches—the ophthalmic, maxillary, and mandibular divisions—the trigeminal nerve transmits tactile, pain, and temperature sensations, forming the anatomical basis for both physiological oral sensation and pathological pain perception. In addition to its sensory functions, the mandibular branch contains motor fibers that innervate the muscles of mastication, including the masseter, temporalis, and pterygoid muscles, which are essential for effective chewing. Dysfunction of the trigeminal nerve may result in severe facial pain, altered oral sensitivity, impaired mastication, and temporomandibular dysfunction, making it central to the diagnosis and management of neurostomatological pain syndromes.

The facial nerve (CN VII) plays a critical role in facial expression and several essential oral functions. It innervates the muscles responsible for facial movement, which are necessary for speech articulation, emotional expression, and maintaining oral competence. In addition, the facial nerve provides parasympathetic innervation to the submandibular and sublingual salivary glands, thereby regulating salivary secretion and contributing to oral homeostasis. Taste perception from the anterior two-thirds of the tongue is also mediated by this nerve. Disorders of the facial nerve can lead to facial asymmetry, difficulty with speech and eating, reduced salivation, and disturbances in taste, all of which significantly affect oral function and patient quality of life.

The glossopharyngeal nerve (CN IX) has both sensory and motor components that are essential for normal oral and pharyngeal activity. It supplies general sensory and taste innervation to the posterior one-third of the tongue and provides sensory input from the oropharynx. This nerve also plays an important role in the swallowing reflex and contributes to the motor control of pharyngeal muscles. Pathology involving the glossopharyngeal nerve may manifest as dysphagia, altered taste sensation, throat pain, or referred otalgia, which can complicate the differential diagnosis of orofacial pain.

The vagus nerve (CN X) has a broad functional scope and is involved in both voluntary and autonomic control. Within the context of neurostomatology, it innervates muscles of the soft palate, pharynx, and larynx, making it essential for speech production and coordinated swallowing. Beyond the head and neck, the vagus nerve provides parasympathetic innervation to the heart, lungs, and gastrointestinal tract, linking oral function with systemic regulation. Dysfunction of this nerve may present with voice changes, swallowing disorders, and autonomic symptoms, underscoring its relevance in complex clinical cases.

The hypoglossal nerve (CN XII) is the primary motor nerve of the tongue and is crucial for precise tongue movements. It innervates both intrinsic and extrinsic tongue muscles, enabling articulation, food manipulation, and the oral phase of swallowing. Damage to the hypoglossal nerve may result in tongue deviation, weakness, atrophy, and impaired speech and swallowing, leading to significant functional limitations.

In summary, while all cranial nerves contribute to the coordinated activity of the head and neck, these selected nerves play a dominant role in neurostomatological physiology and pathology. A detailed understanding of their anatomy and function is essential for accurate diagnosis, differential evaluation, and effective management of neurostomatological diseases.

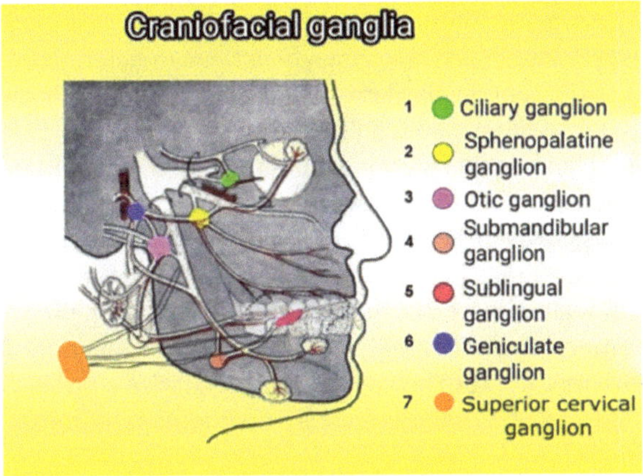

Craniofacial ganglia are clusters of neuronal cell bodies located outside the central nervous system that serve as critical relay and integration centers for sensory and autonomic pathways in the head and neck region. These ganglia play a fundamental role in neurostomatology, as they mediate the transmission of sensory information, regulate autonomic functions, and participate in complex pain mechanisms involving the oral and maxillofacial structures. Understanding the anatomy and physiology of craniofacial ganglia is essential for accurate diagnosis and effective management of neurostomatological disorders.

One of the most important craniofacial ganglia is the trigeminal ganglion, which contains the cell bodies of primary sensory neurons responsible for conveying sensory input from the face, scalp, oral cavity, nasal cavity, and teeth. These neurons form the first-order afferent pathway for somatosensory information and transmit tactile, thermal, proprioceptive, and nociceptive signals from peripheral tissues to the central nervous system. Through its three major divisions—ophthalmic, maxillary, and mandibular—the trigeminal ganglion provides extensive sensory coverage of the craniofacial region, making it a central structure in neurostomatology.

The trigeminal ganglion functions as a critical relay and modulation station, where peripheral sensory signals are organized and transmitted to brainstem nuclei for further processing and integration. This hierarchical transmission allows the nervous system to generate appropriate protective reflexes, such as withdrawal from painful stimuli, as well as adaptive responses involved in mastication, speech, and facial expression. In addition, interactions between sensory and autonomic pathways at the level of the trigeminal system contribute to vascular and inflammatory responses in craniofacial tissues.

In neurostomatological practice, dysfunction or hyperexcitability of the trigeminal ganglion is closely associated with facial pain syndromes, altered dental sensitivity, and neuropathic conditions affecting the oral cavity. Abnormal trigeminal signaling may lead to heightened pain perception, spontaneous pain, or exaggerated responses to normally non-painful stimuli. Such pathophysiological changes underline the importance of the trigeminal ganglion as a key anatomical and functional substrate in the development and persistence of orofacial pain disorders.

In addition to sensory processing, craniofacial ganglia are essential components of the autonomic nervous system, particularly the parasympathetic division. The otic, submandibular, and pterygopalatine ganglia play key roles in the regulation of salivary gland secretion. These ganglia provide parasympathetic innervation to the parotid, submandibular, and sublingual glands, ensuring adequate saliva production. Saliva is vital for maintaining oral homeostasis, as it facilitates mastication and swallowing, supports speech, protects oral tissues, buffers acids, and contributes to antimicrobial defense. Impairment of parasympathetic ganglionic function can result in hyposalivation or xerostomia, which significantly increases the risk of dental caries, periodontal disease, mucosal inflammation, and oral infections. Therefore, proper functioning of these ganglia is indispensable for oral health and overall quality of life.

The superior cervical ganglion (SCG), the largest sympathetic ganglion in the cervical region, also holds considerable significance in neurostomatology. As part of the sympathetic nervous system, the SCG regulates vascular tone, blood flow, and sensory modulation in the head and neck. Sympathetic fibers originating from the SCG innervate the arterioles of the dental pulp, periodontal tissues, and oral mucosa. Through vasoconstriction and vasodilation, these fibers influence pulpal blood flow and tissue metabolism. Dysregulation or irritation of the SCG can alter vascular responses and sensory thresholds, contributing to various forms of orofacial pain.

Clinical observations indicate that sympathetic dysfunction originating in the cervical region may be responsible for certain dental allodynias, particularly cold-sensitive teeth. In such cases, exaggerated sympathetic activity can lower pain thresholds in the dental pulp, causing non-noxious thermal stimuli to be perceived as painful. This phenomenon highlights the close interaction between the sympathetic nervous system and sensory pathways in the craniofacial region. Recognition of this mechanism is crucial for differentiating odontogenic pain from neuropathic or sympathetically maintained pain syndromes.

Craniofacial ganglia and their associated neural pathways also play a pivotal role in pain modulation and chronic pain development. Disruption of normal ganglionic function can lead to persistent nociceptive signaling, central sensitization, and the development of chronic pain conditions. One of the most well-known examples is trigeminal neuralgia, a debilitating disorder characterized by recurrent episodes of severe facial pain. Abnormal activity within the trigeminal ganglion or its central connections is believed to contribute to the pathophysiology of this condition. Chronic ganglionic dysfunction not only causes pain but also affects chewing, speaking, emotional well-being, and social functioning, thereby significantly impairing patients' quality of life.

From a neurostomatological perspective, the craniofacial ganglia represent a key intersection between dentistry, neurology, and pain medicine. Their involvement in sensory transmission, autonomic regulation, and vascular control explains why disorders of these ganglia can present with complex and overlapping clinical symptoms. Effective management of neurostomatological diseases requires a comprehensive understanding of ganglionic anatomy and function, as well as an appreciation of their role in both peripheral and central nervous system interactions.

In summary, craniofacial ganglia are integral to the sensory and autonomic regulation of the head and neck region. They facilitate sensory perception, maintain salivary gland function, regulate vascular tone, and modulate pain pathways. Proper functioning of these ganglia is essential for oral health, while their dysfunction can lead to significant neurostomatological disorders, including chronic orofacial pain and dental hypersensitivity. A detailed understanding of craniofacial ganglia is therefore fundamental for advancing diagnostic accuracy and improving therapeutic strategies in neurostomatology.

3 Brief Oral Anatomy

The region inside the mouth, surrounded by the muscles of the jaw and the adjacent facial and cranial bones, is known as the oral cavity. Although this definition appears simple, the oral cavity represents a highly complex anatomical and functional unit that plays a central role in numerous vital human activities. It serves as the initial segment of the digestive tract and participates directly in ingestion, mastication, deglutition, speech articulation, facial expression, and respiration (see Fig. 3.1). In addition to these mechanical and physiological functions, the oral cavity also contributes significantly to sensory perception, immune defense, and social interaction.

From an anatomical and clinical standpoint, the oral cavity forms an essential interface between the external environment and the internal systems of the body. Because of this unique position, it is frequently exposed to mechanical stress, microbial flora, chemical agents, and temperature variations. Consequently, a thorough understanding of oral anatomy is fundamental for clinicians, particularly dentists, neurologists, maxillofacial surgeons, and otorhinolaryngologists, who must assess and manage a wide range of functional disorders and pathological conditions affecting this region.

3.1 Main Components of Oral Anatomy

The oral cavity is composed of several distinct anatomical structures, each contributing in a specific way to overall oral function and health.

Lips

The lips are two highly mobile, fleshy structures that form the anterior opening of the oral cavity. They consist of skin externally, mucosa internally, and a core of skeletal muscle, primarily the orbicularis oris. The lips play a crucial role in food intake, articulation of speech sounds, facial expression, and maintenance of oral competence.

A. Kalanov, *Neurostomatology*, https://doi.org/10.1007/978-981-92-2552-1_3

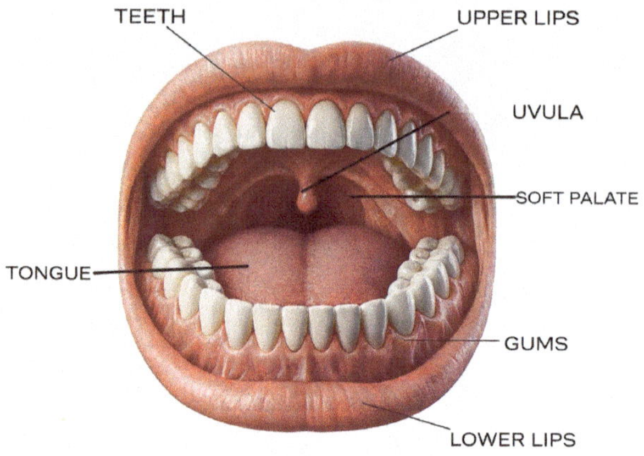

Fig. 3.1 Oral cavity

Their rich sensory innervation makes them highly sensitive to touch, temperature, and pain, which is clinically relevant in neurological and dental assessments.

Salivary Glands

Salivary glands are responsible for the secretion of saliva, a complex biological fluid essential for digestion, speech, lubrication of oral tissues, and maintenance of oral hygiene. The major salivary glands include the parotid, submandibular, and sublingual glands, while numerous minor salivary glands are distributed throughout the oral mucosa. Saliva facilitates enzymatic digestion, buffers oral pH, protects against microbial invasion, and aids in the formation and swallowing of the food bolus.

Teeth

The teeth, collectively referred to as the dentition, are specialized structures embedded in the alveolar bone of the jaws. Their primary function is mastication, allowing mechanical breakdown of food into smaller particles suitable for digestion. Teeth also play an important role in speech articulation, facial aesthetics, and maintenance of vertical facial dimension. The integrity of the dentition is closely linked to overall oral and systemic health.

Tongue

The tongue is a muscular organ located on the floor of the oral cavity. It is essential for taste sensation, manipulation of food during chewing, initiation of swallowing, and articulation of speech. The tongue is richly innervated by both sensory and motor nerves, reflecting its complex functional role. Its surface contains papillae that house taste buds, contributing to gustatory perception.

Hard Palate

The hard palate forms the anterior portion of the roof of the oral cavity and is composed of bone covered by a keratinized mucosa. It separates the oral cavity from the nasal cavity and provides a rigid surface against which the tongue can press food during mastication and speech. The hard palate also contributes to resonance during phonation.

Soft Palate

Posterior to the hard palate lies the soft palate, a movable muscular structure covered with mucosa. The soft palate plays a critical role in swallowing by elevating to close off the nasopharynx, thereby preventing food or liquid from entering the nasal cavity. It also contributes to speech by regulating airflow during phonation.

Uvula

The uvula is a conical projection of soft tissue extending from the posterior margin of the soft palate. Although small, it plays an important role in speech resonance and in assisting the soft palate during swallowing. The uvula also participates in the gag reflex, a protective mechanism that prevents aspiration.

Tonsils

The tonsils are aggregations of lymphoid tissue located at the opening of the oropharynx. They form part of Waldeyer's ring and play a role in immune defense by trapping and responding to pathogens entering through the oral and nasal cavities. Clinically, tonsillar tissue is relevant in infections, inflammatory conditions, and obstructive disorders.

Periodontium

The periodontium refers to the supporting structures of the teeth, including the gingiva, periodontal ligament, alveolar bone, and cementum. These components work together to anchor teeth securely within the jaws while allowing slight physiological movement during mastication. Periodontal health is essential for maintaining tooth stability and preventing tooth loss.

Lower Jaw (Mandible)

The mandible is the only movable bone of the skull. It plays a central role in mastication, speech, and facial aesthetics. Movements of the mandible allow opening and closing of the mouth, as well as lateral and protrusive motions necessary for effective chewing.

Upper Jaw (Maxilla)

The maxilla forms the upper jaw and holds the upper teeth. It also contributes to the formation of the hard palate, nasal cavity, and orbital floor. The maxilla plays a key role in facial structure and the stability of the midface.

Temporomandibular Joint and Articular Disc

The temporomandibular joint (TMJ) is a complex synovial hinge joint that connects the mandible to the temporal bone of the skull. It is located anterior to the external auditory canal and is essential for jaw movement during chewing, speaking, and swallowing (see Fig. 3.2). Unlike most synovial joints, the TMJ allows both rotational and translational movements, making it functionally unique.

Within the TMJ lies an articular disc, a fibrocartilaginous structure positioned between the mandibular condyle and the temporal bone. This disc acts as a cushion, absorbing mechanical stress, distributing pressure evenly, and facilitating smooth joint movement. Proper alignment and function of the disc are critical for painless jaw

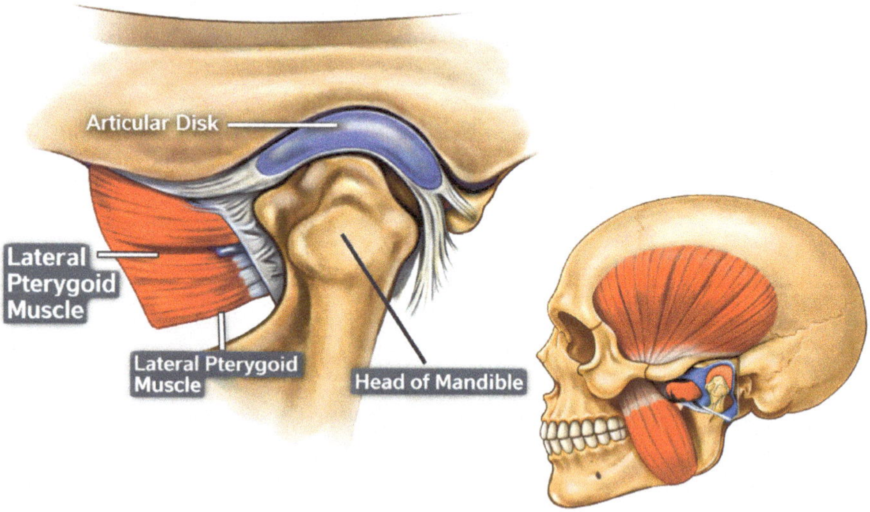

Fig. 3.2 Anatomy of the temporomandibular joint (TMJ)

motion. Dysfunction of the TMJ or displacement of the disc can lead to pain, restricted movement, and referred symptoms affecting the head and neck.

Anatomical Division of the Oral Cavity

Anatomically, the oral cavity is divided into two main regions:

1. Oral Vestibule
   The oral vestibule is the space between the lips and cheeks externally and the teeth and gingiva internally. It includes the buccal and labial gingivae, the superior and inferior labial frenula, and the vestibular surfaces of the lips. This region serves as a transitional space that facilitates food intake and oral hygiene procedures.
2. Oral Cavity Proper
   The oral cavity proper is the space enclosed by the upper and lower dental arches. Its roof is formed by the hard and soft palates, while the floor consists of a muscular diaphragm composed primarily of the mylohyoid muscles, with the geniohyoid muscles situated superiorly. The tongue rests above these muscles and is often considered a functional component of the floor of the oral cavity.

3.2 Clinical Importance of Oral Anatomy

Overall, oral anatomy and its associated structures are essential not only for oral function but also for general health and well-being. Disorders affecting the oral cavity can significantly impair nutrition, communication, social interaction, and quality of life. Furthermore, many systemic diseases manifest early signs within the oral cavity, making it an important diagnostic region.

A detailed understanding of oral anatomy is therefore crucial for dental and medical professionals. It provides the anatomical foundation necessary for accurate assessment, diagnosis, and treatment of a wide range of oral, maxillofacial, and neurostomatological conditions.

Check for updates

Tooth-Related Orofacial Pain

4

Abstract

Tooth-related orofacial pain (TROFP) is a common reason for urgent dental consultation and includes pain arising from the pulp, dentin, periodontium, and periapical tissues. Although often odontogenic, toothache may also reflect neuropathic, neurovascular, idiopathic, or referred pain, making accurate diagnosis essential to avoid unnecessary irreversible procedures. This chapter presents a modern overview of TROFP as a diagnostic framework rather than a simple list of dental disorders. It reviews major causes, including pulpitis, apical periodontitis, cracked tooth syndrome, dentin hypersensitivity, and periodontal pain, and emphasizes the limitations of isolated diagnostic tests. Contemporary evidence supports an integrated approach combining history, pain characteristics, clinical examination, sensibility testing, and imaging. The chapter also highlights mechanism-based management, including minimally invasive care, selective use of vital pulp therapy, nonopioid analgesia, antibiotic stewardship, and timely referral when non-odontogenic pain is suspected.

Keywords

Tooth-related orofacial pain · Odontogenic pain · Pulpitis · Apical periodontitis · Cracked tooth syndrome · Dentin hypersensitivity · Differential diagnosis · Nonopioid analgesia

Tooth-related orofacial pain (TROFP) refers to pain reported in a tooth or dentoalveolar region that arises from dental structures, including the dental pulp, dentin, periodontium, and periapical tissue. Clinically, TROFP is one of the most frequent causes of emergency dental visits and urgent dental consultations globally [1, 2]. Despite this prevalence, it is also one of the most frequently misdiagnosed pain complaints because a number of non-odontogenic illnesses, such as neurovascular orofacial pain, trigeminal

A. Kalanov, *Neurostomatology*, https://doi.org/10.1007/978-981-92-2552-1_4

neuropathic syndromes, and transferred pain from nondental structures, can manifest as "toothache" [3, 4]. Accurate diagnosis is therefore critical, as unnecessary endodontic treatment or tooth extraction may not only fail to resolve pain but can worsen outcomes, particularly in neuropathic and idiopathic dentoalveolar pain disorders [5].

A modern approach to TROFP includes two complementary frameworks:

- Traditional odontogenic pain processes caused by tissue damage, inflammation, or infection, and
- Structured models for phenotyping and classifying orofacial pain are intended to enhance interdisciplinary communication and diagnostic precision.

According to recent research, pain classification is not just academic; rather, it actively influences treatment choices, referral pathways, diagnostic algorithms, and research comparability in otolaryngology, neurology, dentistry, and pain medicine [2, 6].

In many cases, TROFP is clinically straightforward when symptoms and findings are concordant. For example, carious lesions associated with stimulus-dependent thermal pain typically indicate pulpal inflammation; percussion sensitivity accompanied by periapical radiographic changes suggests apical pathology; and exposed cervical dentin producing brief, sharp pain is consistent with dentin hypersensitivity [1, 7, 8].

However, real-world presentations are frequently more complex. Diagnostic interpretation is frequently complicated by mixed symptom patterns, such as broken teeth that present with intermittent biting pain and persistent cold sensitivity [7]. Additionally, pain referral within trigeminal pathways reduces the reliability of patient-reported localization [3, 4].

Two significant clinical facts related to TROFP misdiagnosis have been brought to light by the literature throughout the past 5 years. First, while still necessary, conventional diagnostic methods such as electric pulp testing, cold testing, percussion, and palpation show inconsistent diagnostic accuracy when used alone.

Reversible pulpitis, irreversible pulpitis, and non-odontogenic pain states cannot be accurately distinguished by a single test, according to systematic reviews and clinical research [7, 9]. As a result, contemporary recommendations advocate for a structured diagnostic approach that integrates patient history, pain characteristics, clinical examination, sensibility testing, and imaging findings [6, 9].

Second, there is extensive overlap between odontogenic pain and non-odontogenic orofacial pain, leading to a high likelihood of "false dental positives." While trigeminal neuropathic pain may manifest as persistent tooth-localized pain in the absence of dental pathology, neurovascular orofacial pain may move between teeth, mimic pulpitis, and coexist with headache-like symptoms [3–5].

Persistent idiopathic dentoalveolar pain (PIDAP), previously referred to as atypical odontalgia, exemplifies this challenge. Recent reviews highlight ongoing confusion in terminology and diagnostic criteria but consistently emphasize that repeated irreversible dental procedures in the absence of objective pathology are unlikely to resolve pain and may exacerbate central sensitization [5, 10–12]. Neuroimaging studies further support this view by demonstrating altered central pain processing in patients with persistent dentoalveolar pain [13]. Detailed information is available in the relevant chapter of the PIDAP.

Consequently, TROFP should be viewed and taught as a diagnostic science rather than merely a list of dental disorders. Odontogenic causes, pain processes, structured differential diagnosis using modern categorization models, and evidence-based treatment with a focus on minimally invasive care, nonopioid analgesia, and prompt referral when non-odontogenic pain is suspected must all be covered in a modern framework [2, 6, 10].

Dental pain is often described as biologically "loud" because tooth-related tissues—particularly the dental pulp—are densely innervated and strongly represented within trigeminal nociceptive pathways. Contemporary translational reviews emphasize that acute odontogenic pain is not a single-process phenomenon; rather, it emerges from coordinated activity across peripheral nociceptors, local immune–inflammatory mediators, trigeminal ganglion neuroimmune signaling, and central processing within brainstem and higher centers responsible for sensory-discriminative and affective dimensions of pain. Clinical aspects such as radiation of pain beyond the affected tooth, referred pain to neighboring teeth or craniofacial regions, and inter-individual variations in symptom intensity are all explained by this integrated perspective [14–16].

Inflammation (reversible/irreversible pulpitis; apical periodontitis), mechanical damage (cracks, fractures), dentin exposure (dentin hypersensitivity), or infection-driven tissue damage are the main causes of odontogenic pain. At the peripheral level, inflammatory mediator cascades (e.g., prostaglandins, cytokines, neuropeptides) reduce activation thresholds of nociceptors and improve transmission of heat and mechanical stimuli, creating hyperalgesia and allodynia [14, 17].

Molecular-level analyses of pulp afferent biology further support this model by demonstrating inflammation-associated sensitization of ion channels and receptors that amplify nociceptive signaling within pulpal afferents. When nociceptive input is persistent or intense, central sensitization may develop within trigeminal nociceptive circuits, increasing gain and widening receptive fields. Such changes can contribute to pain amplification and spread, thereby accounting for patient reports of exaggerated pain that appears disproportionate to objective dental findings [14, 17, 18].

A clinically important implication is that symptom quality alone does not establish a diagnosis. Although "sharp" pain is commonly associated with dentin hypersensitivity and "throbbing" pain with infection, overlap is frequent, and descriptors may shift with disease progression. A frequently cited probabilistic trajectory is as follows: reversible pulpitis tends to present with brief, stimulus-evoked sensitivity (often to cold); symptomatic irreversible pulpitis more often produces lingering pain and may become spontaneous; later pulpal necrosis may reduce thermal responsiveness while periapical inflammation increases pain on biting or percussion. Recent reviews stress combining the history with several tests (thermal testing, percussion, palpation, and imaging) and reevaluating the working diagnosis as findings change because these patterns are not definitive. This integrative approach is supported by mechanisms-oriented classification frameworks. Clinical categories (odontogenic, neuropathic, neurovascular, musculoskeletal, and idiopathic) improve diagnostic reasoning by promoting explicit hypothesis testing rather than automatically attributing all pain near a tooth to odontogenic pathology, according to recent analyses of the classification of chronic orofacial pain [18, 19].

This is particularly relevant for post-procedural pain persistence. A subset of patients develops persistent dentoalveolar pain without ongoing dental disease—sometimes after endodontic treatment, extraction, or implant therapy—where current literature describes neuropathic/idiopathic mechanisms and underscores avoidance of unnecessary repeat procedures when clinical tests and imaging do not support active odontogenic disease [20–22]. Evidence-based analgesia is informed by mechanism-based thinking. Current clinical guidance for the management of acute dental pain favors nonopioid regimens, particularly nonsteroidal anti-inflammatory medications (NSAIDs), commonly used in combination with acetaminophen, as first-line therapy for post-extraction pain and the short-term treatment of toothache.

Opioids are only used in rare circumstances where first-line treatments are either inadequate or inappropriate. Both superior comparative efficacy and contemporary safety-focused pain management approaches are reflected in this approach [23, 24].

Major causes of tooth-related orofacial pain include pulpitis and early periapical disease. Among the most prevalent and clinically significant causes of tooth-related orofacial pain (TROFP) are inflammatory and degenerative disorders of the dental pulp and periapical tissues, including reversible pulpitis, symptomatic irreversible pulpitis, pulp necrosis, and early apical periodontitis.

These disorders account for a considerable proportion of urgent dental visits and are typically associated with acute pain that demands emergency care. At the same time, they represent a substantial source of diagnostic mistake since their clinical presentations may overlap with non-odontogenic orofacial pain diseases, particularly neuropathic and neurovascular pain states. For this reason, thorough diagnostic reasoning is vital to avoid unnecessary or inefficient dental procedures [25, 26]. Pulpitis and early apical inflammation share overlapping symptoms, including thermal sensitivity and poorly localized pain, making accurate diagnosis clinically challenging. Patient history often provides the initial diagnostic anchor in suspected pulpal and early periapical disease.

Thermal sensitivity—particularly pain elicited by cold stimuli—is commonly associated with pulpal inflammation, whereas pain triggered by biting, chewing, or percussion suggests involvement of the periapical tissues. Lingering pain following cold stimulation, spontaneous pain episodes, and nocturnal pain are traditionally linked to symptomatic irreversible pulpitis.

However, as patient-reported pain characteristics do not consistently correlate with histopathologic pulpal state, current research warns against a simplistic interpretation of symptom descriptors. Reversible pulpitis, irreversible pulpitis, and early apical illness are not clearly defined entities, but rather exist along a biological continuum, according to recent systematic reviews.

Therefore, comparable histologic findings may present with varied symptom patterns, and identical clinical symptoms may reflect various underlying tissue states. This discrepancy contributes greatly to diagnostic ambiguity and emphasizes the limits of relying exclusively on symptom quality or intensity when evaluating dental pain [25, 27].

Clinical assessment of suspected pulpal and periapical disease typically involves a combination of sensibility tests and mechanical provocation. Commonly employed methods include cold testing, electric pulp testing, percussion, palpation, periodontal probing, and bite tests. Each of these tools provides indirect information about pulpal or periapical status, yet none is diagnostic in isolation.

Recent systematic reviews evaluating diagnostic accuracy in pulpitis demonstrate considerable variability in sensitivity and specificity across testing modalities. Cold testing is widely regarded as the most reliable sensibility test for assessing pulpal vitality, yet it cannot distinguish with certainty between reversible and irreversible inflammation.

Electric pulp testing assesses neural responsiveness rather than vascular supply and may yield false-positive or false-negative results, particularly in traumatized teeth, immature teeth, or teeth with calcified canals. Percussion and palpation are useful indicators of periapical inflammation but may be negative in early disease stages [27, 28].

Current diagnostic recommendations place more emphasis on integrating many data sources than relying just on a single test result due to these constraints. Clinicians are advised to investigate internal consistency across patient history, sensitivity testing, mechanical tests, and imaging data. Instead of taking immediate, irreversible action when the results are inconsistent, a provisional diagnosis should be made and reevaluated over time [26, 28].

In early periapical disease, conventional radiographs may appear normal or show minimal changes, so careful clinical correlation and, when indicated, advanced imaging are essential for accurate diagnosis. Radiographic assessment remains a cornerstone of evaluating suspected periapical pathology. Conventional periapical radiographs are typically the first-line imaging modality, providing information about caries depth, restoration quality, periodontal bone levels, and gross periapical changes.

However, an important diagnostic limitation is that early apical inflammation may be present before radiographically visible bone changes develop. Consequently, a normal periapical radiograph does not exclude early apical periodontitis when clinical signs and symptoms strongly suggest periapical involvement [26, 28].

Cone-beam computed tomography (CBCT) has expanded diagnostic capabilities in selected cases, particularly when conventional imaging is inconclusive. Indications for CBCT include suspected root fractures, complex anatomy, inconsistent clinical findings, or persistent pain without clear radiographic explanation. Nevertheless, recent consensus statements emphasize judicious use of CBCT, balancing diagnostic benefit against radiation exposure and the risk of overinterpretation of incidental findings [28].

Management of acute pain associated with pulpitis and early apical disease has been the subject of extensive research in recent years, particularly with respect to pharmacologic analgesia. Evidence-based guidelines consistently recommend non-opioid regimens as first-line therapy for acute odontogenic pain. Nonsteroidal anti-inflammatory drugs (NSAIDs), either alone or in combination with acetaminophen, demonstrate superior analgesic efficacy compared with opioid-containing regimens for dental pain and are associated with a more favorable safety profile [29].

These instructions apply both to post-extraction pain and to the temporary management of pain emerging from pulpal and periapical inflammation when definitive dental treatment must be delayed. As a result of broader trends toward mechanism-based and safety-oriented pain management, opioids are only used in extreme situations where first-line treatments are ineffective or inappropriate [29].

Definitive management of pulpitis and early apical disease depends on accurate diagnosis, tooth restorability, and patient-specific factors. For symptomatic irreversible pulpitis and acute apical pathology, removal of the inflamed or necrotic pulp via endodontic treatment, or extraction in non-restorable cases, remains the traditional standard of care.

However, recent years have seen renewed interest in vital pulp therapy (VPT) techniques for mature permanent teeth diagnosed with irreversible pulpitis. Contemporary systematic reviews and clinical trials suggest that, under carefully selected conditions, procedures such as partial pulpotomy or full pulpotomy using biocompatible materials may achieve favorable outcomes comparable to root canal treatment. This evolving paradigm reflects a shift toward more conservative, biologically oriented management strategies and underscores the importance of precise diagnosis and case selection [30].

Unclear diagnostic boundaries between pulpal and early apical pathology increase the risk of misattributing symptoms, potentially leading to delayed or inappropriate treatment. A clinically critical issue in TROFP is the recognition that not all pain perceived as tooth-related is odontogenic in origin.

Recent analyses highlight that neurovascular and neuropathic orofacial pain disorders may closely mimic pulpitis, leading to inappropriate dental interventions. Pain that migrates between teeth, involves multiple teeth simultaneously, persists despite appropriate dental treatment, or is accompanied by headache, autonomic features, or sensory disturbances should prompt consideration of non-odontogenic mechanisms.

A recent discussion focusing on symptomatic irreversible pulpitis in the broader context of orofacial pain emphasizes practical clinical warning signs that should raise diagnostic caution. These include equivocal or inconsistent test results, absence of a clear structural correlate, and pain patterns that do not follow expected odontogenic trajectories. In such cases, interdisciplinary evaluation and avoidance of irreversible dental procedures are strongly advised [25, 31].

Cracked tooth syndrome (CTS) represents a classic diagnostic challenge within the spectrum of TROFP. Patients often report sharp, intermittent pain triggered by biting or release of pressure, with variable responses to thermal stimuli. Symptoms may fluctuate over time and may coexist with pulpal inflammation, further complicating diagnosis [32]. Recent narrative and systematic studies identify CTS as a "diagnostic dilemma" due to the lack of a single definitive diagnostic test and the wide diversity in clinical presentation. Prognosis and symptom manifestation are influenced by crack orientation, depth, extent, and pulpal involvement. Superficial enamel cracks may be asymptomatic; however, deeper cracks extending into dentin or pulp can generate significant discomfort and pulpal disease [33].

Clinical assessment of suspected CTS relies on a combination of targeted diagnostic maneuvers. Bite tests using cotton rolls or specialized devices can help localize pain associated with crack propagation. Transillumination, magnification, and careful visual inspection may reveal crack lines not visible to the naked eye. Periodontal probing is particularly important for detecting isolated deep pockets, which may indicate a crack extending subgingivally or vertically along the root surface.

Advanced imaging techniques, including CBCT, may assist in selected cases, although their sensitivity for detecting cracks remains limited. Recent reviews emphasize that many suspected cracks are never definitively visualized, reinforcing the need for cautious, staged management when diagnostic certainty is low [33].

Management of CTS depends on crack severity, tooth restorability, and pulpal status. Conservative measures such as occlusal adjustment and temporary stabilization may be appropriate initial steps when symptoms are mild or diagnosis is uncertain.

Definitive restorative approaches include bonded restorations or cuspal coverage to reduce flexure and prevent crack propagation. Endodontic treatment is indicated when irreversible pulpal involvement is present, while extraction is reserved for non-restorable fractures extending vertically into the root. A consistent message across recent reviews is the importance of staged decision-making. Stabilization followed by symptom reassessment can prevent unnecessary irreversible procedures and allows time for diagnostic clarification.

Dentin hypersensitivity is characterized by brief, sharp pain arising from exposed dentin in response to thermal, tactile, osmotic, or chemical stimuli. Although typically less severe than pulpitis-related pain, it is a common cause of TROFP and a frequent diagnostic consideration in clinical practice.

Recent reviews emphasize the importance of differential diagnosis, as symptoms of hypersensitivity may overlap with those of early caries, cracked teeth, defective restorations, or pulpal inflammation. Etiologic factors include enamel erosion, abrasion, gingival recession, and periodontal therapy.

Management follows a stepped approach, beginning with at-home desensitizing agents and progressing to in-office treatments when necessary. Prevalence estimates for dentin hypersensitivity vary widely across studies, reflecting differences in diagnostic criteria and assessment methods. This variability highlights the need for standardized definitions and reinforces the importance of careful clinical evaluation [26].

Periodontal pain may present as localized tenderness, dull aching, or pain on chewing, particularly in the presence of acute periodontal abscesses. Clinical examination focuses on periodontal probing, mobility assessment, evaluation for food impaction, and identification of local infectious or inflammatory sources. Periodontal and pulpal diseases may coexist, as in endo-periodontal lesions, requiring careful interpretation of probing patterns, vitality testing, and imaging findings. Misclassification can lead to inappropriate treatment sequencing and suboptimal outcomes [28].

Across pulpitis, early apical disease, cracked tooth syndrome, dentin hypersensitivity, and periodontal pain, a unifying diagnostic principle emerges: effective diagnosis relies on pattern recognition combined with active exclusion of alternative explanations. Clinicians must identify features that strongly support a specific odontogenic diagnosis while remaining alert to signs that suggest non-odontogenic mechanisms. The significance of diagnostic flexibility is emphasized in recent classification-oriented studies. Clinicians should be ready to switch from an odontogenic framework to neuropathic or neurovascular models when clinical evidence does not support a dental pain generator. This strategy decreases the danger of overtreatment, protects tooth structures, and improves patient results by aligning therapy with underlying pain processes [25, 31]. Two concurrent and interrelated goals are necessary for the management of tooth-related orofacial pain (TROFP): (1) removing or controlling the odontogenic pain source when it exists, and (2) providing safe, evidence-based analgesia while definitive treatment and diagnostic clarification take place.

According to recent research, these objectives must be pursued concurrently because, while early, irreversible dental treatment based on insufficient diagnostic certainty may result in chronic pain, overtreatment, and iatrogenic harm, poor pain management can mask diagnostic features and worsen patient distress [34–36].

When a clear odontogenic source is identified, management focuses on definitive elimination of the pain generator while preserving tooth structure whenever possible. Inflammatory disorders of the pulp and periapical tissues—such as reversible pulpitis, symptomatic irreversible pulpitis, pulp necrosis, and early apical periodontitis—remain among the most common causes of acute TROFP requiring urgent care [37, 38].

For symptomatic irreversible pulpitis and apical disease, removal of inflammatory or necrotic pulp tissue with endodontic therapy remains the conventional standard of care, with extraction reserved for non-restorable teeth or those with a bad prognosis [38, 39].

Interim procedures, including temporary restorations, sealing of exposed dentin, and selective occlusal correction, may be suitable when urgent definitive care is not available or when diagnostic doubt exists [39]. Recent systematic reviews challenge the rigid dichotomy between reversible and irreversible pulpitis, emphasizing that pulpal disease exists along a biological continuum rather than as discrete entities [37, 40]. Histologic findings often do not correlate consistently with clinical symptoms, which supports a cautious and staged approach to treatment when findings are equivocal. This evolving understanding has contributed to renewed interest in vital pulp therapy (VPT) for mature permanent teeth diagnosed clinically with irreversible pulpitis.

According to recent research, partial or complete pulpotomy utilizing biocompatible materials such calcium silicate-based cements can achieve success rates comparable to traditional root canal therapy under carefully chosen circumstances. A 2022 systematic review and meta-analysis found positive short- and medium-term outcomes for VPT in mature teeth with irreversible pulpitis, encouraging a change toward more biologically conservative management options when suitable [41].

However, these methods necessitate stringent case selection, meticulous bleeding control, and vigilant monitoring, highlighting the significance of accurate diagnosis. When irreversible pulpal involvement is confirmed, endodontic therapy should be used, but extraction should only be used for non-restorable fractures that extend vertically into the root. In cases where conservative management resolves symptoms, early stabilization and diagnostic surveillance may save needless endodontic treatment, according to recent evaluations [32, 42].

Dentin hypersensitivity is another frequent contributor to TROFP and a common source of diagnostic confusion. It is characterized by brief, sharp pain arising from exposed dentin in response to thermal, tactile, osmotic, or chemical stimuli. Although typically less severe than pulpitis-related pain, its high prevalence makes it a frequent consideration in clinical practice.

Management follows a stepwise approach, beginning with at-home desensitizing agents and progressing to professional in-office therapies such as varnishes, sealants, or resin infiltration when necessary [43]. Importantly, clinicians must exclude alternative diagnoses—including early caries, cracked teeth, or pulpal inflammation—before attributing symptoms to hypersensitivity alone.

Periodontal pain may show as localized discomfort, dull aching, or pain on chewing, particularly in the context of severe periodontal abscesses or food impaction. As in endo-periodontal lesions, periodontal and pulpal illnesses may coexist, requiring careful interpretation of imaging results, sensibility tests, and probing patterns to determine the best course of treatment [44]. A common principle that applies to all odontogenic disorders is that conclusive dental treatment should be predicated on concordant findings from imaging, sensitivity testing, clinical examination, and history. Provisional management and reassessment are better than immediate irreversible intervention when such concordance is absent [36, 39]. Pharmacologic management of TROFP has undergone a substantial paradigm shift over the past 5 years, driven by comparative effectiveness research and public health concerns regarding opioid use. Multiple high-quality guidelines now explicitly recommend nonopioid analgesics as first-line therapy for acute dental pain across adolescent, adult, and older adult populations [34, 45].

For acute toothache and postoperative dental pain, nonsteroidal anti-inflammatory medications (NSAIDs) consistently show better analgesic efficacy than opioid-containing regimens, either alone or in combination with acetaminophen. This result is consistent with the underlying biology of the majority of acute odontogenic pain, which is mostly caused by inflammation. NSAIDs diminish prostaglandin-mediated peripheral sensitization, whereas acetaminophen provides parallel central analgesic actions, resulting in additive or synergistic pain alleviation [45, 46]. The 2020 American Dental Association evidence-based clinical practice guideline specifically advises NSAIDs, with or without acetaminophen, as first-line therapy for acute dental pain and reserves opioids for rare cases in which first-line medications are contraindicated or ineffective [34]. Combined ibuprofen–acetaminophen regimens frequently produce analgesia that is on par with or better than that of opioid-containing combinations, without the dangers of reliance, respiratory depression, or abuse, according to subsequent reviews [46, 47].

Local anesthetic strategies may further contribute to pain control, particularly in the perioperative setting. Use of long-acting local anesthetics can reduce early postoperative pain and decrease reliance on systemic analgesics; however, their use should be individualized and balanced against potential adverse effects [34].

A critical component of evidence-based TROFP management is appropriate antibiotic stewardship. Antibiotics are not analgesics and should not be prescribed for pain control in the absence of infection. Their use is indicated only when there is evidence of spreading odontogenic infection, systemic involvement, or specific high-risk conditions, in accordance with established dental infection guidelines. Multiple reviews underline that antibiotics do not treat inflammatory tooth pain without infection and that incorrect administration adds to antimicrobial resistance and unnecessary patient exposure to harmful effects [48, 49]. Clear patient education regarding the role and limitations of antibiotics is therefore crucial, particularly in emergency dentistry settings, where expectations for immediate symptom relief may be high and misconceptions about antibiotic effectiveness in pain management are common. An increasingly acknowledged concern in TROFP is the persistence of tooth-localized pain after technically appropriate dental therapy and absence of continuing disease. Neuropathic or idiopathic processes should be carefully considered in these situations.

When pain persists without clear odontogenic correlates, referral to an orofacial pain specialist, neurologist, or pain medicine clinician is recommended to support accurate diagnosis and targeted management. Current literature emphasizes multidisciplinary management strategies rather than repeated dental interventions, underscoring the value of coordinated, evidence-based care for complex, persistent, or diagnostically ambiguous orofacial pain presentations [11, 12].

Recent studies that directly compare neuropathic and neurovascular orofacial pain syndromes with symptomatic irreversible pulpitis offer useful insights into typical diagnostic errors. These investigations show that some characteristics are unusual for odontogenic pain and should cause diagnostic caution, such as pain migration between teeth, involvement of many teeth, persistence after proper dental treatment, or correlation with headache-like or autonomic symptoms. When considered collectively, current research supports a contemporary concept of tooth-related pain as a diagnostic problem that necessitates organized assessment, receptivity to non-odontogenic processes, and coordinated interdisciplinary management. Limiting overtreatment and coordinating care with underlying pain processes require a focus on diagnostic flexibility and tiered decision-making [50].

References

1. Manfredini D, et al. Epidemiology of dental pain and toothache: a systematic review. J Oral Rehabil. 2021;48(6):709–21. https://doi.org/10.1111/joor.13158.
2. International Classification of Orofacial Pain (ICOP). Orofacial pain classification and diagnostic criteria. Cephalalgia. 2020/2021;41(2):129–221. https://doi.org/10.1177/0333102420976054.
3. Baad-Hansen L. Atypical odontalgia and persistent dentoalveolar pain: clinical features and diagnostic challenges. J Oral Facial Pain Headache. 2021;35(1):7–16. https://doi.org/10.11607/ofph.2626.
4. Zakrzewska JM, Padfield D. Trigeminal neuropathic and neurovascular pain mimicking toothache. Pain. 2022;163(4):e495–502. https://doi.org/10.1097/j.pain.0000000000002407.
5. Renton T, Van der Cruyssen F. Persistent dentoalveolar pain disorder: diagnosis and management. Br Dent J. 2021;230(4):205–11. https://doi.org/10.1038/s41415-021-2594-2.
6. Ohrbach R, Durham J. Orofacial pain assessment and classification: moving toward mechanism-based diagnosis. J Dent Res. 2022;101(3):243–50. https://doi.org/10.1177/00220345211058056.
7. Mejàre I, et al. Diagnostic accuracy of pulp sensibility tests: an updated systematic review. Int Endod J. 2021;54(6):880–902. https://doi.org/10.1111/iej.13476.
8. West NX, Seong J, Davies M. Dentin hypersensitivity: pain mechanisms and management. J Dent. 2021;110:103700. https://doi.org/10.1016/j.jdent.2021.103700.
9. Al-Ani A, et al. Clinical decision-making in pulpitis diagnosis: limits of sensibility testing. J Endod. 2023;49(5):612–20. https://doi.org/10.1016/j.joen.2023.01.012.
10. Nixdorf DR, et al. Outcomes of dental interventions in persistent dentoalveolar pain. J Oral Rehabil. 2022;49(10):1019–28. https://doi.org/10.1111/joor.13342.
11. Zakrzewska JM, Forssell H. Persistent idiopathic dentoalveolar pain. Pain Rep. 2020;5(4):e820. https://doi.org/10.1097/PR9.0000000000000820.

12. Nixdorf DR, Law AS, Lindquist K, et al. Persistent pain after endodontic treatment. J Endod. 2021;47(2):169–76. https://doi.org/10.1016/j.joen.2020.10.012.
13. Ichesco E, Quintero A, Clauw DJ, et al. Altered brain connectivity in persistent dentoalveolar pain. Pain. 2020;161(7):1524–34. https://doi.org/10.1097/j.pain.0000000000001852.
14. Kang Y, Fu Y, Jian K, Liu J, Shao L. Decoding orofacial pain: a translational review of mechanisms and novel therapies. J Headache Pain. 2025;26(1):252. https://doi.org/10.1186/s10194-025-02173-2.
15. Goto T, Kuramoto E, Iwai H, Yamanaka A. Associations with neuropathic pain in the orofacial region. J Oral Biosci. 2024;66(3):485–90. https://doi.org/10.1016/j.job.2024.07.003.
16. Edwards D, Allison JR, Coulter J, Durham J, Beecroft EV. Symptomatic irreversible pulpitis and other orofacial pain: overcoming challenges in diagnosis and management. Br Dent J. 2025;238:517–26. https://doi.org/10.1038/s41415-025-8441-9.
17. Chung M-K, Raman S, Szallasi Á. Tooth pulp afferents and transient receptor potential (TRP) ion channels as key regulators of pulp homeostasis, inflammation, and pain. Int J Mol Sci. 2025;27(1):182. https://doi.org/10.3390/ijms27010182.
18. Ramirez I, et al. Is dental pulp inflammation capable of causing central sensitization? Arch Oral Biol. 2025;177:106320. https://doi.org/10.1016/j.archoralbio.2025.106320.
19. Canfora F, Ottaviani G, Calabria E, Pecoraro G, Leuci S, Coppola N, Sansone M, Biasotto, Adamo D. Advancements in understanding and classifying chronic orofacial pain: key insights from biopsychosocial models and international classifications (ICHD-3, ICD-11, ICOP). Biomedicines. 2023;11(12):3266. https://doi.org/10.3390/biomedicines11123266.
20. Park H-J, Jeong S. Comprehensive management approach to persistent idiopathic dentoalveolar pain: a narrative review. Oral Biol Res. 2024;48(3):69–74. https://doi.org/10.21851/obr.48.03.202409.69.
21. Cervantes-Chavarría AR. Persistent idiopathic dentoalveolar pain: literature review and clinical case treated with intraoral application of botulinum toxin. Odovtos Int J Dent Sci. 2021;23(3):57–65. https://doi.org/10.15517/ijds.2021.46105.
22. Tamura M, et al. Cortical plasticity in patients with persistent idiopathic dentoalveolar pain. Cephalalgia Rep. 2025;8. 10.1177/25158163251359643.
23. Carrasco-Labra A, et al. Evidence-based clinical practice guideline for the pharmacologic management of acute dental pain in adolescents, adults, and older adults. J Am Dent Assoc. 2024;155:102–117.e9.
24. Centers for Disease Control and Prevention (CDC). CDC clinical practice guideline for prescribing opioids for pain—United States, 2022 (Notes ADA recommendations for NSAIDs as first-line for acute dental pain). Morbidity and Mortality Weekly Report (MMWR). 2022. https://www.cdc.gov/mmwr/volumes/71/rr/rr7103a1.htm
25. Nixdorf DR, Moana-Filho EJ, Law AS, McGuire LA, Hodges JS, John MT. Frequency of persistent tooth pain after root canal therapy: a systematic review and meta-analysis. J Endod. 2019;45(2):124–32. https://doi.org/10.1016/j.joen.2018.11.006.
26. Renton T, Durham J, Aggarwal VR. The classification and differential diagnosis of orofacial pain. Br Dent J. 2020;228(9):654–61. https://doi.org/10.1038/s41415-020-1476-6.
27. Mejare IA, Axelsson S, Davidson T, et al. Diagnosis of pulpitis: a systematic review of diagnostic accuracy studies. Int Endod J. 2020;53(5):583–97. https://doi.org/10.1111/iej.13244.
28. Patel S, Brown J, Semper M, Abella F, Mannocci F. European Society of Endodontology position statement: use of CBCT in endodontics. Int Endod J. 2019;52(12):1675–8. https://doi.org/10.1111/iej.13187.
29. Moore PA, Ziegler KM, Lipman RD, et al. Benefits and harms associated with analgesic medications used in the management of acute dental pain. J Am Dent Assoc. 2021;152(1):24–36.e3. https://doi.org/10.1016/j.adaj.2020.08.008.
30. Aguilar P, Linsuwanont P. Vital pulp therapy in vital permanent teeth with irreversible pulpitis: a systematic review. J Endod. 2021;47(5):681–94. https://doi.org/10.1016/j.joen.2020.12.018.
31. Zakrzewska JM, Padfield D, Brennan PA. Symptomatic irreversible pulpitis versus non-odontogenic pain: diagnostic pitfalls. Oral Surg Oral Med Oral Pathol Oral Radiol. 2025;139(2):123–31. https://doi.org/10.1016/j.oooo.2024.10.003.

32. Banerji S, Mehta SB, Millar BJ. Cracked tooth syndrome: a review. Br Dent J. 2021;230(8):499–504. https://doi.org/10.1038/s41415-021-2930-8.
33. Hilton TJ, Ferracane JL, Broome JC. Summit on cracked tooth syndrome: review of current diagnostic and treatment concepts. J Am Dent Assoc. 2020;151(6):456–66. https://doi.org/10.1016/j.adaj.2020.02.013.
34. American Dental Association. Evidence-based clinical practice guideline on the pharmacologic management of acute dental pain. J Am Dent Assoc. 2020;151(11):867–76. https://doi.org/10.1016/j.adaj.2020.08.020.
35. Renton T, Durham J, Aggarwal VR. Chronic orofacial pain: a multidisciplinary problem. Oral Surg Oral Med Oral Pathol Oral Radiol. 2021;131(3):252–63. https://doi.org/10.1016/j.oooo.2020.10.009.
36. Baad-Hansen L, Benoliel R. Neuropathic orofacial pain: facts and fiction. Cephalalgia. 2021;41(6):670–9. https://doi.org/10.1177/0333102420980149.
37. Mejàre IA, Axelsson S, Davidson T, et al. Diagnosis of the condition of the dental pulp: a systematic review. Int Endod J. 2021;54(2):163–83. https://doi.org/10.1111/iej.13408.
38. Abbott PV, Yu C. A clinical classification of pulp and apical disease. Int Endod J. 2022;55(4):295–306. https://doi.org/10.1111/iej.13674.
39. Duncan HF, Chong BS, Pitt Ford TR. The diagnostic challenges of pulpal pain. Br Dent J. 2021;230(9):573–9. https://doi.org/10.1038/s41415-021-2946-0.
40. Ricucci D, Siqueira JF, Loghin S. Histologic correlates of pulpal pain. J Endod. 2020;46(2):188–98. https://doi.org/10.1016/j.joen.2019.11.009.
41. Duncan HF, Cooper PR, Smith AJ. Vital pulp therapy for mature permanent teeth with irreversible pulpitis: a systematic review. Int Endod J. 2022;55(5):469–82. https://doi.org/10.1111/iej.13695.
42. Opdam NJM, Frankenberger R, Loomans BAC. Longevity of restorations for cracked teeth. J Dent. 2020;99:103388. https://doi.org/10.1016/j.jdent.2020.103388.
43. West NX, Seong J, Davies M. Dentin hypersensitivity: a review of its diagnosis and management. J Dent. 2021;104:103558. https://doi.org/10.1016/j.jdent.2020.103558.
44. Herrera D, Sanz M, Kebschull M. Treatment of periodontal abscesses. J Clin Periodontol. 2022;49(Suppl 24):186–97. https://doi.org/10.1111/jcpe.13620.
45. Moore PA, Hersh EV. Combining ibuprofen and acetaminophen for acute pain management after dental surgery. J Am Dent Assoc. 2021;152(10):847–57. https://doi.org/10.1016/j.adaj.2021.05.009.
46. Hersh EV, Kane WT, O'Neil MG, et al. Prescribing recommendations for acute dental pain management. Compend Contin Educ Dent. 2020;41(3):155–64.
47. Denisco RC, Kenna GA, O'Neil MG, et al. Prevention of opioid misuse in dental pain management. J Am Dent Assoc. 2021;152(10):803–11. https://doi.org/10.1016/j.adaj.2021.06.001.
48. Cope AL, Francis NA, Wood F, Chestnutt IG. Antibiotic prescribing in dentistry. Br Dent J. 2021;230(1):25–9. https://doi.org/10.1038/s41415-020-2509-9.
49. Thompson W, Williams D, Pavitt S. Antibiotics and dental pain: an evidence-based approach. Br J Gen Pract. 2020;70(695):e328–36. https://doi.org/10.3399/bjgp20X708029.
50. Benoliel R, Gaul C. Persistent orofacial pain—diagnostic challenges. Cephalalgia. 2022;42(3):247–58. https://doi.org/10.1177/03331024211047005.

Trigeminal Neuralgia

5

Abstract

Trigeminal neuralgia (TN) is a severe neuropathic pain disorder characterized by recurrent, brief, unilateral paroxysms of electric shock–like facial pain within one or more divisions of the trigeminal nerve. Typically triggered by innocuous stimuli such as chewing, speaking, toothbrushing, or light facial touch, TN has major diagnostic relevance in neurostomatology because it is frequently mistaken for odontogenic or temporomandibular pain, leading to unnecessary dental procedures. This chapter reviews the contemporary understanding of TN, including its definition, epidemiology, etiopathogenesis, clinical phenotype, differential diagnosis, and evidence-based management. Particular emphasis is placed on the central role of neurovascular compression and focal demyelination in classical TN, while also addressing secondary causes such as multiple sclerosis, tumors, trauma, metabolic disorders, and iatrogenic injury. The clinical presentation, including trigger zones, pain distribution, refractory periods, continuous background pain, and mild cranial autonomic features, is discussed in relation to diagnostic accuracy and phenotypic variation. Current treatment strategies are summarized, encompassing first-line pharmacotherapy with carbamazepine and oxcarbazepine, second-line and adjunctive agents, botulinum toxin therapy, and procedural interventions such as microvascular decompression, stereotactic radiosurgery, and peripheral neurectomy. The chapter also examines emerging concepts regarding occlusal modulation, disclusion time reduction, and neuronal–glial interactions as potential contributors to symptom expression and future therapeutic innovation. A structured, mechanism-based, and interdisciplinary approach is essential for timely diagnosis, avoidance of overtreatment, and optimization of long-term outcomes in patients with TN.

A. Kalanov, *Neurostomatology*, https://doi.org/10.1007/978-981-92-2552-1_5

Keywords

Trigeminal neuralgia · Neuropathic facial pain · Neurostomatology · Neurovascular compression · Trigger zones · Microvascular decompression · Stereotactic radiosurgery · Carbamazepine · Facial pain diagnosis · Occlusal modulation · DTR

5.1 Definition

Trigeminal neuralgia is a chronic neuropathic pain disorder characterized by recurrent episodes of severe, unilateral facial pain arising from dysfunction of the trigeminal nerve (cranial nerve V). The trigeminal nerve is the principal sensory nerve of the face and oral cavity, transmitting tactile, nociceptive, and thermal information from the skin, mucosa, teeth, and deep facial structures to the brainstem.

Irritation or injury to this nerve results in the hallmark pain of TN, which is typically described by patients as sudden, sharp, stabbing, or electric shock–like in quality. These paroxysmal attacks are brief, lasting from a fraction of a second to 2 min, yet may occur repeatedly throughout the day and are often separated by pain-free intervals.

The existence of distinct trigger zones—discrete regions of the face or oral cavity where even slight stimulation can cause an attack—is a hallmark clinical characteristic of TN. Light touch, chewing, speaking, brushing teeth, shaving, and even air movement across the face are common triggers.

These triggers are especially important in neurostomatological practice since pain can be mistakenly diagnosed as odontogenic, which often results in needless dental operations before a proper diagnosis is made. The unpredictable and powerful character of TN attacks generally results in significant anxiety, avoidance behaviors, and marked decline in quality of life.

Epidemiologically, TN is a relatively uncommon condition, with an annual incidence ranging from 4 to 13 cases per 100,000 individuals. Women are affected more frequently than men, and in patients without multiple sclerosis, disease onset most commonly occurs between the fifth and sixth decades of life [1].

Although TN is typically unilateral, bilateral involvement may occur, particularly in patients with demyelinating diseases such as multiple sclerosis. The condition is classified as either classical TN, most often associated with neurovascular compression at the trigeminal root entry zone, or secondary TN, resulting from identifiable structural or neurological causes including tumors, demyelinating plaques, or brainstem lesions [2]. Focal demyelination of trigeminal afferent fibers is a key factor in the production of pain from a pathological standpoint. Ephaptic transmission, or aberrant cross-talk between neighboring nerve fibers, is made easier by this demyelination, which causes hyperexcitability and spontaneous ectopic discharges. Consequently, harmless sensory inputs are mistakenly perceived as excruciating pain.

The capacity to detect neurovascular conflict and secondary causes has improved thanks to advancements in neuroimaging, especially high-resolution magnetic resonance imaging, which helps with treatment planning and diagnostic precision [3].

Trigeminal neuralgia occupies a critical position at the interface of neurology, dentistry, and pain medicine. Accurate recognition of TN is essential in neurostomatology, as its symptoms frequently overlap with dental and temporomandibular disorders.

A structured diagnostic approach, grounded in validated clinical criteria and supported by imaging when appropriate, is fundamental to preventing misdiagnosis and overtreatment and to guiding evidence-based therapeutic decision-making [4]. Given its profound functional and psychological impact, TN demands timely diagnosis and comprehensive management to restore patient quality of life and minimize chronic disability.

5.2 Etiopathogenesis of Trigeminal Neuralgia

The complex interaction of anatomical, neurovascular, inflammatory, metabolic, and genetic factors that ultimately result in trigeminal nerve dysfunction is reflected in the multifactorial etiopathogenesis of trigeminal neuralgia (see Fig. 5.1). The exact source of trigeminal nerve irritation or injury cannot always be determined in every

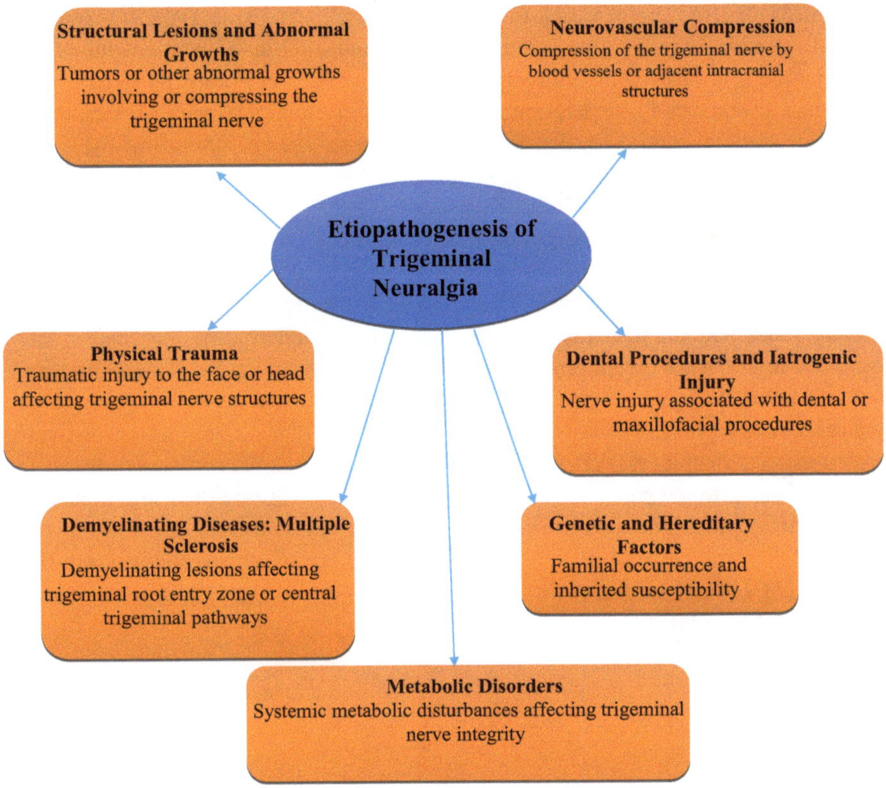

Fig. 5.1 Etiopathogenesis of trigeminal neuralgia

patient, despite the well-defined clinical signs of TN. Advances in neuroimaging, neurophysiology, and molecular neuroscience have considerably increased understanding of the mechanisms underlying TN; however, in a fraction of instances, the disorder remains idiopathic.

Neurovascular Compression

The most widely accepted etiological mechanism in classical trigeminal neuralgia is neurovascular compression at the root entry zone of the trigeminal nerve, where central myelin transitions to peripheral myelin. This region is particularly vulnerable to mechanical stress due to its anatomical configuration and repetitive functional loading. Pulsatile compression by an aberrant or ectatic artery—most commonly the superior cerebellar artery—or less frequently a vein results in focal demyelination of trigeminal afferent fibers [3–5].

Demyelination causes ectopic impulse production, ephaptic transmission between neighboring fibers, and disruption of normal saltatory conduction. Consequently, nociceptive Aδ fibers may be inappropriately activated by non-nociceptive tactile stimuli transmitted by large-diameter Aβ fibers, resulting in paroxysmal pain in response to harmless stimuli. The characteristic trigger zones and stimulus-dependent attacks seen in clinical settings are explained by this process. In addition to causing structural and functional alterations in the trigeminal nucleus, chronic neurovascular compression also contributes to central sensitization and the persistence of symptoms [4].

Structural Lesions and Abnormal Growths

Identifiable structural abnormalities impacting the trigeminal nerve along its intracranial or extracranial path are the cause of secondary trigeminal neuralgia. Cerebellopontine angle meningiomas, schwannomas, epidermoid cysts, and metastatic lesions are examples of tumors that can compress or infiltrate the trigeminal nerve, resulting in pain that resembles typical TN but is frequently accompanied by sensory impairments [6].

In these circumstances, pain is often less entirely paroxysmal and may progress into a continuous or mixed pain pattern. Brainstem lesions involving the trigeminal nuclei or intrapontine fibers can also compromise normal trigeminal signaling. Finding these lesions is essential since treatment approaches for them are very different from those for classical TN and frequently call for surgical or oncological intervention rather than just symptomatic care [6].

Demyelinating Diseases: Multiple Sclerosis

Multiple sclerosis (MS) represents a well-established etiological factor in secondary trigeminal neuralgia. TN occurs in approximately 2%–5% of patients with MS and may be bilateral or recurrent [7]. In MS-associated TN, demyelinating plaques typically involve the trigeminal root entry zone or intrapontine trigeminal pathways, leading to impaired impulse conduction and neuronal hyperexcitability. Unlike neurovascular compression, demyelination in MS is immune-mediated and often multifocal. The coexistence of central demyelination and peripheral neurovascular compression has been demonstrated in some patients, suggesting that MS may lower the threshold for TN development. Clinically, TN may precede other neurological symptoms of MS, emphasizing the importance of neuroimaging in younger patients or those with atypical features [7].

Dental Procedures and Iatrogenic Injury

Dental and maxillofacial procedures have long been implicated as potential triggers or contributing factors in the development of TN. Procedures such as tooth extraction, root canal treatment, dental implant placement, and mandibular nerve blocks may result in direct nerve trauma, ischemic injury, or inflammatory changes affecting branches of the trigeminal nerve [8]. Although true causation is difficult to establish, iatrogenic injury may induce peripheral sensitization and alter afferent input to central trigeminal pathways. In susceptible individuals, this may unmask latent TN or precipitate chronic neuropathic pain syndromes. Importantly, misdiagnosis of TN as odontogenic pain can lead to repeated dental interventions that exacerbate nerve injury without addressing the underlying neuropathic process [8].

Physical Trauma

The trigeminal nerve or its branches may be harmed by trauma to the face or head, such as blunt injuries, facial skeleton fractures, or surgical trauma. Post-traumatic trigeminal neuropathy might appear with persistent pain, paroxysmal bouts, or a mixed pain phenotype resembling TN. Traumatic injury may produce axonal damage, demyelination, or compression from post-traumatic fibrosis. Central plastic alterations following peripheral nerve damage further contribute to pain persistence. Although post-traumatic cases are frequently categorized independently from classical TN, overlap in clinical presentation emphasizes the common mechanisms of trigeminal hyperexcitability and maladaptive neuronal remodeling [9].

Genetic and Hereditary Factors

Genetic predisposition is increasingly acknowledged as a contributing element in trigeminal neuralgia. Familial clustering of TN has been described, and mutations affecting ion channels involved in neuronal excitability—such as sodium and calcium channels—have been hypothesized. These genetic differences could reduce pain thresholds, increase vulnerability to ectopic firing, or hinder remyelination after nerve damage. Genetic variables probably combine with environmental and structural cues to determine disease manifestation, even though no single gene mutation has been conclusively linked to TN.

Variability in age of beginning, severity, and responsiveness to pharmacological therapy may be explained by an understanding of hereditary factors [10].

Metabolic Disorders

Metabolic disturbances, although less common, represent an important etiological consideration in TN. Conditions such as diabetes mellitus can cause microvascular ischemia and peripheral neuropathy affecting trigeminal nerve fibers. Thyroid dysfunction and other systemic metabolic disorders may alter neuronal metabolism, ion channel function, and inflammatory signaling, thereby contributing to nerve vulnerability [11]. Chronic metabolic imbalance may not directly cause TN but can exacerbate nerve dysfunction and impair recovery following injury.

In neurostomatological practice, evaluation of metabolic status is particularly relevant in patients with atypical pain patterns or poor response to standard treatment.

Integrative Pathophysiological Model

Current evidence supports an integrative pathophysiological model in which trigeminal neuralgia arises from the interaction of peripheral and central nervous system mechanisms. Initial peripheral events, such as focal demyelination of trigeminal afferent fibers caused by neurovascular compression, inflammatory processes, metabolic stress, or traumatic injury, lead to abnormal impulse generation and impaired signal conduction. These peripheral changes promote ectopic discharges and aberrant transmission of sensory information. Sustained abnormal afferent input subsequently induces maladaptive plasticity within central trigeminal pathways, resulting in increased neuronal excitability and central sensitization. As a consequence, normally innocuous stimuli are perceived as painful, and pain attacks may become more frequent or severe. This multidimensional etiopathogenesis accounts for the clinical heterogeneity of trigeminal neuralgia and highlights the necessity for individualized diagnostic evaluation and mechanism-based therapeutic strategies.

5.3 Clinical Features of TN

Trigeminal neuralgia is a chronic craniofacial pain disorder characterized by distinctive episodic pain patterns and well-defined clinical features. The pain is severe, unilateral, and recurrent.

Core Pain Phenotype

The hallmark clinical manifestation of trigeminal neuralgia (TN) is the presence of recurrent pain paroxysms, which are defined by sudden, severe, and brief episodes of unilateral facial pain [12, 13]. These paroxysms are not merely intense pain attacks but represent a distinctive temporal and qualitative pattern that is central to the diagnosis of the disorder. The term paroxysm captures the abrupt onset, extreme intensity, short duration, and unpredictable recurrence of the pain episodes, all of which are considered defining clinical characteristics of TN [12, 13]. Patients characteristically describe the pain using vivid and striking descriptors such as electric shock–like, stabbing, shooting, or lancinating sensations. This phenomenology is highly suggestive of a neuropathic pain mechanism and reflects abnormal excitability within the trigeminal sensory pathways [12, 14]. The pain typically reaches maximal intensity almost instantaneously, often without any prodromal warning, and resolves rapidly, usually within a few seconds. Only rarely does an individual pain episode persist beyond 2 min, a temporal profile that is critical in differentiating TN from other facial pain syndromes [12, 13].

The explosive onset and transient nature of TN pain clearly distinguish it from facial pain disorders that are more continuous, aching, burning, or dull in character, such as persistent idiopathic facial pain or temporomandibular disorders [12, 13]. During particularly severe attacks, patients may abruptly stop speaking, freeze in place, grimace, or clutch the affected side of the face. These behaviors reflect both the suddenness and the incapacitating severity of the pain and are often observed during clinical interviews or examinations [13, 15]. Patients with trigeminal neuralgia often exhibit characteristic pain behavior during paroxysms, including marked facial grimacing and protective hand postures corresponding to sudden, electric shock–like facial pain (see Fig. 5.2).

Although TN is thought to originate at or near the trigeminal nerve root entry zone, pain is perceived peripherally within one or more divisions of the trigeminal nerve. This creates a central–peripheral dissociation between the anatomical site of pathology and the subjective localization of pain, which can complicate clinical interpretation [12, 16].

As a consequence, this mismatch frequently leads to misdiagnosis, most notably as odontogenic disease, and may result in unnecessary dental procedures before the correct neurological diagnosis is established [13, 16].

Mild cranial autonomic features, such as lacrimation or slight nasal congestion, may occasionally accompany TN attacks. However, these symptoms are generally subtle and are considered secondary phenomena related to pain intensity rather than evidence of primary autonomic activation, further helping to distinguish TN from trigeminal autonomic cephalalgias [17].

Tic douloureux is the classical clinical descriptor of trigeminal neuralgia, denoting a syndrome of sudden, brief, and recurrent paroxysms of unilateral facial pain with an electric shock–like, stabbing, or lancinating quality, typically confined to one or more divisions of the trigeminal nerve and frequently precipitated by innocuous mechanical or sensory stimuli such as light touch, chewing, speaking, or brushing the teeth, with complete pain-free intervals between attacks and an otherwise normal neurological examination [12].

The term remains clinically and historically relevant because it encapsulates the reflexive facial grimacing and abrupt pain behavior characteristic of classical trigeminal neuralgia, most commonly associated with neurovascular compression at the trigeminal root entry zone, and continues to be used in scholarly literature

Fig. 5.2 Typical facial pain behavior during a trigeminal neuralgia paroxysm

to distinguish this disorder from continuous facial pain syndromes and secondary trigeminal neuropathies [18].

Pain Distribution and Laterality

Trigeminal neuralgia (TN) pain is characteristically unilateral, affecting one side of the face in the vast majority of patients. True unilateral involvement remains a defining clinical feature and reflects the focal nature of trigeminal nerve pathology. The pain most commonly involves the second (maxillary) and third (mandibular) divisions of the trigeminal nerve, either in isolation or in combination [12, 15]. These divisions supply extensive sensory territories of the midface and lower face, which explains the frequent localization of pain to the cheek, upper or lower jaw, lips, and chin. In contrast, involvement of the ophthalmic (first) division is less frequent and, when present, often occurs in conjunction with maxillary or mandibular division involvement rather than as an isolated presentation [19]. Pain distribution in TN may be perceived both extraorally and intraorally. Intraoral pain can involve the teeth, gingiva, hard or soft palate, floor of

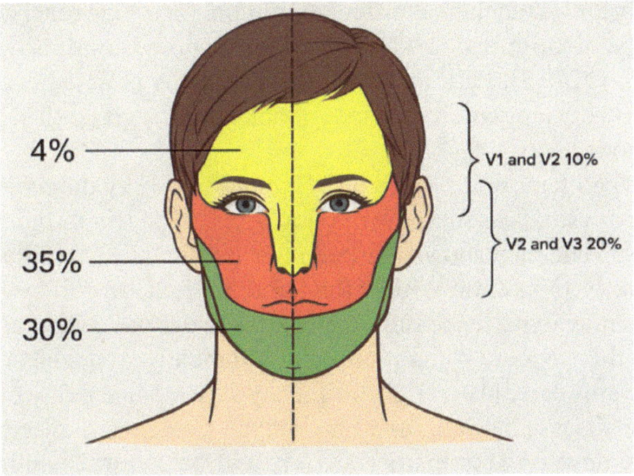

Fig. 5.3 Distribution of trigeminal neuralgia across trigeminal nerve divisions [15]

the mouth, or tongue, closely mimicking odontogenic or periodontal disease. This overlap with dental pain syndromes represents a major reason for misdiagnosis and inappropriate dental interventions, including unnecessary tooth extractions, prior to recognition of the neurological origin of symptoms [20]. As a result, many patients experience significant diagnostic delay before receiving appropriate neurological evaluation and management. Accurate clinical assessment therefore requires careful neuroanatomical mapping of pain distribution in relation to trigeminal nerve divisions, with attention to trigger zones, radiation patterns, and consistency with known trigeminal sensory territories [19]. Bilateral pain, simultaneous involvement of both sides of the face, or alternating laterality over time are atypical features for classical TN and should raise suspicion for secondary causes. Such presentations warrant thorough investigation for underlying conditions, including demyelinating disease, space-occupying lesions, or other structural abnormalities affecting the trigeminal system [21].

This schematic representation demonstrates the typical topographic distribution of pain in trigeminal neuralgia. The maxillary (V2) and mandibular (V3) divisions are most frequently affected, either individually or in combination, whereas isolated ophthalmic (V1) involvement is uncommon. The predominance of V2 and V3 involvement explains the frequent localization of pain to the cheek, upper and lower jaw, lips, and chin, while frontal and periorbital pain is less often reported. The figure highlights the clinical relevance of trigeminal division mapping in diagnosis, differential assessment, and treatment planning for trigeminal neuralgia [22, 23] (see Fig. 5.3).

Duration, Frequency, Refractory Period

Individual trigeminal neuralgia (TN) attacks are defined by sudden, severe paroxysms of pain that typically last from a fraction of a second up to 2 min. These attacks arise abruptly, reach peak intensity almost instantaneously, and terminate just as suddenly, distinguishing TN from other forms of continuous or inflammatory facial pain [12, 15]. Pain episodes may occur as isolated events or present in rapid clusters, during which multiple attacks are triggered in close succession over short periods.

The stereotyped temporal profile of TN attacks is a key diagnostic feature and reflects the paroxysmal discharge behavior of affected trigeminal afferent fibers [12, 24]. Attack frequency shows marked interindividual variability and may fluctuate significantly over time within the same patient. In mild or early disease stages, patients may experience only a few attacks per day, whereas during severe exacerbations, the frequency can increase dramatically to dozens or even several hundred paroxysms daily [19]. High-frequency attack patterns are often associated with heightened disease burden, functional impairment, and reduced quality of life, as patients may develop anticipatory anxiety and behavioral avoidance of known trigger activities such as chewing, speaking, or facial hygiene [19, 25]. The wide variability in attack frequency further complicates clinical assessment and underscores the importance of longitudinal symptom evaluation rather than reliance on isolated clinical snapshots.

A defining and diagnostically valuable characteristic of TN is the presence of a refractory period following an attack. During this interval, which may last seconds to minutes, repeated stimulation of the same trigger zone fails to provoke another pain paroxysm [12]. This refractory phenomenon clearly distinguishes TN from other neuropathic and nociplastic facial pain conditions and provides important insight into the underlying pathophysiology. Experimental and clinical evidence suggests that the refractory period reflects transient inactivation of hyperexcitable trigeminal neurons, likely mediated by altered sodium channel kinetics and impaired impulse propagation within demyelinated afferent fibers [24, 26]. The consistent presence of a refractory period strongly supports abnormal neuronal excitability and dysfunctional impulse generation as core mechanisms in TN.

The natural history of TN is additionally characterized by spontaneous remissions, during which patients may remain entirely pain-free for months or even years without treatment. These remission phases are unpredictable in duration and frequency and are often followed by relapses of varying severity [12]. Over time, remissions may shorten, and attacks may become more frequent or refractory to therapy, contributing to the progressive clinical course observed in a subset of patients [25]. The alternation between active disease phases and remission further complicates disease management and highlights the importance of early recognition and individualized treatment strategies.

Together, the brief duration of attacks, extreme variability in frequency, presence of a refractory period, and fluctuating course with spontaneous remissions form a distinctive clinical signature of TN (see Fig. 5.4). Recognition of these temporal features is essential for accurate diagnosis, differentiation from other orofacial pain disorders, and appropriate therapeutic decision-making [12, 24, 25].

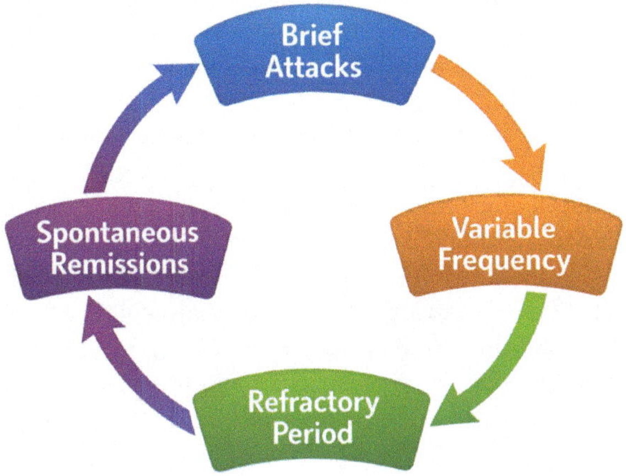

Fig. 5.4 Cyclical clinical characteristics of paroxysmal pain disorders in TN

Continuous Background Pain and Sex Differences

Nevertheless, growing clinical and epidemiological evidence indicates that a substantial proportion of patients—estimated to include up to approximately half of all cases—also experience concomitant continuous background pain affecting the same trigeminal distribution [12, 27]. This persistent pain component occurs independently of the paroxysmal attacks and represents an important deviation from the classical, purely episodic description of TN.

The interparoxysmal pain is most commonly characterized by patients as dull, burning, or aching in quality and may remain present between acute paroxysmal exacerbations, thereby contributing to a mixed pain phenotype that combines continuous and paroxysmal elements [15]. Although less intense than the characteristic lancinating attacks, this background pain can be persistent and distressing, significantly influencing daily functioning and overall pain perception.

From a phenotypic perspective, the presence of continuous background pain has emerged as a clinically meaningful modifier of TN. This phenotype has been reported more frequently in female patients and is consistently associated with a greater overall disease burden. Affected individuals often demonstrate higher levels of pain-related disability, increased psychological impact, and diminished quality of life, as well as reduced responsiveness to both pharmacological therapy and surgical interventions when compared with patients exhibiting purely paroxysmal TN [28]. These observations suggest that continuous pain may reflect a more complex or advanced disease state.

Supporting this concept, neuroimaging studies have identified structural and functional differences in patients with continuous background pain, indicating the potential involvement of distinct or additional central and peripheral mechanisms beyond those implicated in purely paroxysmal TN [27]. Such findings reinforce the view that continuous pain represents a marker of disease complexity rather than a separate diagnostic entity.

Importantly, contemporary classification systems and expert consensus guidelines emphasize that the presence of continuous background pain does not exclude the diagnosis of TN, provided that the defining clinical feature of characteristic paroxysmal attacks is clearly present [12]. Awareness and recognition of this phenotype are therefore essential for accurate clinical characterization, appropriate prognostic assessment, and the development of individualized, mechanism-informed therapeutic strategies.

Trigger Zones and Provoking Factors

Trigger zones are small, precisely localized areas of skin or mucosa where minimal, otherwise innocuous stimulation can reliably precipitate a full trigeminal neuralgia (TN) pain attack. These zones are typically situated in regions innervated by the second and third divisions of the trigeminal nerve and are most commonly identified around the nose, lips, cheeks, perioral skin, or within the intraoral mucosa [13, 15]. Mechanical stimulation of these areas, even when light and non-noxious, may result in abrupt, severe paroxysmal pain characteristic of TN.

A wide range of everyday activities may act as provoking factors by inadvertently stimulating trigger zones. Commonly reported triggers include brushing teeth, chewing, speaking, smiling, swallowing, shaving, and exposure to cold air or wind [13, 29] (see Fig. 5.5). These stimuli are typically brief and low-intensity, yet they can elicit intense pain disproportionate to the physical input, reflecting the underlying pathophysiological hyperexcitability of trigeminal afferent fibers.

As a consequence of this predictable stimulus–pain relationship, many patients develop pronounced avoidance behaviors. These may include reduced oral hygiene practices, reluctance to eat or drink, limitation of speech, and social withdrawal due to fear of provoking pain episodes [16, 30]. Over time, such behaviors may contribute to secondary complications such as poor dental health, weight loss, and diminished quality of life.

In contrast to musculoskeletal or myofascial facial pain syndromes, trigeminal neuralgia rarely improves with conservative physical measures such as massage or application of heat. In fact, manipulation of trigger zones may exacerbate symptoms and provoke acute pain attacks rather than provide relief [13, 30]. This distinguishing feature is clinically important and aids in differentiating TN from other causes of chronic facial pain during diagnostic evaluation.

Cumulative evidence derived from detailed anatomical studies and corroborated by clinical investigations supports the concept that facial sensory innervation follows a highly organized somatotopic pattern corresponding to the three major

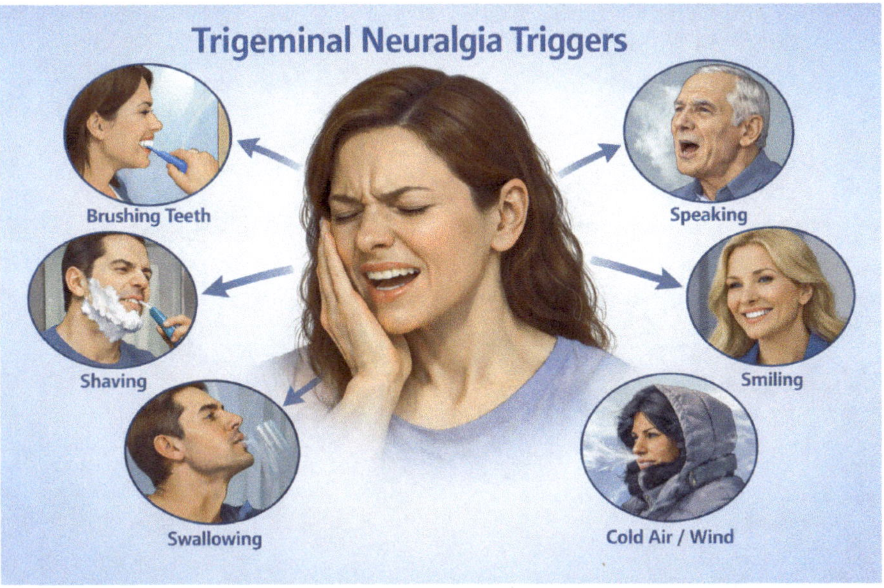

Fig. 5.5 Typical trigger factors provoking pain paroxysms in trigeminal neuralgia

divisions of the trigeminal nerve, namely the ophthalmic (V1), maxillary (V2), and mandibular (V3) branches. This orderly distribution underlies the conventional subdivision of the face into distinct and clinically relevant sensory zones, which provides a reliable anatomical framework for the localization and evaluation of pain in trigeminal neuralgia and related trigeminal neuropathic conditions, thereby facilitating accurate diagnosis and consistent clinical assessment across patient populations [31, 32] (see Fig. 5.6).

Although the pathological process in classical trigeminal neuralgia is generally considered to arise proximally at the trigeminal root entry zone or within the trigeminal ganglion, patients most commonly perceive pain distally within the cutaneous and mucosal territories of the maxillary and mandibular branches, a clinical phenomenon that is clearly reflected by the midfacial and lower facial regions emphasized in this distribution model [13, 33, 34].

The systematic subdivision of facial sensory territories according to trigeminal branch innervation is of particular diagnostic value, as it assists clinicians in differentiating trigeminal neuralgia from odontogenic pain, sinus disease, temporomandibular disorders, and primary headache syndromes, which typically demonstrate non-dermatomal, diffuse, or overlapping pain patterns rather than strict trigeminal territorial confinement [35, 36].

Beyond its diagnostic utility, this anatomical representation plays a critical role in therapeutic decision-making, guiding branch-specific pharmacologic strategies as

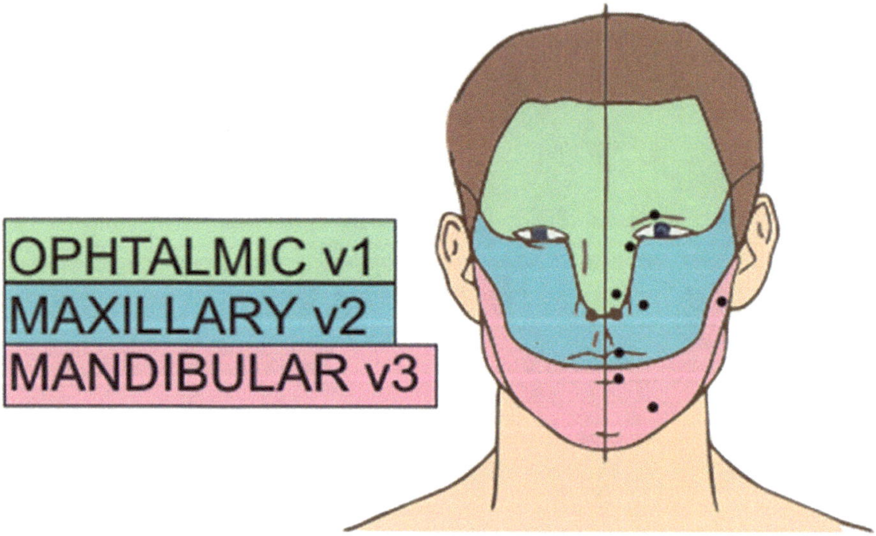

Fig. 5.6 A template showing the six locations based on the trigeminal branch innervation

well as interventional procedures such as peripheral nerve blocks, radiofrequency thermocoagulation, and neurosurgical approaches including microvascular decompression, where precise identification of the symptomatic trigeminal division is directly associated with treatment efficacy and risk reduction [37–39].

Accordingly, standardized visualization of trigeminal sensory territories, as exemplified by this six-zone facial template, remains an essential tool in both clinical practice and academic reporting, promoting anatomical clarity, reproducibility, and consistency across clinical trials and observational studies investigating the mechanisms, diagnosis, and management of trigeminal neuralgia [40, 41].

Although mild cranial autonomic features, such as lacrimation, conjunctival injection, or nasal secretion, may occasionally accompany attacks, these symptoms are generally subtle, transient, and inconsistent [21, 22]. Importantly, autonomic manifestations in trigeminal neuralgia lack the intensity, persistence, and stereotyped presentation observed in trigeminal autonomic cephalalgias. Their presence should therefore be interpreted with caution and does not, in isolation, suggest a primary autonomic cephalalgia. In most cases, autonomic features in trigeminal neuralgia are considered secondary phenomena related to nociceptive trigeminal activation rather than defining components of the disorder.

Differentiation from trigeminal autonomic cephalalgias relies on careful recognition of the characteristic clinical profile of trigeminal neuralgia, including the extremely brief duration of pain paroxysms, the reproducible triggerability by innocuous stimuli, and the presence of a refractory period following attacks [12, 13]. In contrast to autonomic cephalalgias, trigeminal neuralgia lacks prolonged attacks, circadian or circannual periodicity, and prominent ipsilateral autonomic activation. A structured assessment of attack duration, trigger dependence, and post-paroxysmal refractoriness remains central to accurate diagnosis and helps prevent misclassification, inappropriate investigations, and suboptimal treatment strategies.

5.4 Management of TN

Management of trigeminal neuralgia should initially focus on the identification and modification of pain-triggering stimuli. Patients are commonly advised to recognize and avoid specific mechanical, thermal, or functional factors known to precipitate paroxysmal attacks. This adaptive behavioral approach represents an essential early component of care, enabling patients to adjust daily activities in response to the episodic and stimulus-dependent nature of the disorder while pharmacological and interventional strategies are being optimized. Common daily activities such as toothbrushing and shaving are frequently reported precipitants of paroxysmal pain attacks in patients with trigeminal neuralgia (TN). A characteristic clinical feature of TN is the presence of well-defined trigger zones, typically located within the perioral and midfacial territories, including the lips, nasal ala, cheek, chin, and gingiva. Stimulation of these areas by innocuous tactile input may reliably evoke stereotyped pain paroxysms. This phenomenon reflects mechanically evoked allodynia, whereby activation of low-threshold mechanoreceptors results in disproportionate nociceptive discharge within trigeminal pathways [42, 43].

Toothbrushing constitutes a compound mechanical stimulus, combining light tactile contact, vibratory input, and coordinated orofacial movement involving the lips, cheeks, and gingival tissues. Shaving similarly applies repetitive low-intensity mechanical stimulation across facial regions commonly associated with trigger zones. Both activities are consistently identified as frequent precipitants in systematic reviews and consensus documents addressing the clinical characteristics of TN [44, 45]. Systematic evaluations of patient-reported trigger factors indicate that facial touch, speaking, chewing, and toothbrushing are among the most commonly identified precipitants of TN attacks, reported by a substantial subset of affected individuals. Although not universally present, such triggers are considered characteristic of the disorder and are routinely emphasized in diagnostic criteria and clinical teaching. The occurrence of pain paroxysms induced by brushing or shaving is therefore consistent with established clinical descriptions of TN [45, 46].

The following measures represent pragmatic strategies aimed at reducing mechanical stimulation of trigeminal trigger zones during daily hygiene practices. These approaches are adjunctive in nature and do not substitute for pharmacological or surgical treatment. The use of extra-soft toothbrushes and minimal applied pressure is recommended to limit low-threshold mechanical input to sensitive perioral and gingival regions. Smaller brush heads may further reduce inadvertent contact with trigger zones. Warm water is generally preferred, as cold thermal stimuli are frequently reported as additional precipitants. Oral hygiene should be interrupted promptly upon the emergence of prodromal sensations. When flossing provokes symptoms, alternative interdental cleaning methods may be considered, provided they are tolerated [45, 47].

Mechanical stimulation during shaving may be attenuated through the use of electric trimmers, which typically generate less skin drag than manual blades. Shaving following exposure to warmth, such as after a shower, may further reduce cutaneous resistance. The application of adequate lubrication and minimal pressure is advised. During periods of increased symptom activity, avoidance of close shaving may reduce trigger-zone activation [45, 47].

Because persistent severe attacks significantly affect nutrition, sleep, mood, and overall function, management strives to: [48] quickly suppress paroxysms; [49] minimize adverse effects and drug–drug interactions; and [12] escalate early to procedural options when drugs are unsuccessful or not tolerated [12, 48]. Since etiology affects treatment response and the timing/choice of operation, modern frameworks place a strong emphasis on phenotyping (classical, secondary, idiopathic) and identifying possibly treatable structural causes on MRI [12, 13, 48].

Pharmacological therapy remains the cornerstone of initial management for trigeminal neuralgia (TN), particularly in patients with classical forms of the disorder. Rapid suppression of paroxysmal pain episodes, prevention of recurrence, and reduction of side effects are the main objectives of medication therapy. According to recent guideline pronouncements, current recommendations stress a progressive strategy, moving from first-line sodium channel blockers to second-line anticonvulsants and, if needed, adjunctive or third-line therapy in refractory cases [50, 51].

First-Line Therapy

Carbamazepine (CBZ) and oxcarbazepine remain the only pharmacological agents supported by high-quality evidence and are therefore recommended as first-line treatment for classical trigeminal neuralgia [50, 51]. Both drugs exert their principal analgesic effect through blockade of voltage-gated sodium channels, thereby stabilizing hyperexcitable neuronal membranes and reducing ectopic discharges within the trigeminal nerve. This mechanism is particularly relevant in classical TN, where neurovascular compression and focal demyelination are believed to facilitate ephaptic transmission and abnormal impulse generation [51].

In treatment-naïve patients, CBZ and oxcarbazepine demonstrate robust efficacy, with initial response rates of approximately 80%–90%, often resulting in rapid and substantial reduction of paroxysmal pain attacks [50–52]. Randomized controlled trials, observational research, and long-term clinical experience have all continuously shown this high level of efficacy, making these drugs the pharmacological standard by which alternative treatments are evaluated. Nevertheless, both medications have a substantial load of side effects despite their demonstrated effectiveness. In addition to more significant side effects such as hyponatremia, hepatic enzyme increase, leukopenia, and, in rare cases, severe dermatological reactions, common dose-dependent side effects include dizziness, somnolence, ataxia, diplopia, and gastrointestinal complaints [50, 52].

As a consequence of these adverse effects, treatment discontinuation is required in nearly one-third of patients, either due to intolerance or the emergence of medical contraindications during long-term use [50, 52]. Oxcarbazepine is often preferred over CBZ in clinical practice because of its improved pharmacokinetic profile and lower risk of drug–drug interactions; however, the incidence of hyponatremia may be higher, particularly in elderly patients or those receiving concomitant diuretic therapy [51]. These limitations frequently necessitate the consideration of alternative pharmacological strategies, either as replacement monotherapy or as adjunctive treatment.

Second-Line Therapy

When first-line sodium channel blockers are ineffective, poorly tolerated, or contraindicated, several agents may be considered as second-line options, although the supporting evidence is limited and heterogeneous. Gabapentin has been among the most extensively studied alternatives. Multiple small clinical trials and meta-analyses have evaluated gabapentin in patients with TN, reporting analgesic efficacy that appears broadly comparable to CBZ, particularly with respect to reduction in attack frequency and pain intensity [53, 54]. Importantly, gabapentin is generally associated with a more favorable tolerability profile, with fewer serious adverse effects and minimal drug–drug interactions, making it an attractive option in older patients or those with significant comorbidities.

Nevertheless, interpretation of the gabapentin literature is constrained by several methodological shortcomings. Many studies are limited by small sample sizes, variable diagnostic criteria, short follow-up periods, and lack of adequate blinding or randomization [51]. In addition, heterogeneity in dosing regimens and outcome measures further complicates direct comparison with first-line therapies. As a result, while gabapentin may be considered in patients who cannot tolerate CBZ or oxcarbazepine, its overall level of evidence remains inferior, and it cannot be regarded as an equivalent alternative [50, 51].

Pregabalin, a structurally related $\alpha 2\delta$ calcium-channel ligand, has attracted increasing interest over the past 5 years as a potential therapeutic option for TN. Unlike sodium channel blockers, pregabalin modulates calcium-dependent neurotransmitter release and may exert broader effects on central pain processing. Prospective observational studies and open-label trials suggest that pregabalin can reduce pain intensity and attack frequency, particularly in patients with concomitant continuous pain, mixed neuropathic phenotypes, or features overlapping with painful trigeminal neuropathy [55–57]. These characteristics may make pregabalin especially useful in patients who exhibit partial response to CBZ or oxcarbazepine but continue to experience background pain between paroxysms.

Despite these promising findings, the absence of large, high-quality randomized controlled trials precludes strong recommendations regarding pregabalin use. Current evidence supports its consideration as a second-line agent, either as monotherapy in selected cases or, more commonly, as add-on therapy to first-line sodium channel blockers [50,56]. Dose-limiting adverse effects such as dizziness, somnolence, peripheral edema, and weight gain must be taken into account, particularly during long-term treatment.

Topiramate has also been investigated as an alternative anticonvulsant for TN. Its proposed mechanisms include voltage-gated sodium channel inhibition, enhancement of γ-aminobutyric acid (GABA)–mediated inhibition, and modulation of glutamatergic transmission. Limited clinical studies and small randomized trials have suggested potential analgesic efficacy in TN, particularly in refractory cases [58, 59]. However, the clinical utility of topiramate is constrained by a high incidence of adverse effects, including cognitive slowing, paresthesias, fatigue, and mood disturbances, which frequently limit dose escalation and long-term adherence. As a result, current evidence supports only a weak recommendation for topiramate, and its use should be restricted to carefully selected patients who have failed or cannot tolerate other pharmacological options [50].

Adjunctive and Third-Line Therapies

Baclofen, a γ-aminobutyric acid ($GABA_B$) receptor agonist, represents one of the earliest pharmacological alternatives investigated for TN. Its analgesic effect is thought to result from inhibition of excitatory neurotransmitter release at the level of the spinal trigeminal nucleus and higher central pain pathways, thereby reducing synaptic transmission, neuronal hyperexcitability, and central sensitization associated with chronic orofacial pain states. Baclofen has demonstrated efficacy in a small placebo-controlled trial, followed by open-label extension studies that confirmed its potential benefit [60]. Importantly, combination therapy with CBZ has been associated with higher response rates than baclofen monotherapy, supporting its role as an adjunctive agent in patients with partial response to first-line treatment [60, 61].

Despite these advantages, the long-term effectiveness of baclofen often diminishes over time, and dose escalation is frequently limited by central nervous system adverse effects, including sedation, dizziness, muscle weakness, and impaired coordination [50, 61]. These factors restrict its widespread use and necessitate careful titration and monitoring, particularly in elderly patients or those with pre-existing neuromuscular disorders.

Pimozide, a dopamine D2 receptor antagonist, demonstrated analgesic efficacy in early comparative trials against CBZ and was historically considered a potential alternative for TN [62]. However, subsequent clinical experience revealed a high risk of serious adverse effects, including extrapyramidal symptoms, tardive dyskinesia, QT prolongation, and potentially life-threatening cardiac arrhythmias. As a result, contemporary guidelines strongly discourage its use, and pimozide should be considered only in exceptional circumstances when all other pharmacological options have failed and close cardiac monitoring is feasible [50, 51]. In modern clinical practice, its role is therefore largely historical.

OnabotulinumtoxinA has emerged as an important adjunctive and third-line option for patients with refractory TN. Unlike systemic pharmacotherapies, onabotulinumtoxinA exerts its effects locally by inhibiting the release of nociceptive neurotransmitters and neuropeptides from peripheral nerve terminals, thereby attenuating both peripheral and central sensitization. This localized mechanism is particularly valuable in patients who are unable to tolerate higher doses of oral medications or who continue to experience disabling pain despite optimized conventional therapy. Randomized controlled trials and meta-analyses published within the last 5 years have demonstrated significant reductions in pain intensity, frequency of paroxysmal attacks, and dependence on rescue medication when compared with placebo [63–65]. Importantly, the safety and tolerability profile of onabotulinumtoxinA remains favorable, with most adverse effects reported as mild, transient, and confined to the injection site, supporting its role in carefully selected patients.

According to EAN guidance, onabotulinumtoxinA may be considered in patients who are refractory to or intolerant of oral pharmacotherapy, particularly when systemic adverse effects limit further dose escalation [50]. However, optimal dosing regimens, precise injection sites, and ideal treatment intervals have not yet been fully standardized across clinical practice. Additional high-quality, large-scale studies are therefore required to define more clearly the precise role of onabotulinumtoxinA within the pharmacological treatment algorithm for trigeminal neuralgia. See Tables 5.1 and 5.2 for a summary of currently reported treatment approaches.

Trigeminal neuralgia is treated pharmacologically using established dose ranges that focus safety and slow titration to achieve sufficient pain control. Carbamazepine and oxcarbazepine are examples of first-line medications that are started at modest dosages and gradually escalated to the generally recognized therapeutic ranges of 400–1200 mg/day and up to 1800 mg/day, respectively. Baclofen, lamotrigine, gabapentin, pregabalin, and topiramate are examples of adjunctive or second-line drugs that are used within approved dosage limits, with dose modifications determined by tolerability and individual response. It is typically advised to prevent escalation beyond conventional maximum doses due to the elevated risk of negative systemic and neurological consequences. Table 5.3 provides a summary of these dosage recommendations, which are in line with accepted clinical practice [15].

Table 5.1 Pharmacological treatment options for trigeminal neuralgia—first-line therapies

Therapy class	Drug	Mechanism of action	Level of evidence/ recommendation	Clinical role	Key limitations
First-line therapy	Carbamazepine	Voltage-gated sodium channel blockade	High-quality evidence; strong recommendation [12, 14, 15]	First-line monotherapy for classical TN	Frequent adverse effects; drug–drug interactions; ~30% discontinuation
	Oxcarbazepine	Voltage-gated sodium channel blockade	High-quality evidence; strong recommendation [12, 14, 15]	First-line alternative to carbamazepine	Hyponatremia; dose-limiting CNS effects

Table 5.2 Pharmacological treatment options for trigeminal neuralgia—second-line therapies, adjunctive, and third-line therapies

Therapy category	Drug	Mechanism of action	Evidence and recommendation	Clinical use	Considerations/ Adverse effects
Second-line therapy	Gabapentin	$\alpha 2\delta$ calcium-channel modulation	Low–moderate evidence; weak recommendation [13, 16]	Second-line monotherapy or add-on	Methodological limitations of studies; variable efficacy
	Pregabalin	$\alpha 2\delta$ calcium-channel modulation	Low–moderate evidence; weak recommendation [18–20]	Second-line monotherapy or add-on, especially with continuous pain	Lack of RCTs; sedation, edema, weight gain
	Topiramate	Sodium-channel blockade; glutamate inhibition; GABA facilitation	Low evidence; weak recommendation [21, 22]	Selected refractory cases	Cognitive adverse effects; poor tolerability
Adjunctive therapy	Baclofen	$GABA_B$ receptor agonist	Low evidence; adjunctive recommendation [23, 26]	Add-on to carbamazepine	Sedation; diminishing long-term efficacy
Third-line/discouraged	Pimozide	Dopamine D2 receptor antagonism	Historical evidence only; not recommended [12, 14, 24]	Exceptional refractory cases only	Extrapyramidal symptoms; QT prolongation
Adjunctive/non-systemic	Onabotulinum toxinA	Inhibition of peripheral neurotransmitter release	Moderate evidence; conditional recommendation [25, 27, 28]	Refractory TN or intolerance to oral drugs	Injection technique not standardized

Table 5.3. Commonly used pharmacological agents and dosage ranges in trigeminal neuralgia

Drug	Dose	Side Effects	Caution
Carbamazepine	400 mg to 1200 mg/day in 2–3 divided doses. Start: 100 mg–200 mg twice daily with meals. Increase by 100 mg every other day until pain relief is attained.	Somnolence, nystagmus, nausea, dizziness, rash, hyponatremia. Blood disorders (neutropenia, thrombocytopenia, rare aplastic anemia). Severe: Stevens–Johnson syndrome, erythema multiforme.	Liver enzyme elevation—monitor liver function.
Oxcarbazepine	150 mg twice daily, up to 1800 mg/day.	Milder side effects than carbamazepine. Dizziness, fatigue, drowsiness, diplopia. Hyponatremia.	Monitor sodium levels.
Lamotrigine	Dose not provided.	Gizziness, headache, vertigo, excessive sleep. Risk of Stevens–Johnson syndrome if increased too quickly.	Increase dose gradually to reduce risk of severe rash.
Baclofen	25 mg daily (first 2 weeks). Increase to 50 mg daily (25 mg twice daily) for the next 2 weeks. Maintenance: 200–400 mg/day in 2 divided doses. Maximum: 60–80 mg/day in 3–4 divided doses.	Drowsiness, dizziness, fatigue, nausea, hypotension.	Abrupt discontinuation may cause hallucinations.
Pimozide	1–2 mg/day in divided doses.	Arrhythmias, extrapyramidal symptoms, parkinsonism.	Requires cardiac monitoring.
Gabapentin	Start: 300 mg/day. Increase by 300 mg every 2–3 days. Maximum: 1800 mg/day in divided doses.	Mild dizziness, nausea, ankle edema. Hyperlipidemia.	Monitor lipid profile in long-term use.
Pregabalin	150 mg to 600 mg/day.	Drowsiness, dizziness, dry mouth, weight gain, edema.	Monitor for potential weight gain and fluid retention.
Topiramate	100 mg to 400 mg/day in divided doses.	Dizziness, cognitive impairment, weight loss, somnolence.	Cognitive effects may impact daily activities.

Patients with trigeminal neuralgia who were either intolerant of or resistant to carbamazepine medication have reportedly experienced long-lasting and clinically significant pain relief from a combination therapy utilizing pregabalin and amitriptyline. This pharmaceutical approach was used in the reported case series in response to carbamazepine's insufficient response or side effects, and it significantly reduced or eliminated paroxysmal face discomfort. Crucially, the therapeutic effect persisted during lengthy follow-up periods of 10–16 months, indicating both short-term and long-term symptom management. In certain situations of medically refractory trigeminal neuralgia, this combination may be a reasonable alternative because the positive results were obtained with comparatively low drug dosages and were linked to few, reversible side effects [66].

Surgical and Radiosurgical Management of Trigeminal Neuralgia

Surgical and radiosurgical interventions play a pivotal role in the management of trigeminal neuralgia for patients who are refractory to pharmacological therapy or experience intolerable adverse effects. These approaches aim to provide durable pain relief by targeting the underlying neurovascular or neural pathology while balancing efficacy, invasiveness, and long-term safety.

Microvascular Decompression (MVD)

In patients with a confirmed diagnosis of classical trigeminal neuralgia and appropriate anatomical findings, microvascular decompression (MVD) remains the most effective surgical treatment, offering the highest rates of long-term pain relief while preserving trigeminal nerve function. Contemporary outcome studies consistently demonstrate that MVD provides superior durability of pain control compared with ablative and radiosurgical techniques, particularly in patients with neurovascular compression confirmed on high-resolution magnetic resonance imaging (MRI) [5, 67, 68] (see Fig. 5.7).

Recent prospective and large cohort studies published within the last 5 years have strengthened the evidence base for MVD by incorporating standardized outcome measures, longer follow-up intervals, and improved patient phenotyping. In modern series, immediate postoperative pain freedom rates range from 80% to 95%, with sustained pain relief at 5 years reported in approximately 70%–80% of patients [67, 68]. Importantly, these favorable outcomes are most pronounced in patients with classical TN caused by arterial compression of the trigeminal root entry zone, reinforcing the importance of precise preoperative imaging and patient selection [5].

Despite its proven efficacy, the current literature still highlights methodological limitations, including variability in outcome definitions, inconsistent use of validated pain and quality-of-life instruments, and heterogeneity in follow-up duration. Recent consensus statements therefore emphasize the need for future prospective studies employing psychometrically validated pain scales, standardized quality-of-life assessments, and independent postoperative evaluators to enable meaningful comparison across surgical modalities [69]. Such methodological rigor is essential to accurately quantify the true long-term benefit of MVD and to refine patient selection criteria.

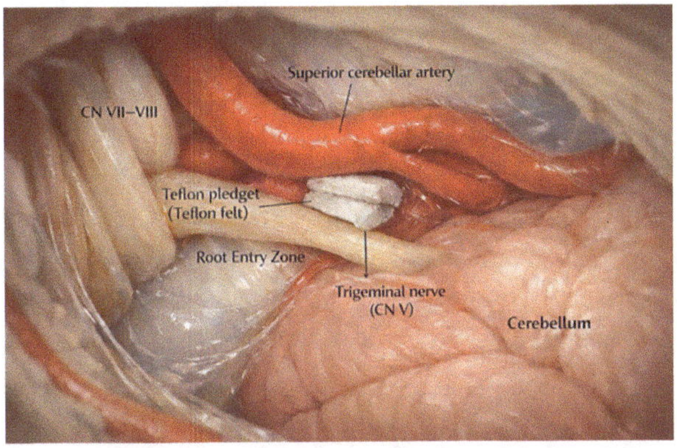

Fig. 5.7 Intraoperative microsurgical view of microvascular decompression showing the trigeminal nerve (CN V) at the cerebellopontine angle, compressed at the root entry zone by a loop of the superior cerebellar artery and separated by interposed Teflon felt via a right retrosigmoid approach

Peripheral Neurectomy

Peripheral neurectomy refers to the surgical division or avulsion of distal branches of the trigeminal nerve and represents one of the oldest surgical treatments for trigeminal neuralgia. Although largely supplanted by more definitive neurosurgical and radiosurgical approaches, peripheral neurectomy continues to play a role in carefully selected patients. This procedure is most commonly considered in individuals who have failed pharmacological therapy and are either medically unfit for intracranial surgery or lack access to specialized neurosurgical facilities [70] (see Fig. 5.8).

The technique can be performed as an outpatient procedure under local or regional anesthesia and targets specific peripheral branches corresponding to the patient's pain distribution. Reported targets include the supraorbital, supratrochlear, infratrochlear, lacrimal, infraorbital, inferior alveolar, lingual, and mental nerves [70]. The relative simplicity of the procedure, minimal anesthetic requirements, and low direct cost make peripheral neurectomy an appealing option in geographically or socioeconomically constrained settings. However, the long-term efficacy of peripheral neurectomy is inferior to that of MVD. Pain recurrence is common, typically occurring within 1–3 years due to nerve regeneration or collateral sprouting. Systematic reviews and critical appraisals published in recent years report wide variability in pain-free intervals, with recurrence rates exceeding 50% at 3 years in many series [70]. Furthermore, sensory loss in the distribution of the treated nerve is unavoidable and may be associated with dysesthesia or anesthesia dolorosa in a minority of patients.

Despite these limitations, peripheral neurectomy retains value as a palliative intervention in patients with limited life expectancy, severe medical comorbidities, or those residing in regions without access to advanced neurosurgical or radiosurgical

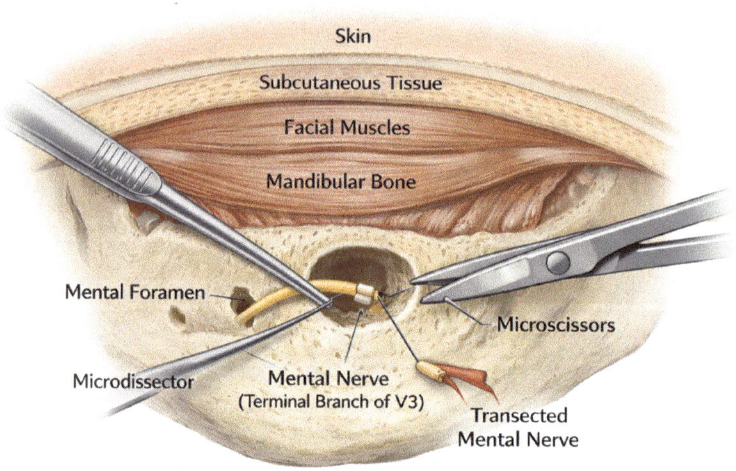

Fig. 5.8 Peripheral neurectomy of the mental nerve, a terminal branch of the mandibular division (V3), performed at the level of the mental foramen. The nerve is isolated and transected using microsurgical instruments, with adjacent soft tissues and mandibular bone shown as key anatomical landmarks

infrastructure. In such contexts, it provides meaningful short-term pain relief with an acceptable risk profile and remains a relevant component of the global therapeutic armamentarium for trigeminal neuralgia [70].

Stereotactic Radiosurgery

Stereotactic radiosurgery (SRS) has emerged as a minimally invasive alternative for patients with trigeminal neuralgia who are either refractory to medical therapy or decline open surgical intervention.

Radiosurgery is typically considered after failure of first-line pharmacological treatment and may be offered earlier in the disease course for patients who are poor candidates for MVD due to advanced age, medical comorbidities, or unfavorable anatomical considerations [71] (see Fig. 5.9).

Multiple systematic reviews and meta-analyses published within the last decade confirm that SRS provides effective pain relief in a substantial proportion of patients, although long-term outcomes remain inferior to those achieved with MVD [71]. Initial pain control rates following radiosurgery range from 70% to 85%, but durable pain freedom at 5 years declines to approximately 45%–60% [7]. The delayed onset of pain relief, which may take weeks to months following treatment, represents an additional limitation compared with the immediate postoperative benefit observed after MVD.

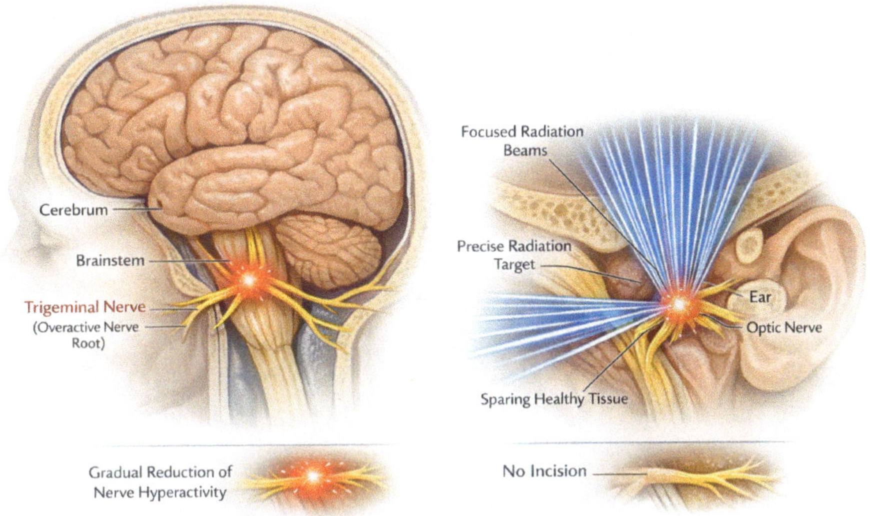

Fig. 5.9 Stereotactic radiosurgery for trigeminal neuralgia illustrating precise, non-invasive focused radiation targeting the trigeminal nerve root entry zone, sparing adjacent neural structures and resulting in gradual reduction of neural hyperactivity

Patient selection is critical when considering radiosurgical therapy. Factors such as age, duration of symptoms, trigeminal nerve anatomy, and prior surgical interventions significantly influence outcomes. Older patients and those with shorter disease duration tend to experience more favorable responses, whereas individuals with long-standing TN or multiple prior procedures exhibit higher recurrence rates [71].

Frameless CyberKnife Radiosurgery

Among radiosurgical techniques, frameless robotic radiosurgery using the CyberKnife system has gained increasing attention due to its precision, patient comfort, and favorable safety profile. A large contemporary cohort study conducted at Centro Diagnostico Italiano (CDI) in Milan evaluated 527 patients with trigeminal neuralgia treated using CyberKnife radiosurgery [72]. This study represents one of the largest single-center experiences with frameless radiosurgery for TN and provides valuable long-term outcome data.

Using carefully defined constraints for radiation dose, nerve volume, and exposure to critical structures such as the brainstem and Gasserian ganglion, the authors reported durable pain control in approximately 76% of patients [72]. Importantly, the incidence of bothersome sensory complications was low, with the majority of patients experiencing either no sensory disturbance or mild, non-disabling hypoesthesia. These

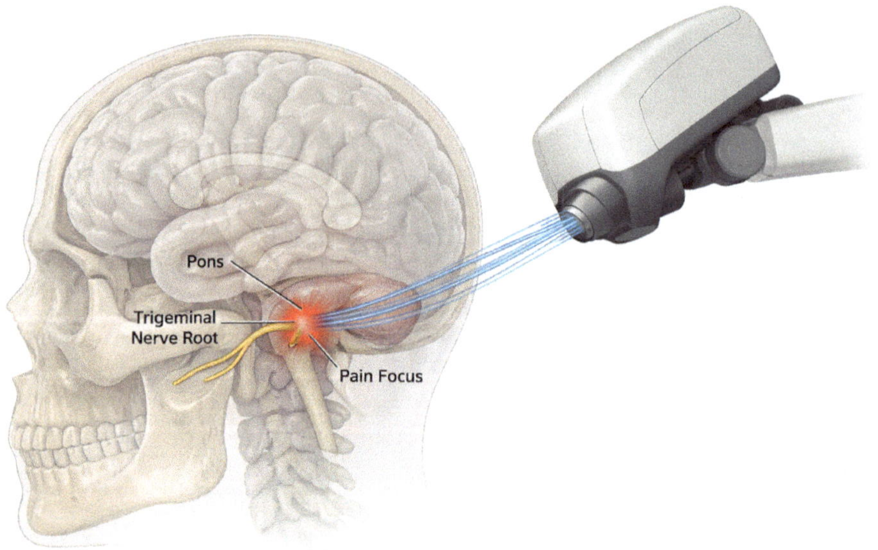

Fig. 5.10 Schematic illustration of frameless CyberKnife stereotactic radiosurgery showing selective irradiation of the trigeminal nerve root entry zone at the pontine level in trigeminal neuralgia

findings underscore the importance of meticulous treatment planning in minimizing adverse effects while preserving efficacy. Compared with frame-based radiosurgical techniques, frameless CyberKnife radiosurgery offers several advantages, including improved patient tolerance, enhanced targeting accuracy through real-time image guidance, and the ability to deliver highly conformal dose distributions [72]. These features contribute to its favorable safety profile and make it particularly attractive for elderly patients or those with significant medical comorbidities (see Fig. 5.10).

Nevertheless, as with other radiosurgical modalities, CyberKnife treatment does not match the long-term pain relief rates of MVD, particularly in younger patients with classical TN and clear neurovascular compression. Consequently, current evidence supports frameless radiosurgery as an effective, low-morbidity alternative rather than a replacement for MVD in appropriately selected patients. The choice among MVD, peripheral neurectomy, and radiosurgical techniques must be individualized and guided by a comprehensive assessment of patient-specific factors, including age, comorbidities, anatomical findings, disease duration, prior treatments, and patient preference. Contemporary treatment algorithms increasingly emphasize a stratified approach, reserving MVD for medically fit patients with classical TN and demonstrable neurovascular compression, while offering radiosurgery or peripheral procedures to those for whom open surgery carries unacceptable risk [5, 71].

From a health-systems perspective, the availability of specialized surgical expertise and technological infrastructure also plays a critical role in determining treatment pathways. In high-resource settings, MVD remains the gold standard for long-term disease control, whereas in low-resource environments, peripheral neurectomy and radiosurgery provide viable alternatives that can substantially reduce suffering despite their inherent limitations [70–72].

The strongest evidence for microvascular decompression (MVD) comes mostly from large retrospective cohort studies and current prospective observational series. The overall level of evidence supporting surgical and radiosurgical therapies for trigeminal neuralgia is moderate. The lack of randomized controlled trials highlights the ethical, practical, and methodological difficulties that come with conducting surgical research on this group, even if these studies repeatedly show that MVD provides better long-term pain control than ablative and radiosurgical procedures. As a result, it is still difficult to compare treatment approaches directly.

The variety of outcome measures, including inconsistent definitions of pain alleviation, heterogeneous application of validated pain and quality-of-life instruments, and irregular follow-up intervals, represents a significant methodological limitation that complicates comparison, synthesis, and interpretation of findings across the literature and undermines the reliability of conclusions. Potential reporting bias is introduced by the fact that many studies use surgeon-reported outcomes rather than evaluations conducted by impartial observers. Furthermore, patient populations are frequently not adequately phenotyped in terms of neurovascular compression features, trigeminal neuralgia subtype, and previous treatment history, which could complicate the interpretation of results.

Comparative efficacy and durability of surgical and radiosurgical treatment modalities are summarized in Table 5.4, while detailed mechanisms, patient selection criteria, and complication profiles are provided in Table 5.5.

Relatively shorter follow-up periods and variations in treatment protocols, such as radiation dosage, target selection, and surgical technique, further limit the evidence for radiosurgical and peripheral treatments. Future multicenter trials should standardize diagnostics, outcomes, and long-term follow-up.

Occlusal Disclusion Time Reduction and Modulation of Trigeminal Neuralgia

While most authors do not consider occlusal factors to be a primary etiological cause of trigeminal neuralgia (TN), a growing body of clinical and neuro-occlusal research suggests that occlusion may act as a significant peripheral modulator of trigeminal nociceptive activity (see Fig. 5.11).

Several studies employing combined biometric technologies—specifically computerized occlusal analysis (T-Scan 10) and surface electromyography (BioEMG™ III)—have investigated the role of occlusal dynamics in patients with TN and related craniofacial pain disorders. These investigations have demonstrated that targeted computer-assisted occlusal adjustment can be associated with meaningful reductions in pain intensity and attack frequency, alongside improvements in patient-reported quality of life [73–76]. Emerging evidence indicates that prolonged occlusal disclusion time

Table 5.4 Comparative efficacy of surgical and radiosurgical treatment modalities for trigeminal neuralgia

Treatment modality	Initial pain relief	Long-term pain control	Onset of pain relief	Evidence base
Microvascular decompression (MVD)	80%–95%	~70%–80% at 5 years	Immediate	Prospective observational cohorts; large retrospective series [1–4]
Peripheral neurectomy	High (short-term)	Poor durability; >50% recurrence in 1–3 years	Immediate	Retrospective case series; critical reviews [6]
Stereotactic radiosurgery (SRS)	70%–85%	~45%–60% at 5 years	Delayed (weeks to months)	Systematic reviews; cohort studies [7]
Frameless CyberKnife radiosurgery	Comparable to SRS	~70%–76% in large cohorts	Delayed (weeks to months)	Large single-center cohort studies [8]

Table 5.5 Mechanisms, patient selection, and complications of surgical and radiosurgical treatments for trigeminal neuralgia

Treatment modality	Mechanism of action	Best candidate profile	Typical complications
Microvascular decompression (MVD)	Microsurgical separation of offending vessel from trigeminal root entry zone	Classical TN; MRI-confirmed arterial neurovascular compression; medically fit patients	CSF leak, hearing loss, facial numbness, cerebellar injury (low incidence in experienced centers)
Peripheral neurectomy	Division or avulsion of distal trigeminal nerve branches	Pharmacoresistant TN; patients unfit for intracranial surgery; limited-resource or palliative settings	Permanent sensory loss, dysesthesia, anesthesia dolorosa (rare)
Stereotactic radiosurgery (SRS)	Focal radiation-induced lesioning of trigeminal nerve root	Pharmacoresistant TN; older patients or those declining MVD	Facial numbness, dysesthesia, delayed recurrence
Frameless CyberKnife radiosurgery	Image-guided robotic radiosurgery (technical variant of SRS)	Elderly or comorbid patients; preference for noninvasive treatment	Mild sensory disturbance; bothersome complications uncommon

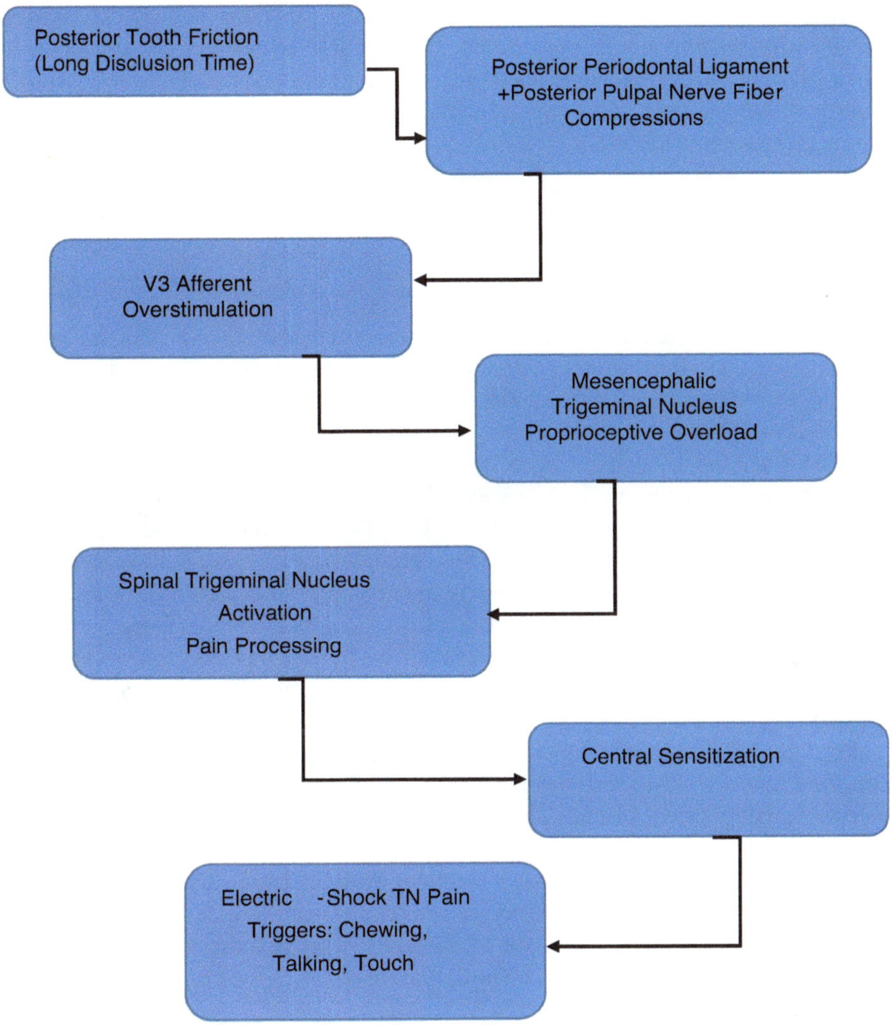

Fig. 5.11 Neuro-mechanistic pathway of disclusion time reduction–related modulation of trigeminal nociceptive processing

(DT), defined as the duration of posterior tooth contact during mandibular excursive movements, represents a critical biomechanical parameter. Excessive DT—typically exceeding 0.5 s—results in sustained posterior tooth friction during lateral or protrusive excursions, producing prolonged posterior tooth periodontal ligament and pulpal nerve fiber compressions and heightened activation of periodontal and pulpal mechanoreceptors. This peripheral stimulation is primarily conveyed via the mandibular (V3 from the mandibular posterior teeth + V2 from the maxillary posterior teeth) divisions of the trigeminal nerve and is accompanied by elevated electromyographic activity in the masseter and temporalis muscles, reflecting excessive jaw function muscle recruitment [73, 74].

Repeated and prolonged afferent input from periodontal, pulpal, and masticatory structures is believed to contribute to sensitization of central trigeminal nuclei, particularly the mesencephalic trigeminal nucleus involved in proprioceptive processing and the spinal trigeminal nucleus, which serves as a key relay center for facial nociception and for repetitive swallowing function.

At the neurochemical level, excessive afferent bombardment may enhance excitatory neurotransmission through mediators such as glutamate and substance P, thereby lowering neuronal activation thresholds and promoting a state of trigeminal hyperexcitability. In this context, normally innocuous stimuli—chewing with teeth frictionally engaging, speaking, light facial touch, or exposure to cold air—may become sufficient to trigger the characteristic paroxysmal, electric shock–like pain of TN [75, 76]. Posterior excursive contacts during prolonged DT have also been conceptualized as repetitive micro-trauma generators. Each sustained contact episode produces localized periodontal ligament and pulpal nerve fiber compressions, jaw muscle splinting, high-amplitude electromyographic bursts, and transient inflammatory responses within the myofascial tissues. In patients with established TN, such repeated peripheral insults may further exacerbate central sensitization, reinforcing hyperresponsiveness within trigeminal trigger zones [73, 75].

Disclusion Time Reduction (DTR) therapy, performed using the Immediate Complete Anterior Guidance Development (ICAGD) procedure, is designed to eliminate these pathological stimuli. The ICAGD protocol employs highly precise, computer-guided occlusal adjustments to shorten DT—often to less than 0.5 s—by rapidly disengaging posterior teeth during excursive movements and establishing effective anterior guidance. As a result, posterior periodontal and pulpal compressions are drastically minimized, and excessive co-contraction of the masseter and temporalis muscles and the swallow mechanism musculature during lateral excursions is markedly reduced [77, 78].

Objective neuromuscular measurements have demonstrated that DTR is associated with an 80%–98% reduction in excursive electromyographic activity, elimination of frictional posterior contacts, and a substantial decrease in abnormal periodontal and pulpal mechanoreceptor firing. By removing the chronic mechanical driver of excessive trigeminal stimulation, DTR significantly reduces the volume and intensity of afferent sensory input entering the central nervous system [74–76].

This reduction in peripheral sensory input facilitates partial reversal of central sensitization. Decreased afferent bombardment lowers nociceptive gain within the spinal trigeminal nucleus, promotes downregulation of hyperactive interneuronal circuits, and diminishes convergence between periodontal and pulpal proprioceptive inputs and facial trigger-zone pathways. In parallel, normalization of jaw-muscle reflex loops reduces exaggerated responses to mandibular movement and facial touch, thereby lowering susceptibility to pain paroxysms [75, 76, 78] (see Fig. 5.12).

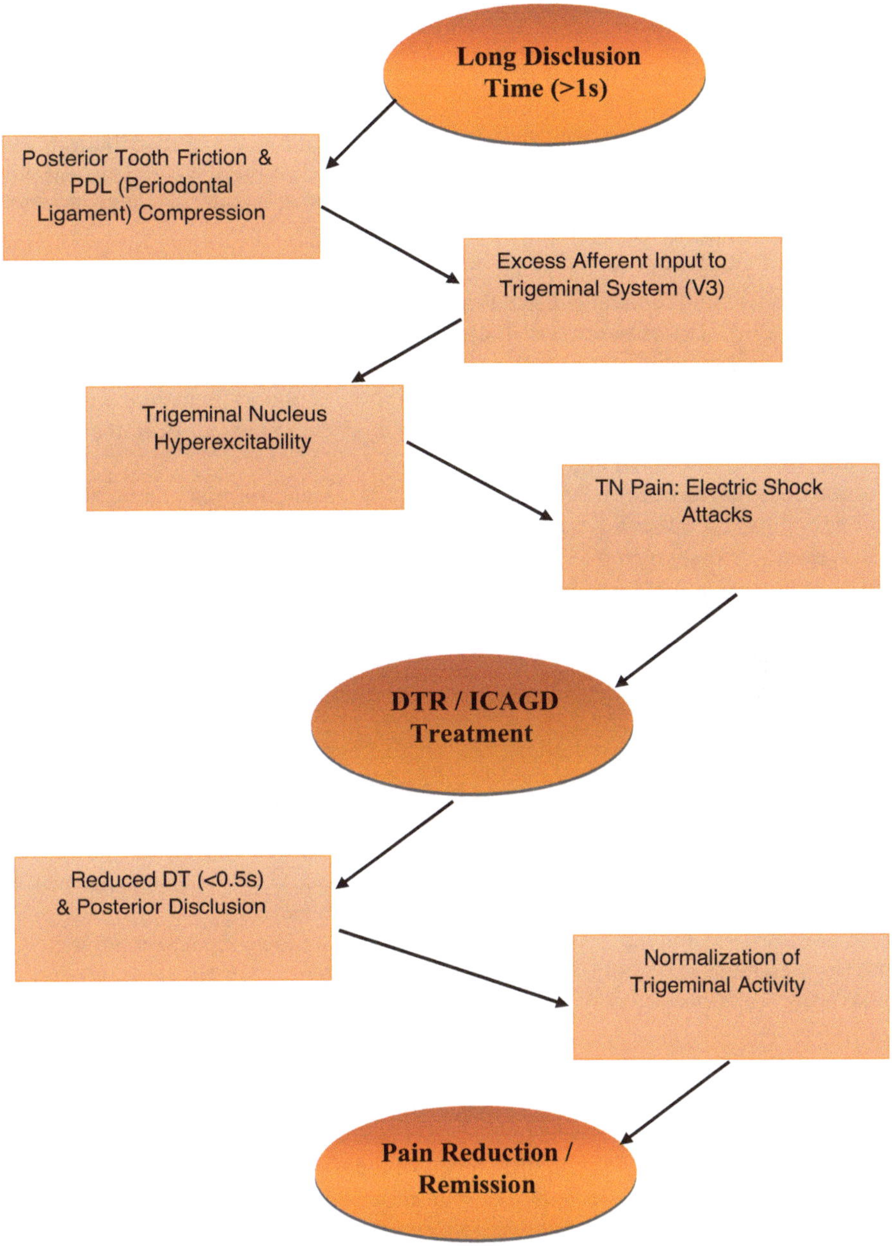

Fig. 5.12 Schematic illustration of the proposed occlusion-related mechanism of trigeminal neuralgia and its modulation by Disclusion Time Reduction (DTR/ICAGD) therapy, showing how prolonged disclusion time leads to trigeminal hyperexcitability and electric shock–like pain, while high-precision computer-guided occlusal adjustment normalizes trigeminal activity and reduces or abolishes pain

Clinically, these neurophysiological changes correspond to improvements in TN symptomatology. Following DTR therapy, patients have reported reductions in pain frequency and intensity, as well as diminished sensitivity during routine activities such as chewing, speaking, tooth brushing, shaving, and exposure to wind. Objective findings—including shortened DT values, normalized bite-force distribution, and sustained suppression of abnormal masticatory muscle activity—closely parallel these symptomatic improvements [74–76]. Importantly, the relevance of DTR in TN extends beyond its established role in temporomandibular disorders.

Although TN is frequently associated with central etiologies such as neurovascular compression or demyelinating disease, peripheral triggers play a critical modulatory role in the initiation of paroxysmal attacks. By reducing posterior tooth friction, minimizing periodontal and pulpal compressions, and restoring more physiologic trigeminal sensory input, DTR may decrease the likelihood that routine orofacial activities will precipitate paroxysmal trigeminal discharge, even when the primary pathology is central in origin [73–76].

In summary, while occlusion is not thought to represent a primary cause of trigeminal neuralgia, Disclusion Time Reduction therapy appears capable of modulating TN symptom expression by eliminating posterior tooth friction that chronically overstimulates mandibular trigeminal afferents, thereby reducing trigeminal nucleus hyperexcitability and decreasing the sensitivity of TN trigger zones. The current evidence supporting Disclusion Time Reduction therapy in trigeminal neuralgia is primarily derived from retrospective cohort studies, case series, and mechanistic biometric electromyographic analyses, corresponding to a moderate level of clinical evidence. Most available studies lack randomized controlled designs and are limited by small sample sizes, heterogeneous diagnostic criteria for trigeminal neuralgia, and potential selection bias. Consequently, while the observed clinical improvements and objective neuromuscular changes are biologically plausible, well-designed prospective controlled trials are required to establish causality and define the precise role of occlusal modulation in trigeminal neuralgia management.

A more precise and integrative understanding of the molecular and cellular mechanisms underlying trigeminal neuralgia (TN) is essential for the development of innovative, mechanism-based therapeutic strategies. Current treatments largely target neuronal excitability at a symptomatic level and do not adequately address the complex pathophysiological processes that sustain the disorder. Increasing evidence suggests that TN is not solely a disorder of the trigeminal neuron itself but rather involves dynamic interactions between neurons, glial cells, and the surrounding microenvironment at both peripheral and central levels of the trigeminal system.

In particular, maladaptive neuronal–glial interactions have emerged as a critical contributor to ectopic impulse generation, central sensitization, and the maintenance of paroxysmal pain. Schwann cells, satellite glial cells in the trigeminal ganglion, and central astrocytes and microglia may modulate ion channel expression, neurotransmitter release, inflammatory signaling, and myelin integrity, thereby amplifying nociceptive transmission. Neurovascular compression–induced focal demyelination further facilitates pathological cross-talk between axons and glial elements, reinforcing abnormal signal propagation.

Despite these advances, the precise molecular pathways linking neuronal dysfunction and glial activation in trigeminal neuralgia remain incompletely defined. Additional experimental and translational research is therefore required to clarify these interactions, identify disease-specific biomarkers, and uncover novel pharmacological targets. A deeper mechanistic insight into neuronal–glial signaling networks may ultimately enable the development of more effective, targeted, and durable therapies that go beyond symptomatic pain control and modify the underlying disease process [79–81].

References

1. Chakraborty AR, Sunshine K, Miller JP, Sweet JA. Current applications of ablative therapies for trigeminal neuralgia. Neurosurg Clin N Am. 2023;34(2):285–90. https://doi.org/10.1016/j.nec.2022.12.005.
2. Bendtsen L, Zakrzewska JM, Abbott J, et al. European Academy of Neurology guideline on trigeminal neuralgia. Eur J Neurol. 2019;26(6):831–49. https://doi.org/10.1111/ene.13950.
3. Cruccu G, Finnerup NB, Jensen TS, et al. Trigeminal neuralgia: new classification and diagnostic grading for practice and research. Neurology. 2016;87(2):220–8. https://doi.org/10.1212/WNL.0000000000002840.
4. Maarbjerg S, Di Stefano G, Bendtsen L, Cruccu G. Trigeminal neuralgia – diagnosis and treatment. Cephalalgia. 2021;41(6):648–57. https://doi.org/10.1177/0333102420987062.
5. Love S, Coakham HB. Trigeminal neuralgia: pathology and pathogenesis. Brain. 2001;124(12):2347–60. https://doi.org/10.1093/brain/124.12.2347.
6. Bick SKB, Eskandar EN. Surgical treatment of trigeminal neuralgia. Neurosurg Clin N Am. 2017;28(3):429–38. https://doi.org/10.1016/j.nec.2017.02.005.
7. Zakrzewska JM, Wu J, Mon-Williams M, Phillips N, Pavitt SH. Trigeminal neuralgia and multiple sclerosis. Pain. 2019;160(8):1741–50. https://doi.org/10.1097/j.pain.0000000000001569.
8. Renton T. Dental (odontogenic) pain and trigeminal neuralgia. Oral Surg Oral Med Oral Pathol Oral Radiol. 2019;128(2):110–8. https://doi.org/10.1016/j.oooo.2019.02.011.
9. Benoliel R, Gaul C. Persistent orofacial muscle pain. Cephalalgia. 2017;37(7):680–94. https://doi.org/10.1177/0333102417696930.
10. Di Stefano G, Truini A. Pharmacological treatment of trigeminal neuralgia. Expert Rev Neurother. 2017;17(10):1003–11. https://doi.org/10.1080/14737175.2017.1373022.
11. Truini A, Garcia-Larrea L, Cruccu G. Reappraising neuropathic pain in humans—how symptoms help disclose mechanisms. Nat Rev Neurol. 2013;9(10):572–82. https://doi.org/10.1038/nrneurol.2013.180.
12. Bendtsen L, Zakrzewska JM, Heinskou TB, Hodaie M, Leal PRL, Nurmikko T, Cruccu G. Advances in diagnosis, classification, pathophysiology, and management of trigeminal neuralgia. Lancet Neurol. 2020;19(9):784–96. https://doi.org/10.1016/S1474-4422(20)30233-7.
13. Lambru G, Zakrzewska JM, Matharu MS, Al-Kaisy A. Trigeminal neuralgia: a practical guide. Pract Neurol. 2021;21(5):392–402. https://doi.org/10.1136/practneurol-2020-002782.
14. Abd-Elsayed CMA. Trigeminal nerve pain. Springer; 2021. https://doi.org/10.1007/978-3-030-60687-9
15. Ovallath S. Textbook of trigeminal neuralgia. Nova Science Publishers; 2020.
16. Kikkeri NS, Nagalli S, Kaur A. Trigeminal neuralgia. In StatPearls. StatPearls Publishing; 2024. https://doi.org/10.55235/statpearls.trigeminalneuralgia
17. Zakrzewska JM, Linskey ME. Autonomic features in trigeminal neuralgia and differential diagnosis. J Oral Facial Pain Headache. 2024;38(1):12–20. https://doi.org/10.22514/jofph.2024.023.

18. Di Stefano G, Maarbjerg S, Nurmikko T, Truini A, Cruccu G. Triggering trigeminal neuralgia. Cephalalgia. 2019;39(6):806–13. https://doi.org/10.1177/0333102418773016.
19. Lambru G, Zakrzewska JM, Matharu MS. Trigeminal neuralgia: a practical guide. Pract Neurol. 2021;21(5):392–402. https://doi.org/10.1136/practneurol-2020-002782.
20. Hassan AF, El-Sayed AA, El-Sharkawy AA. Misdiagnosis-driven dental extractions in patients with trigeminal neuralgia: a retrospective analysis. J Oral Rehabil. 2025;51:785. https://doi.org/10.1111/joor.13652
21. Kolakowski L, Michels L, Wiest R. Interdisciplinary strategies for diagnosis and treatment of trigeminal neuralgia. Swiss Medical Weekly. 2024;154:w3460 https://doi.org/10.4414/smw.2024.w3460.
22. Cruccu G, Finnerup NB, Jensen TS, Scholz J, Sindou M, Svensson P, Treede RD, Zakrzewska JM, Nurmikko T. Trigeminal neuralgia: new classification and diagnostic grading. Eur J Neurol. 2016;23(10):1406–16. https://doi.org/10.1111/ene.12831.
23. Elzohry AAM. Trigeminal neuralgia: update in our knowledges. Arch Neurol Neuro Disord. 2019;2(2):24–30.
24. Cruccu G, Di Stefano G, Truini A. Trigeminal neuralgia. N Engl J Med. 2020;383(8):754–62. https://doi.org/10.1056/NEJMra1914484.
25. Maarbjerg S, Gozalov A, Olesen J, Bendtsen L. Trigeminal neuralgia — diagnosis and treatment. Cephalalgia. 2017;37(7):648–57. https://doi.org/10.1177/0333102416687280.
26. Chen QL, Xu Y, Zhang Y. Molecular mechanisms underlying trigeminal neuralgia. Int J Mol Sci. 2022;23(7):3604 https://doi.org/10.3390/ijms23073604.
27. Li S, Chen M, Wang Y, Liu X. Neuroimaging correlates of continuous pain in classical trigeminal neuralgia. Front Neurol. 2023;14:1277654 https://doi.org/10.3389/fneur.2023.1277654.
28. Headache Medicine Editorial Board. Clinical phenotype variations in trigeminal neuralgia. Headache Med. 2025;16(2):101–9. https://doi.org/10.48208/HeadacheMed.2025.1375.
29. Chiotoroiu AL, Popescu CD, Dumitru M. Therapeutic and clinical considerations in trigeminal neuralgia. Rom J Neurol. 2025;24(3):145–52. https://doi.org/10.37897/RJN.2025.3.11.
30. Maarbjerg S, Gozalov A, Olesen J, Bendtsen L. Trigeminal neuralgia – diagnosis and treatment. Cephalalgia. 2021;41(6):648–57. https://doi.org/10.1177/0333102420987863.
31. Bendtsen L, Zakrzewska JM, Abbott J, Braschinsky M, Di Stefano G, Donnet A, Cruccu G. European Academy of Neurology guideline on trigeminal neuralgia. Eur J Neurol. 2020;27(6):831–49. https://doi.org/10.1111/ene.14193.
32. Cruccu G, Di Stefano G, Truini A. Trigeminal neuralgia. N Engl J Med. 2020;383(8):754–62. https://doi.org/10.1056/NEJMra1914484.
33. Maarbjerg S, Gozalov A, Olesen J, Bendtsen L. Trigeminal neuralgia – a prospective systematic study of clinical characteristics in 158 patients. Headache. 2017;57(1):153–64. https://doi.org/10.1111/head.12991.
34. Di Stefano G, Truini A, Cruccu G. Current and innovative pharmacological options to treat trigeminal neuralgia. Drugs. 2020;80(14):1437–52. https://doi.org/10.1007/s40265-020-01374-6.
35. Renton T, Durham J, Aggarwal VR, Whyman R. Orofacial pain and trigeminal neuralgia: diagnostic challenges. Br Dent J. 2020;228(9):659–66. https://doi.org/10.1038/s41415-020-1538-4.
36. Benoliel R, Svensson P. Persistent orofacial muscle pain and trigeminal neuropathic pain. Cephalalgia. 2021;41(13):1334–47. https://doi.org/10.1177/03331024211018054.
37. Gronseth G, Cruccu G, Alksne J, Argoff C, Brainin M, Burchiel K, Zakrzewska JM. Practice guideline update summary: trigeminal neuralgia. Neurology. 2018;90(17):784–9. https://doi.org/10.1212/WNL.0000000000005353.
38. Tuleasca C, Régis J, Sahgal A, De Salles A, Hayashi M, Ma L, Levivier M. Stereotactic radiosurgery for trigeminal neuralgia: a systematic review. J Neurosurg. 2019;130(3):733–57. https://doi.org/10.3171/2017.12.JNS172276.
39. Maarbjerg S, Wolfram F, Gozalov A, Olesen J, Bendtsen L. Significance of neurovascular contact in classical trigeminal neuralgia. Brain. 2014;137(12):3110–7. https://doi.org/10.1093/brain/awu241.

40. Zakrzewska JM, Linskey ME, Trigeminal Neuralgia Consensus Group. Trigeminal neuralgia: consensus on classification and diagnosis. Pain. 2020;161(S1):S1–9. https://doi.org/10.1097/j.pain.0000000000001862.
41. Di Stefano G, Maarbjerg S, Nurmikko T, Truini A, Cruccu G. Triggering trigeminal neuralgia. Cephalalgia. 2022;42(9):867–75. https://doi.org/10.1177/03331024221083502.
42. Maarbjerg S, et al. Trigeminal neuralgia: diagnosis and treatment. Lancet Neurol. 2020;19(9):784–96. https://doi.org/10.1016/S1474-4422(20)30233-7.
43. Bendtsen L, et al. Trigeminal neuralgia – an update on diagnosis and pharmacotherapy. Ther Adv Neurol Disord. 2020;13:1–14. https://doi.org/10.1177/1756286420932031.
44. Zakrzewska JM, et al. Patient-reported triggers and quality-of-life impact in trigeminal neuralgia: a systematic review. Pain. 2021;162(8):2244–56. https://doi.org/10.1097/j.pain.0000000000002206.
45. Cruccu G, et al. AAN–EFNS guidelines on trigeminal neuralgia: recent updates and clinical implications. Neurology. 2020;95(10):e1454–63. https://doi.org/10.1212/WNL.0000000000010267.
46. Obermann M, et al. Clinical characteristics and trigger mechanisms in trigeminal neuralgia: systematic review. Cephalalgia. 2022;42(6):515–27. https://doi.org/10.1177/03331024211070892.
47. Bendtsen L, Zakrzewska JM. Non-pharmacological considerations and patient self-management in trigeminal neuralgia. Pain Rep. 2023;8(1):e1058 https://doi.org/10.1097/PR9.0000000000001058.
48. Ashina S, Robertson CE, Srikiatkhachorn A, et al. Trigeminal neuralgia. Nat Rev Dis Primers. 2024;10(1):39 https://doi.org/10.1038/s41572-024-00523-z.
49. Pergolizzi JV Jr., et al. An update on pharmacotherapy for trigeminal neuralgia. Expert Rev Neurother. 2024;24(8):773–86. https://doi.org/10.1080/14737175.2024.2365946.
50. Cruccu G, et al. European Academy of Neurology guideline on trigeminal neuralgia. Eur J Neurol. 2020;27:e70–90. https://doi.org/10.1111/ene.14471.
51. Bendtsen L, et al. Trigeminal neuralgia: diagnosis and treatment. Lancet Neurol. 2020;19:784–96. https://doi.org/10.1016/S1474-4422(20)30233-7.
52. Di Stefano G, et al. Pharmacological treatment of trigeminal neuralgia. Expert Rev Neurother. 2018;18:289–300. https://doi.org/10.1080/14737175.2018.1458612.
53. Zhang H, et al. Gabapentin versus carbamazepine in trigeminal neuralgia: a meta-analysis. Pain Pract. 2016;16:1083–91. https://doi.org/10.1111/papr.12414.
54. Wiffen PJ, et al. Gabapentin for chronic neuropathic pain in adults. Cochrane Database Syst Rev. 2017;6:CD007938 https://doi.org/10.1002/14651858.CD007938.pub4.
55. Di Stefano G, et al. Pregabalin add-on therapy in refractory trigeminal neuralgia. J Headache Pain. 2021;22:60. https://doi.org/10.1186/s10194-021-01274-6.
56. Maarbjerg S, et al. Trigeminal neuralgia—recent advances in management. Cephalalgia. 2021;41:648–57. https://doi.org/10.1177/0333102421992367.
57. Obermann M. Recent advances in pharmacotherapy of trigeminal neuralgia. Ther Adv Neurol Disord. 2022;15:1–12. https://doi.org/10.1177/17562864221096744.
58. Gilron I, et al. Topiramate in trigeminal neuralgia: a randomized, double-blind trial. Neurology. 2001;56:1304–7. https://doi.org/10.1212/WNL.56.10.1304.
59. Zakrzewska JM, et al. Systematic review of drug therapy for trigeminal neuralgia. Pain. 2017;158:985–95. https://doi.org/10.1097/j.pain.0000000000000850.
60. Fromm GH, et al. Baclofen in the treatment of trigeminal neuralgia: a double-blind study. Neurology. 1984;34:960–2. https://doi.org/10.1212/WNL.34.7.960.
61. Cruccu G, Truini A. Refractory trigeminal neuralgia: non-surgical treatment options. Cephalalgia. 2013;33:399–406. https://doi.org/10.1177/0333102412472465.
62. Lechin F, et al. Pimozide versus carbamazepine in trigeminal neuralgia. Arch Neurol. 1989;46:960–3. https://doi.org/10.1001/archneur.1989.00520460046015.
63. Wu CJ, et al. Botulinum toxin type A for trigeminal neuralgia: systematic review and meta-analysis. J Headache Pain. 2020;21:83. https://doi.org/10.1186/s10194-020-01147-1.

64. Zhang H, et al. Efficacy and safety of botulinum toxin A in trigeminal neuralgia. Neurology. 2019;92:e2688–97. https://doi.org/10.1212/WNL.0000000000007586.
65. Li S, et al. Long-term efficacy of botulinum toxin A in refractory trigeminal neuralgia. Pain Physician. 2022;25:E353–62. https://doi.org/10.36076/ppj.2022.e353.
66. Lawal AO, Adisa AO, Akinyamoju AO, Kolude B. Management of trigeminal neuralgia using amitriptyline and pregabalin combination therapy: a case report. Afr J Biomed Res. 2012;15(3):201–3.
67. Broggi G, Ferroli P, Panciani PP, et al. Microvascular decompression for trigeminal neuralgia: long-term outcomes and prognostic factors in a contemporary series. Neurosurgery. 2020;87(4):761–9. https://doi.org/10.1093/neuros/nyaa148.
68. Bick SKB, Huie D, Eskandar EN. Long-term outcomes of microvascular decompression for trigeminal neuralgia: a modern cohort study. J Neurosurg. 2021;134(6):1803–10. https://doi.org/10.3171/2020.4.JNS20318.
69. Zakrzewska JM, Linskey ME. Trigeminal neuralgia: current concepts and future directions. Lancet Neurol. 2021;20(11):950–60. https://doi.org/10.1016/S1474-4422(21)00272-3.
70. Yuvaraj V, Krishnan B, Therese BA, Balaji TS. Efficacy of neurectomy of peripheral branches of the trigeminal nerve in trigeminal neuralgia: a critical review of the literature. J Maxillofac Oral Surg. 2019;18(1):15–22. https://doi.org/10.1007/s12663-018-1108-1.
71. Tuleasca C, Régis J, Sahgal A, et al. Stereotactic radiosurgery for trigeminal neuralgia: a systematic review. J Neurosurg. 2019;130(3):733–57. https://doi.org/10.3171/2017.9.JNS1754.
72. Romanelli P, Conti A, Redaelli I, et al. CyberKnife radiosurgery for trigeminal neuralgia: setting and study design. Cureus. 2019;11(10):e6014 https://doi.org/10.7759/cureus.6014.
73. Kerstein RB (ed). Handbook of research on t-scan technology applications in dental medicine. IGI Global; 2024. https://doi.org/10.4018/978-1-6684-9313-7
74. Sutter B, Teragawa S, Radke J. Trigeminal neuralgia patients treated with disclusion time reduction (DTR): a retrospective cohort study. Adv Dent Tech. 2020;90–105.
75. Thumati P, Thumati RP, Kerstein RB, Radke J. Trigeminal neuralgia improvements after disclusion time reduction (DTR). Adv Dent Tech. 2022;66–75.
76. Sutter BA. Complex medical diagnoses with an underlying dental etiology: case reviews. In: Kerstein RB (ed). Handbook of research on clinical applications of computerized occlusal analysis in dental medicine. IGI Global; 2019. p. 1243–1315.
77. Thumati PP, Thumati RP, Thumati P, Kerstein RB, Radke J. Efficacy of disclusion time reduction occlusal therapy for headache and migraine. J Adv Dent Tech. 2021;3(2):42–55.
78. Sutter BA. Complex medical conditions with an underlying occlusal etiology successfully resolved by disclusion time reduction (DTR). In: Kerstein RB (ed). Handbook of research on t-scan technology applications in dental medicine. IGI Global; 2025. p. 1607–1722. https://doi.org/10.4018/978-1-6684-9313-7
79. Ashina S, Bendtsen L, Buse DC, Lyngberg AC, Lipton RB, Ashina M. Trigeminal neuralgia. Nat Rev Dis Primers. 2021;7:63 https://doi.org/10.1038/s41572-021-00292-3.
80. Zakrzewska JM, Linskey ME. Autonomic features in trigeminal neuralgia and differential diagnosis. J Oral Facial Pain Headache. 2024;38(1):12–20. https://doi.org/10.22514/jofph.2024.023.
81. Evers S, Afra J. Trigemino-autonomic activation in facial pain disorders. Cephalalgia. 2025;45(2):0333102424123456 https://doi.org/10.1177/0333102424123456.

6 Temporomandibular Dysfunction

Abstract

Temporomandibular disorders (TMDs) are a heterogeneous group of musculoskeletal and neuromuscular conditions affecting the temporomandibular joint, masticatory muscles, and associated craniofacial structures. They represent a major cause of nondental orofacial pain and commonly manifest with pain, joint sounds, restricted mandibular movement, and functional limitation. Contemporary evidence supports a multifactorial biopsychosocial model in which structural abnormalities, trauma, parafunctional activity, muscular dysfunction, psychosocial distress, and altered pain modulation interact in the onset and persistence of disease. This chapter reviews the current understanding of TMDs, including their epidemiology, etiopathogenesis, clinical manifestations, and diagnostic principles, with particular emphasis on the Diagnostic Criteria for Temporomandibular Disorders (DC/TMD). It further summarizes contemporary management strategies, highlighting conservative, reversible, and evidence-based interventions as the foundation of care, while reserving minimally invasive and surgical procedures for selected refractory cases with clear structural indications. A structured and multidimensional approach is essential for accurate diagnosis, prevention of overtreatment, and individualized patient management.

Keywords

Temporomandibular disorders · Temporomandibular joint · Orofacial pain · DC/TMD · Biopsychosocial model · Myofascial pain · TMD management

A. Kalanov, *Neurostomatology*, https://doi.org/10.1007/978-981-92-2552-1_6

6.1 Definition

Temporomandibular disorders (TMDs), also referred to as temporomandibular dysfunctions, encompass a heterogeneous group of musculoskeletal and neuromuscular conditions affecting the temporomandibular joint (TMJ) complex, the masticatory muscles, and associated osseous and soft-tissue structures. These disorders are characterized by pain, functional limitation, and biomechanical alterations involving the jaw, craniofacial musculature, and related neural pathways.

Historically, TMDs have been conceptualized within the broader framework of orofacial pain disorders, with early authoritative descriptions emphasizing their multifactorial origin and close interaction between joint mechanics, muscular function, and neural modulation of pain [1] (see Table 6.1).

In contemporary clinical practice, TMDs are recognized as one of the most common causes of nondental pain in the orofacial region, frequently presenting with symptoms that mimic odontogenic, otologic, or neurological conditions. This diagnostic overlap contributes to delayed recognition and, in some cases, unnecessary dental or surgical interventions. Modern classification systems, particularly the Diagnostic Criteria for Temporomandibular Disorders (DC/TMD), have refined the diagnostic approach by integrating clinical examination with validated psychosocial assessment tools, reinforcing the view of TMDs as disorders arising from the interaction of biological, psychological, and social factors [2–4]. Within this framework, TMDs are broadly categorized into pain-related disorders (such as myalgia, myofascial pain, and arthralgia) and intra-articular disorders (including disc displacement and degenerative joint disease), acknowledging both peripheral and central mechanisms of pain generation.

From an epidemiological perspective, TMDs represent a substantial global health burden. Population-based studies consistently report a high prevalence of TMD signs and symptoms, although estimates vary widely depending on diagnostic criteria, assessment methods, and demographic characteristics of the studied populations. Recent large-scale analyses indicate that approximately one-third of adults worldwide

Table 6.1 Classification of temporomandibular disorders with representative clinical examples

Category	Subcategory	Examples
Disorders of the joints	Joint issues	- Joint pain (*arthralgia*) - Disc displacement (*disc is not in its normal position*) - Bone destruction (*degenerative joint disease*)
Disorders of the muscles used for chewing	Muscle-related pain	- Localized pain (*myalgia*)—worsens with pressure - Referred pain (*myofascial pain with/without referral*) - Spreads beyond origin
Headaches associated with TMD	TMD-related headaches	- Any type of headache occurring with a painful TMD

Adapted from National Institute of Dental and Craniofacial Research (NIDCR), National Institutes of Health, *Temporomandibular Disorders (TMDs)* (2023), https://www.nidcr.nih.gov/health-info/tmd

experience at least one symptom consistent with TMD during their lifetime, while a smaller but clinically significant proportion develops chronic and functionally limiting disease [2, 5]. Importantly, TMDs account for a major share of healthcare utilization related to orofacial pain, underscoring their relevance not only for dental professionals but also for neurologists, pain specialists, otolaryngologists, and primary care physicians.

Geographical location appears to exert a measurable influence on the epidemiology of temporomandibular disorders. A recent meta-analysis synthesizing global prevalence data demonstrated marked regional variability, with the highest prevalence reported in South America (approximately 47%), followed by Asia (33%) and Europe (29%) [2] (see Fig. 6.1). African epidemiological evidence exists but is dominated by single-country, nonrepresentative designs—most commonly student-based or clinic-based samples—which limits the validity of continent-level pooled prevalence estimates. Egyptian cross-sectional studies among postgraduate students using Fonseca Anamnestic Index report relatively high TMD prevalence, whereas Nigerian data are largely derived from clinical populations and yield lower or highly variable estimates depending on diagnostic criteria and recruitment setting [6–8]. In Australia, national survey analyses suggest TMD experience in ~10% of adults, whereas university cohorts report markedly higher symptom prevalence—highlighting how sampling frame and case definition strongly influence estimates [9, 10]. In North America, epidemiological data indicate that temporomandibular disorders are common but highly dependent on case definition and methodology, with pooled prevalence estimates from clinically

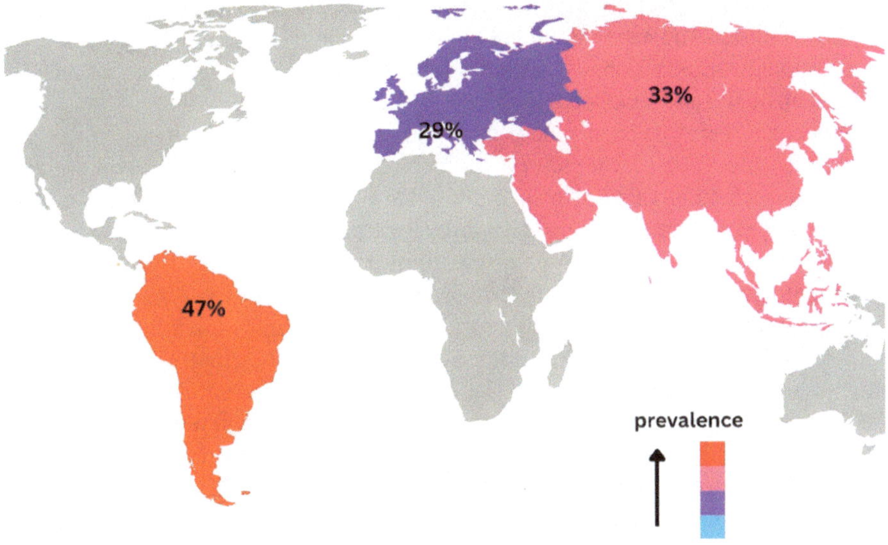

Fig. 6.1 Geographic distribution of prevalence and demographic characteristics of temporomandibular disorders (TMD) across major world regions

examined populations of approximately 15%–20%, population-based diagnostic prevalence around 9%–10%, and prospective cohort studies demonstrating an annual incidence of first-onset painful TMD of about 4% in adults [11–13].

These differences likely reflect a complex interplay of genetic background, cultural norms related to pain reporting, healthcare accessibility, diagnostic awareness, and methodological heterogeneity across studies. For instance, populations with greater awareness of TMD symptoms and better access to specialized care may report higher prevalence rates, whereas underdiagnosis may contribute to lower estimates in certain regions.

Age distribution further modifies the epidemiological profile of TMDs. Although TMD symptoms can occur across the lifespan, prevalence peaks in young and middle-aged adults, particularly between 18 and 60 years of age. This pattern has been attributed to increased exposure to psychosocial stressors, parafunctional behaviors such as bruxism, and cumulative biomechanical loading of the temporomandibular joint during these decades [2, 14].

In older populations, degenerative joint changes may persist, yet pain intensity and symptom reporting often decrease, possibly reflecting adaptive coping mechanisms, altered nociceptive processing, or age-related changes in pain perception and reporting behavior. Across virtually all geographical regions and age groups, sex-related differences in TMD prevalence are consistently observed. Epidemiological studies uniformly report a higher occurrence of TMDs in females compared with males, with female-to-male ratios ranging from near parity in certain European cohorts to pronounced female predominance in South American populations [2]. In the aforementioned meta-analysis, the average proportion of affected females exceeded that of males by 9%–56%, depending on the continent, with the highest female-to-male ratio observed in South America (1.56) and the lowest in Europe (1.09) [2].

These findings suggest that while female predominance is a global phenomenon, its magnitude varies regionally, possibly reflecting sociocultural influences and differences in health-seeking behavior.

Gender has therefore been widely discussed as a potential risk factor—or more precisely, a risk modifier—for temporomandibular disorders. Biological, psychological, and social mechanisms have all been proposed to explain the observed sex differences. From a biological standpoint, hormonal influences, particularly fluctuations in estrogen levels, have been implicated in pain sensitivity, connective tissue metabolism, and inflammatory responses within the temporomandibular joint [15–17].

Experimental and clinical studies suggest that estrogen may modulate nociceptive processing both peripherally and centrally, potentially contributing to heightened pain sensitivity in women with TMDs. Psychological factors further differentiate the experience and expression of TMD-related pain between genders. Women with TMDs are more likely to report higher pain intensity, longer pain duration, and greater functional impairment compared with men, even when objective clinical findings are similar [3, 18]. This difference has been linked to variations in pain appraisal, coping strategies, and emotional processing. Women tend to interpret pain through a more affective-emotional lens, whereas men more often emphasize the physical and mechanical aspects of pain. Such differences may influence symptom reporting, healthcare utilization, and response to treatment.

Social and behavioral factors also play a critical role. Women generally demonstrate greater vigilance toward bodily symptoms and are more likely to seek medical attention for pain-related conditions. In the context of TMDs, heightened concern about pain may interfere with daily activities, including work performance and social interactions, thereby amplifying perceived disability [3, 18].

Conversely, men may underreport symptoms or delay seeking care, potentially contributing to an apparent gender gap in prevalence rather than a true difference in disease incidence. Importantly, contemporary research emphasizes that gender alone should not be viewed as an independent causal factor for TMDs. Rather, it represents a complex modifier that interacts with psychosocial stress, parafunctional behaviors, sleep disturbances, and central pain processing mechanisms [4, 14, 19]. Longitudinal studies have demonstrated that psychosocial variables such as anxiety, depression, catastrophizing, and perceived stress are among the strongest predictors of TMD onset and chronification, often outweighing purely biomechanical factors [20, 21].

These findings reinforce the biopsychosocial model that underpins current diagnostic and therapeutic strategies for TMDs. Ongoing research continues to refine the understanding of risk factors associated with temporomandibular disorders, with increasing emphasis on central sensitization, altered pain modulation, and overlap with other chronic pain conditions such as fibromyalgia, migraine, and tension-type headache [19, 22]. Advances in neuroimaging and quantitative sensory testing have provided further evidence that a subset of patients with TMDs exhibit widespread pain sensitivity and dysfunctional central pain processing, supporting the classification of TMDs within the spectrum of chronic overlapping pain conditions. In summary, temporomandibular disorders constitute a diverse group of musculoskeletal and neuromuscular conditions that represent a major source of nondental orofacial pain. Their prevalence varies substantially across geographical regions, with consistently higher rates observed in females across all continents. While gender-related biological, psychological, and social factors contribute to differences in pain perception and disease expression, current evidence supports a multifactorial etiology in which gender functions as a modifying variable rather than a sole determinant. Continued investigation into the epidemiology and pathophysiology of TMDs is essential to refine diagnostic criteria, optimize individualized treatment strategies, and improve long-term outcomes for affected patients.

6.2 Etiopathogenesis of Temporomandibular Disorders

Temporomandibular disorders represent a heterogeneous group of musculoskeletal and neuromuscular conditions affecting the temporomandibular joint, masticatory muscles, and associated osseous, soft-tissue, and functionally related craniofacial structures (see Fig. 6.2).

The etiopathogenesis of TMDs is widely accepted as multifactorial, involving a complex interaction between structural, functional, neurophysiological, genetic, and

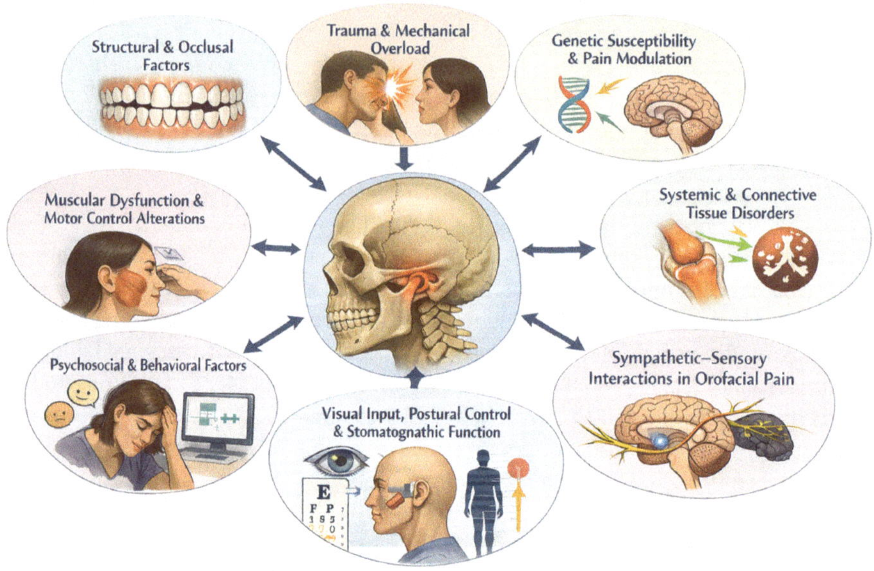

Fig. 6.2 Multifactorial etiopathogenesis of temporomandibular disorders

psychosocial factors rather than a single causative mechanism. This multifactorial framework helps explain the marked variability in clinical presentation, symptom severity, disease progression, and response to treatment observed among affected individuals across different clinical settings and populations.

Structural and Occlusal Factors

Abnormalities in the structure or alignment of the temporomandibular joint can contribute to the development and perpetuation of TMDs. These include internal derangements of the TMJ, altered condyle–disc relationships, and developmental or acquired asymmetries that affect mandibular function. Malocclusion, altered occlusal contacts, and discrepancies between centric relation and maximal intercuspation may increase mechanical loading on the TMJ and associated musculature, particularly in susceptible individuals. Excessive parafunctional activity, such as bruxism or clenching, can further exacerbate joint loading and accelerate degenerative or inflammatory processes within the joint complex [23]. Pediatric and adolescent populations may be particularly vulnerable to the effects of internal derangements on craniofacial development. Imaging-based studies have demonstrated that early TMJ disc displacement can influence mandibular growth patterns, facial symmetry, and long-term joint stability, underscoring the importance of structural integrity in maintaining normal stomatognathic function [24].

Trauma and Mechanical Overload

Trauma to the TMJ, whether acute or repetitive, represents a well-recognized risk factor for TMD onset. Direct blows to the jaw, whiplash injuries, intubation-related strain, and excessive mouth opening during dental or medical procedures can result in capsular injury, disc displacement, or intra-articular inflammation. Even in the absence of overt structural damage, trauma may initiate maladaptive neuromuscular responses and central sensitization processes that contribute to persistent pain and dysfunction [25]. Chronic microtrauma associated with parafunctional habits can similarly induce cumulative mechanical stress on the joint and surrounding tissues. Over time, such stress may lead to synovial inflammation, cartilage degradation, and altered joint biomechanics, particularly when combined with psychosocial stressors or pre-existing anatomical vulnerabilities [26].

Muscular Dysfunction and Motor Control Alterations

Dysfunction of the masticatory muscles plays a central role in many forms of TMD, particularly myogenous and mixed presentations. Muscle overuse, sustained hyperactivity, impaired coordination, and reduced adaptive capacity can all contribute to pain, fatigue, and restricted mandibular movement. Habits such as clenching and grinding, often associated with emotional stress, may perpetuate increased muscle tone and ischemia, thereby promoting nociceptive input from muscle tissues [27]. Diagnostic studies comparing classification systems have highlighted the importance of distinguishing primary muscular disorders from joint-based pathology, as the underlying mechanisms and therapeutic approaches may differ substantially. Alterations in motor control and muscle recruitment patterns have been identified in patients with TMD, suggesting that central nervous system involvement extends beyond peripheral tissue injury [28].

Psychosocial and Behavioral Factors

Psychosocial factors are strongly implicated in both the development and chronicity of TMDs. Stress, anxiety, depression, and maladaptive coping strategies can influence pain perception, muscle activity, and behavioral patterns such as parafunction. Experimental and clinical studies have demonstrated heightened electromyographic activity in masticatory muscles during stress exposure among individuals with chronic TMD pain, supporting a bidirectional relationship between psychological state and neuromuscular function [29]. Contemporary biopsychosocial models emphasize that psychological distress does not merely coexist with TMDs but actively modulates symptom expression, pain intensity, and treatment outcomes. These findings align with broader chronic pain literature, which highlights the role of affective and cognitive processes in pain amplification and persistence [30].

Systemic and Connective Tissue Disorders

Systemic conditions affecting connective tissue integrity and joint stability may predispose individuals to TMDs or exacerbate existing symptoms. Generalized joint hypermobility, inflammatory arthropathies, and degenerative joint diseases can involve the TMJ as part of a wider musculoskeletal disorder. In such cases, inflammation, ligamentous laxity, or cartilage degeneration may compromise joint mechanics and increase susceptibility to pain and dysfunction [31]. Although not all systemic conditions directly target the TMJ, their presence may reduce adaptive capacity and impair tissue repair, thereby contributing to chronicity. Recognition of these systemic contributors is essential for comprehensive diagnosis and individualized management planning, enabling more accurate phenotyping, targeted interventions, and improved patient-centered clinical outcomes.

Genetic Susceptibility and Pain Modulation

Accumulating evidence suggests that genetic factors influence individual vulnerability to TMDs, particularly through mechanisms related to pain perception and modulation. Variations in genes involved in catecholamine metabolism, inflammatory signaling, and neural plasticity have been associated with altered pain sensitivity and increased risk of developing chronic orofacial pain conditions [32]. Large cohort studies have demonstrated that genetic predisposition interacts with environmental and psychosocial stressors to shape pain trajectories over time. These findings support a model in which genetic factors do not act in isolation but modify responses to mechanical load, trauma, and emotional stress [33].

Sympathetic–Sensory Interactions in Orofacial Pain

The role of the sympathetic nervous system (SNS) in orofacial pain has gained increasing attention in recent years. Although traditionally considered distinct from somatosensory pathways, growing evidence indicates that sympathetic–sensory interactions may influence nociceptive processing in the trigeminal system. Experimental studies have demonstrated that sympathetic activity can modulate jaw reflexes, muscle tone, and sensory transmission, suggesting a potential contribution to pain amplification and dysregulation in TMDs [34]. Peripheral sympathetic mechanisms have been implicated in various chronic orofacial pain conditions, including atypical facial pain, stomatodynia, and atypical odontalgia. However, the precise mechanisms through which the SNS influences TMJ-related pain remain incompletely understood. Further elucidation of these interactions may open new avenues for targeted therapeutic strategies aimed at modulating autonomic–sensory coupling [35].

Visual Input, Postural Control, and Stomatognathic Function

Emerging research has highlighted a functional connection between the stomatognathic system, visual input, and postural control mechanisms. Investigations into the vestibulo-ocular reflex (VOR) and visual correction in patients with TMD have revealed that alterations in visual input can influence electromyographic activity of pericranial and masticatory muscles. In adults with chronic TMD, standard optical correction has been shown to increase EMG activity in pericranial muscles, potentially exacerbating muscle tension and pain [36]. Further studies in pediatric populations with functional mandibular lateral deviation have suggested that conventional corrective lenses may adversely affect muscle activity, particularly in the anterior temporal muscles involved in mandibular posture. In contrast, EMG-guided visual correction strategies appear to improve muscle balance and may support orthodontic and functional rehabilitation approaches [37]. These findings reinforce the concept of TMDs as disorders of sensorimotor integration rather than purely localized joint or muscle pathology, emphasizing central nervous system involvement, altered pain modulation, and dysfunctional neuromuscular control mechanisms.

Integrative Perspective

Taken together, the etiopathogenesis of TMDs reflects a dynamic and nonlinear interplay between biomechanical loading, altered neuromuscular control, psychological and behavioral modulation, genetic susceptibility, and multisensory integration within the central nervous system. In many patients, no single etiological factor can be isolated; rather, TMDs develop from the cumulative effect of multiple predisposing and perpetuating influences acting across different biological levels over time. Changes in peripheral tissue may interact with cerebral systems of pain regulation, resulting in chronification and duration of symptoms. This intrinsic complexity highlights the value of thorough, customized evaluation and provides compelling evidence for modern multimodal management approaches that concurrently address the structural, functional, and psychological causes of pain and dysfunction.

6.3 Clinical Features of Temporomandibular Disorders

Temporomandibular disorders comprise a heterogeneous group of musculoskeletal and neuromuscular conditions involving the temporomandibular joint (TMJ), the masticatory muscles, and associated structures. As reflected in contemporary diagnostic frameworks, particularly the Diagnostic Criteria for Temporomandibular Disorders (DC/TMD), the clinical presentation of TMDs is multidimensional and varies according to the specific subtype, chronicity, and degree of peripheral and central pain modulation involved [38, 39]. In most patients, symptoms do not arise from a single

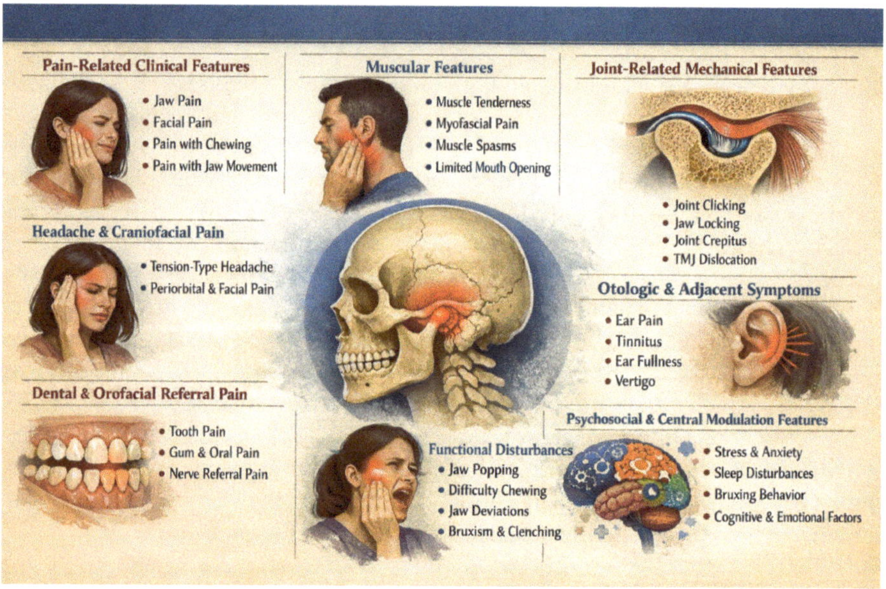

Fig. 6.3 Clinical manifestations of temporomandibular disorders

anatomical structure but rather reflect dysfunction across multiple components of the stomatognathic system, including muscles, joints, ligaments, connective tissues, and neural elements [40] (see Fig. 6.3).

Pain-Related Clinical Features

Pain represents the core clinical feature of most TMDs and may originate from muscular, articular, or mixed sources. Patients commonly report pain localized to the preauricular region, jaw muscles, or TMJ, which may be unilateral or bilateral and can present as intermittent or persistent [41, 42]. The quality of pain varies widely, ranging from dull, aching discomfort typical of myogenous TMDs to sharper or pressure-like pain associated with intra-articular disorders. Importantly, pain intensity and temporal patterns are influenced not only by peripheral nociceptive input but also by central sensitization mechanisms, particularly in chronic TMDs [43].

Pain frequently worsens during jaw function, such as chewing, yawning, speaking, or prolonged mouth opening, reflecting functional loading of affected tissues [44]. Palpation of the masticatory muscles or TMJ often reproduces the patient's familiar pain, a key diagnostic feature emphasized in DC/TMD-based examinations [38]. However, spontaneous pain at rest is also commonly reported in chronic cases and is associated with poorer prognosis and greater psychosocial burden [45].

Muscular Features

Myogenous TMDs are characterized by tenderness, stiffness, fatigue, or soreness of the masticatory muscles, most commonly involving the masseter, temporalis, and medial pterygoid muscles [46]. Patients may describe a sensation of jaw heaviness or muscular exhaustion, particularly after sustained functional activities. Electromyographic and imaging studies suggest altered muscle recruitment patterns and impaired motor control rather than primary muscle pathology in many individuals [47]. Muscle-related pain frequently extends beyond the immediate orofacial region, contributing to regional pain patterns involving the temples, neck, and shoulders [48]. This overlap reflects shared neuromuscular and central processing pathways and supports the classification of chronic myogenous TMD as part of a broader spectrum of chronic overlapping pain conditions [43, 49].

Joint-Related Mechanical Features

Mechanical symptoms of the TMJ are common in patients with intra-articular TMDs. Joint sounds, including clicking, popping, or crepitus, are frequently reported during mandibular opening, closing, or lateral movements [38]. Clicking sounds are most often associated with disc displacement with reduction, whereas crepitus is more suggestive of degenerative joint disease; however, joint sounds alone are not diagnostic and may occur in asymptomatic individuals [50, 51]. Consequently, current guidelines emphasize correlating joint sounds with pain, functional limitation, and imaging findings when clinically indicated [38, 39].

Limitation or deviation of mandibular movement is another key clinical feature. Reduced maximal mouth opening, asymmetrical opening patterns, or jaw locking may occur, particularly in disc displacement without reduction or acute inflammatory joint conditions [44, 52]. Functional limitation is often accompanied by pain, but painless restriction can also be observed, underscoring the need for careful clinical differentiation.

Functional Disturbances

Functional impairment is a hallmark of clinically significant TMDs. Patients frequently report difficulty chewing hard or chewy foods, discomfort during prolonged speaking, or a sensation of bite instability [53]. Subjective occlusal discomfort or perceived changes in bite position are common but should not be interpreted as evidence of primary occlusal pathology. Contemporary evidence demonstrates that occlusal discrepancies are more often a consequence rather than a cause of TMD-related neuromuscular adaptation [54, 55]. Jaw function limitations are typically assessed using validated instruments such as the Jaw Functional Limitation Scale (JFLS), which provides a standardized measure of disability and treatment response [56].

These tools highlight the close relationship between pain severity, functional restriction, and quality-of-life impairment in TMD populations.

Headache and Craniofacial Pain Overlap

Headache is a frequent comorbidity in patients with TMDs, particularly tension-type headache and headache attributed to TMD as defined in the International Classification of Headache Disorders [57]. Pain may be localized to the temporal region or present as diffuse craniofacial discomfort and is often exacerbated by jaw activity or stress. Large cohort studies, including the OPPERA project, have demonstrated shared risk factors and overlapping pain processing mechanisms between TMDs and primary headache disorders [43, 58]. The recognition of headache as a potential manifestation of TMD has important diagnostic implications, as misattribution may delay appropriate management. Current guidelines recommend screening for TMD in patients presenting with recurrent temporal or preauricular headaches, particularly when associated with jaw dysfunction [57].

Otologic and Adjacent Symptoms

Symptoms mimicking otologic disorders are commonly reported by patients with TMDs and include ear pain, fullness, tinnitus, and, less frequently, subjective hearing changes [59]. These manifestations are thought to arise from shared embryologic origin, close anatomical proximity, and convergence of trigeminal and auditory pathways rather than direct ear pathology [60]. Importantly, otologic symptoms in TMDs often occur in the absence of identifiable middle or inner ear disease, despite normal clinical and audiological findings. A multidisciplinary evaluation is therefore strongly recommended to prevent unnecessary otologic interventions, enhance diagnostic accuracy, and ensure appropriate, patient-centered management strategies across specialties [59, 61].

Dental and Orofacial Referral Pain

TMD-related pain may be perceived as toothache or dental hypersensitivity, leading patients to seek dental treatment despite the absence of odontogenic pathology [62]. Referred pain patterns from masticatory muscles or the TMJ can closely mimic pulpal or periodontal pain, posing a significant diagnostic challenge. Structured diagnostic workflows emphasizing exclusion of dental causes and reproduction of pain with muscle or joint palpation are essential to prevent overtreatment [63].

Psychosocial and Central Modulation Features

Psychosocial factors, including stress, anxiety, depression, and pain catastrophizing, play a substantial role in the clinical expression of TMDs, particularly in chronic forms [45, 64]. Patients with elevated psychosocial distress often report greater pain intensity, wider pain distribution, and increased functional limitation. These features reflect altered central pain modulation rather than increased peripheral tissue damage [43, 49]. Contemporary models conceptualize chronic TMDs within a biopsychosocial framework, emphasizing the interaction between biological vulnerability, psychological state, and environmental stressors. Recognition of these dimensions is critical for accurate diagnosis, prognostic assessment, and individualized management planning [38, 65]. Given the substantial overlap between temporomandibular disorder symptoms and those of other orofacial, neurologic, and otologic conditions, individual clinical signs or symptoms should never be interpreted in isolation. Reliance on single features risks misdiagnosis and inappropriate management. A comprehensive clinical evaluation grounded in standardized diagnostic frameworks—most notably the Diagnostic Criteria for Temporomandibular Disorders (DC/TMD)—is therefore essential to ensure diagnostic accuracy, internal consistency, and reproducibility across both clinical practice and research settings [38, 39]. The coexistence of multiple features, including pain intensity and quality, functional limitation, joint sounds, and associated psychosocial distress, should prompt a structured, multidimensional assessment. Such an approach moves beyond a reductionist, structure-based diagnosis and instead acknowledges the complex biopsychosocial nature of TMD, supporting more precise classification and individualized management strategies.

6.4 Management of Temporomandibular Disorders

Management of temporomandibular disorders represents a complex clinical challenge due to the heterogeneity of etiological mechanisms, symptom presentations, and psychosocial modifiers involved in disease expression. Contemporary treatment paradigms emphasize that pain relief and restoration of function should constitute the primary therapeutic objectives during the initial stages of care, with interventions tailored to individual patient profiles and guided by a stepwise, evidence-based framework. Management strategies for temporomandibular disorders are commonly organized within stepped-care frameworks, emphasizing conservative, pharmacological, procedural, and surgical approaches (see Table 6.2).

Principles of Conservative Management

Conservative and reversible therapies form the cornerstone of TMD management and are recommended as first-line interventions by multiple professional organizations and

Table 6.2 Stepped-Care Management Strategies for Temporomandibular Disorders: Conservative, Pharmacological, Procedural, and Surgical Approaches.

Category	Description	Examples
✅ **Self-Help (Start Here)**	Easy, non-invasive treatments to try first.	- Soft foods 🥣 - Heat/cold therapy ❄️🔥 - Jaw exercises 👤 - OTC NSAIDs 💊 - Self-management ☐
🔷 **Doctor-Prescribed (Next Step)**	If self-help isn't enough, seek professional guidance.	- Behavioral therapy 🧠 - Physical therapy 💪 - Prescription medications 💊 Intraoral appliances 🦷 - Complementary treatments 🤝
⚠ **Procedures (Use Caution)**	Involve minor interventions with needles or instruments.	- Prolotherapy 💉 - Arthrocentesis 🦠 - Arthroscopy 🔬
🚨 **Surgery (Last Resort)**	For severe cases only, as recommended by a specialist.	- TMJ implants 🦿
❌ **What's NOT Recommended**	Interventions lacking strong evidence or potentially harmful.	- Botulinum toxin (Botox) 💉 → No strong evidence it hclps TMD

consensus panels. These approaches aim to reduce pain, limit functional impairment, and address contributing biomechanical and behavioral factors while minimizing the risk of iatrogenic harm [12, 66, 67]. Self-management strategies—including patient education, activity modification, avoidance of parafunctional habits, application of heat or cold therapy, and structured jaw exercises—have demonstrated benefit in reducing symptom severity and improving patient-reported outcomes, particularly in early or mild disease [68].

Occlusal appliances, such as stabilization splints, are frequently employed to redistribute occlusal forces, reduce muscle hyperactivity, and protect joint structures during nocturnal parafunction. Although their precise mechanism of action remains incompletely understood, systematic reviews suggest that splint therapy may provide short- to medium-term pain relief in selected patients when combined with education and behavioral counseling [69]. Importantly, irreversible occlusal adjustments are no longer recommended as routine therapy, given the lack of robust evidence supporting long-term benefit and the potential for permanent alteration of dental structures [70].

Physical therapy interventions play a central role in conservative care, particularly for patients with predominant myogenous pain or restricted mandibular mobility. Techniques such as manual therapy, stretching, postural correction, and neuromuscular re-education aim to restore normal movement patterns and reduce muscular overload. Randomized controlled trials have demonstrated that individualized physiotherapy programs can significantly improve pain intensity, jaw function, and quality of life, especially when integrated into interdisciplinary treatment models [71].

This approach reflects a shift away from irreversible or invasive therapies toward conservative, multimodal strategies grounded in biological plausibility, clinical effectiveness, and patient safety.

Pharmacological Treatment Patterns and Evidence

Pharmacological therapy is commonly employed as an adjunct to conservative measures, particularly during periods of acute exacerbation or persistent pain. Data from large observational cohorts provide valuable insight into real-world prescribing patterns and patient exposure to various medication classes.

A National Institute of Dental and Craniofacial Research (NIDCR)–supported registry and related cohort studies involving more than 1500 individuals with chronic TMD symptoms revealed extensive use of pharmacological agents across multiple classes. Within this population, 73% of patients reported the use of anti-inflammatory drugs, 56% nonprescription analgesics, 50% antidepressants, 48% opioids, 41% anxiolytics, and 40% muscle relaxants [66] (see Table 6.3).

These findings highlight both the substantial symptom burden associated with chronic TMDs and the frequent reliance on systemic medications to achieve pain control. NSAIDs and acetaminophen remain the most commonly recommended agents for short-term management of inflammatory pain, supported by moderate evidence of efficacy in acute musculoskeletal conditions [72]. However, their long-term use is limited by gastrointestinal, renal, and cardiovascular risks, particularly in older or medically complex patients.

Antidepressants, including tricyclic antidepressants and serotonin–norepinephrine reuptake inhibitors, are frequently prescribed to address chronic pain mechanisms, sleep disturbance, and comorbid mood disorders. Their analgesic effects are thought to result from modulation of descending inhibitory pain pathways rather than direct anti-inflammatory action [73]. Similarly, anxiolytic agents may provide short-term benefit in patients with significant anxiety-related muscle tension, although their use should be carefully monitored due to risks of dependence and cognitive impairment [74]. Opioid analgesics, despite historically high utilization rates, are increasingly recognized as suboptimal for chronic TMD pain due to limited evidence of sustained benefit and substantial risk of adverse outcomes. Contemporary guidelines strongly discourage routine opioid use for TMDs, emphasizing that these agents should be reserved, if at all, for short-term rescue therapy in exceptional circumstances [75]. The observed prevalence of opioid exposure in registry-based cohorts underscores the importance of continued efforts to align clinical practice with evolving pain-management standards.

Interdisciplinary and Biopsychosocial Approaches

Growing recognition of the biopsychosocial nature of TMDs has driven the adoption of interdisciplinary treatment models that integrate dental, medical, psychological,

Table 6.3 Pharmacological treatment patterns and clinical implications in chronic temporomandibular disorders [66]

Medication class	Proportion of patients (%)	Clinical role in chronic TMD	Key considerations and limitations
Anti-inflammatory drugs (NSAIDs)	73	First-line agents for short-term control of inflammatory and musculoskeletal pain	Moderate evidence for acute pain relief; long-term use is limited by gastrointestinal, renal, and cardiovascular adverse effects, especially in older or medically complex patients.
Non-prescription analgesics (e.g., acetaminophen)	56	Adjunctive or initial therapy for mild to moderate pain	Generally well tolerated at therapeutic doses; limited efficacy in chronic pain states when used alone.
Antidepressants	50	Management of centralized pain, comorbid mood disorders, and sleep disturbance	Reflects high psychosocial and central sensitization burden; delayed onset of analgesic benefit.
Opioids	48	Short-term or rescue analgesia in severe or refractory pain	High reliance despite limited evidence for chronic TMD; risks include dependence, tolerance, and opioid-induced hyperalgesia.
Anxiolytics	41	Reduction of anxiety-related pain amplification and muscle tension	Potential for sedation, cognitive effects, and dependency with prolonged use.
Muscle relaxants	40	Treatment of masticatory muscle hyperactivity and myofascial pain	Evidence is mixed; typically recommended for short-term use due to central nervous system adverse effects.

Treatment categories and examples are adapted from clinical guidelines and patient-reported treatment utilization data supported by the National Institute of Dental and Craniofacial Research (NIDCR). Percentages reflect medication exposure reported in large registry-based cohorts of patients with chronic temporomandibular disorders and are intended to illustrate real-world treatment patterns rather than comparative therapeutic efficacy [66]

and rehabilitative care. Psychological factors—including stress, catastrophizing, depression, and maladaptive coping—are well-established contributors to pain persistence and disability in TMD populations [76]. Cognitive-behavioral therapy (CBT) and other psychotherapeutic interventions have demonstrated efficacy in reducing pain-related distress and improving functional outcomes, particularly when combined with physical and dental therapies [77]. Interdisciplinary care models aim to address peripheral nociceptive drivers while simultaneously targeting central sensitization and psychosocial vulnerability.

Longitudinal studies suggest that patients managed within such frameworks experience greater symptom improvement and reduced healthcare utilization compared with those receiving isolated, discipline-specific interventions [78]. These findings reinforce the importance of coordinated care pathways and early identification of psychosocial risk factors.

Imaging in Diagnosis and Treatment Planning

Imaging plays a critical role in the diagnostic evaluation of patients with suspected intra-articular TMDs, particularly when symptoms are refractory to conservative management or when structural pathology is suspected. MRI is widely regarded as the reference standard for assessing soft-tissue components of the temporomandibular joint, including disc position, disc morphology, joint effusion, and synovial inflammation [79]. CBCT, in contrast, provides superior visualization of osseous structures, allowing detailed assessment of condylar remodeling, cortical erosion, osteophyte formation, and joint space narrowing [80]. Recent advances in image fusion techniques have enabled the integration of MRI and CBCT data, providing a comprehensive representation of both soft- and hard-tissue pathology. Studies evaluating CBCT–MRI fused images demonstrate improved diagnostic accuracy for anterior disc displacement, degenerative joint disease, and combined pathologies compared with either modality alone [73, 74]. This integrated approach facilitates more precise treatment planning and may reduce unnecessary invasive interventions by improving diagnostic confidence.

Minimally Invasive and Surgical Interventions

When conservative and pharmacological therapies fail to provide adequate relief, minimally invasive procedures such as arthrocentesis and arthroscopy may be considered. Arthrocentesis involves lavage of the joint space to remove inflammatory mediators and improve joint mobility, while arthroscopy allows direct visualization and limited surgical intervention within the joint. Systematic reviews suggest that these procedures can provide symptom relief in selected patients, particularly those with internal derangement or inflammatory joint conditions [75, 81]. Open surgical procedures—including disc repositioning, meniscectomy, condylotomy, and total joint replacement—are reserved for severe, refractory cases characterized by advanced structural pathology and significant functional impairment. Importantly, evidence indicates that invasive surgical techniques may worsen symptoms or lead to irreversible joint damage if applied indiscriminately [77]. As a result, the overall frequency of open TMJ surgery has declined, with greater emphasis placed on conservative management and careful patient selection.

Professional guidelines consistently recommend a conservative, nonsurgical approach as the foundation of TMD management. The American Association of Oral and Maxillofacial Surgeons and related expert bodies advise that surgical intervention should be considered only after exhaustive nonsurgical measures have failed and when clear structural indications are present [78]. Epidemiological estimates suggest that no more than 20% of individuals with TMDs ultimately require surgical treatment, underscoring the effectiveness of conservative and minimally invasive strategies for the majority of patients [66, 82].

References

1. Sharav Y, Benoliel R. Orofacial pain and headache. Elsevier; 2003. https://doi.org/10.1016/B978-0-443-07201-7.50017-5.
2. Zieliński G, Pająk-Zielińska B, Ginszt M. A meta-analysis of the global prevalence of temporomandibular disorders. J Clin Med. 2024;13(5):1365 https://doi.org/10.3390/jcm13051365.
3. Maspero C, Segù M, Abate A, Lanteri V. Correlation between temporomandibular pain and gender. Appl Sci. 2023;13(5):3077 https://doi.org/10.3390/app13053077.
4. Ohrbach R, Dworkin SF. The evolution of TMD diagnosis: past, present, future. J Dent Res. 2022;101(8):907–14. https://doi.org/10.1177/00220345221088646.
5. Manfredini D, Guarda-Nardini L, Winocur E, Piccotti F, Ahlberg J, Lobbezoo F. Research diagnostic criteria for temporomandibular disorders: a systematic review. J Oral Rehabil. 2020;47(2):190–201. https://doi.org/10.1111/joor.12888.
6. Elsayed ADA, Taema MA, Elkhashab MA, Aziz EA, Radi IAW. A cross-sectional study of temporomandibular disorders among postgraduate students. Int J Prosthodont. 2024;37(5):495–504. https://doi.org/10.11607/ijp.8716.
7. Adegbiji WA, et al. Otological manifestation of temporomandibular joint disorder in Ekiti, a sub-Saharan African country. J Int Med Res. 2021;49(2):03000605219965l7. https://doi.org/10.1177/03000605219965l7.
8. Diab A. Prevalence and severity of temporomandibular disorders among Egyptian postgraduate students: a cross-sectional study. Adv Dent J. 2023;5(1):166–73
9. Hanna K, et al. Temporomandibular dysfunction experience is associated with oral health-related quality of life: an Australian national study. BMC Oral Health. 2021;21(1):432. https://doi.org/10.1186/s12903-021-01773-z.
10. Lung J, Bell L, et al. Prevalence of temporomandibular disorders among a cohort of university undergraduates in Australia. J Investig Clin Dent. 2018;9:e12341 https://doi.org/10.1111/jicd.12341.
11. Manfredini D, Lombardo L, Siciliani G. Temporomandibular disorders and dental occlusion. A systematic review and meta-analysis of epidemiological studies. J Oral Rehabil. 2025;52(1):45–62. https://doi.org/10.1111/joor.13492.
12. National Academies of Sciences, Engineering, and Medicine. Temporomandibular disorders: priorities for research and care. Washington, DC: National Academies Press; 2020. https://doi.org/10.17226/25652.
13. Slade GD, Ohrbach R, Greenspan JD, et al. Painful temporomandibular disorder: incidence, natural history, and risk factors in the OPPERA study. J Pain. 2013;14(12 Suppl):T1–20. https://doi.org/10.1016/j.jpain.2013.09.009.
14. Slade GD, Ohrbach R, Greenspan JD, et al. Painful temporomandibular disorder: decade of discovery from OPPERA studies. J Dent Res. 2021;100(12):1233–42. https://doi.org/10.1177/00220345211018420.
15. Fillingim RB, King CD, Ribeiro-Dasilva MC, Rahim-Williams B, Riley JL. Sex, gender, and pain: a review of recent clinical and experimental findings. J Pain. 2021;22(1):1–16. https://doi.org/10.1016/j.jpain.2020.08.001.
16. Warren MP, Fried JL. Temporomandibular disorders and hormones in women. Cells Tissues Organs. 2020;209(1):68–78. https://doi.org/10.1159/000505450.
17. Poveda-Roda R, Bagan JV, Jiménez-Soriano Y, Gallud-Ruiz L. Review of temporomandibular joint pathology. Med Oral Patol Oral Cir Bucal. 2022;27(2):e125–35. https://doi.org/10.4317/medoral.25114.

18. Velly AM, Mohit S. Epidemiology of pain and pain-related disability in temporomandibular disorders. J Orofacial Pain Headache. 2021;35(1):10–9. https://doi.org/10.11607/ofph.2563.
19. Ashina S, Bendtsen L, Lyngberg AC, Lipton RB, Jensen R. Chronic overlapping pain conditions. Nat Rev Dis Primers. 2024;10:12 https://doi.org/10.1038/s41572-024-00471-0.
20. Ohrbach R, Slade GD, Bair E, et al. Psychological factors and TMD onset. Pain. 2020;161(6):1183–93. https://doi.org/10.1097/j.pain.0000000000001824.
21. Greenspan JD, Slade GD, Bair E, et al. Pain sensitivity and TMD risk. Pain. 2020;161(8):1729–40. https://doi.org/10.1097/j.pain.0000000000001870.
22. Manfredini D, Winocur E, Guarda-Nardini L, Paesani D, Lobbezoo F. Epidemiology of bruxism and its association with TMD. J Oral Rehabil. 2021;48(2):161–9. https://doi.org/10.1111/joor.13095.
23. Schellhas KP, Pollei SR, Wilkes CH. Pediatric internal derangements of the temporomandibular joint: effect on facial development. Am J Orthod Dentofac Orthoped. 1993;104(1):51–9. https://doi.org/10.1016/0889-5406(93)70027-L.
24. Wilkes CH. Internal derangements of the temporomandibular joint: pathologic variations. Oral Maxillofac Surg Clinics North Am. 2020;32(1):1–12. https://doi.org/10.1016/j.coms.2019.08.002.
25. Attanasio R. Management of temporomandibular disorders and occlusion. Elsevier; 2003. https://doi.org/10.1016/S1059-941X(03)00131-1.
26. Manfredini D, Winocur E, Guarda-Nardini L, Lobbezoo F, Selfe TK. Epidemiology of bruxism in adults: a systematic review. J Oral Rehabil. 2021;48(5):537–49. https://doi.org/10.1111/joor.13143.
27. Reiter S, Goldsmith C, Emodi-Perlman A, Friedman-Rubin P, Winocur E. Masticatory muscle disorders diagnostic criteria: a comparison of classification systems. J Oral Rehabil. 2012;39(12):941–7. https://doi.org/10.1111/j.1365-2842.2012.02337.x.
28. Peck CC, Murray GM, Gerzina TM. How does pain affect jaw muscle activity?. J Oral Rehabil. 2020;47(10):1225–34. https://doi.org/10.1111/joor.13017.
29. Flor H, Birbaumer N, Schulte W, Roos R. Stress-related electromyographic responses in patients with chronic temporomandibular pain. Pain. 1991;46(2):145–52. https://doi.org/10.1016/0304-3959(91)90069-A.
30. Fillingim RB, Slade GD, Diatchenko L, Dubner R, Greenspan JD, Knott C, et al. Summary of OPPERA findings. J Pain. 2018;19(5):450–64. https://doi.org/10.1016/j.jpain.2017.12.008.
31. Pasinato F, Souza JA, Corrêa ECR, da Silva AMT. Temporomandibular disorder and generalized joint hypermobility. Braz J Otorhinolaryngol. 2011;77(4):418–25. https://doi.org/10.1590/S1808-86942011000400003.
32. Diatchenko L, Slade GD, Nackley AG, et al. Genetic basis for individual variations in pain perception. Hum Mol Gen. 2005;14(1):135–43. https://doi.org/10.1093/hmg/ddi013.
33. Smith SB, Maixner DW, Greenspan JD, et al. Potential genetic risk factors for chronic TMD. Pain. 2019;160(6):1284–96. https://doi.org/10.1097/j.pain.0000000000001517.
34. Passatore M, Roatta S. Modulation operated by the sympathetic nervous system on jaw reflexes. Arch Oral Biol. 2007;52(4):343–6. https://doi.org/10.1016/j.archoralbio.2006.11.013.
35. Fan W, Zhu X, He Y, et al. Peripheral sympathetic mechanisms in orofacial pain. J Pain Res. 2018;11:2425–31. https://doi.org/10.2147/JPR.S179327.
36. Monaco A, Ortu E, Giannoni M, et al. Standard correction of vision worsens EMG activity of pericranial muscles in chronic TMD subjects. Pain Res Manage. 2020;2020:3932476 https://doi.org/10.1155/2020/3932476.
37. Bachetti T, Segù M, Collesano V, Dalessandri D, Lanteri V, Migliorati M. Ocular correction effects on EMG activity of stomatognathic muscles in children. Eur J Paediatr Dent. 2006;7(2):81–5.
38. Schiffman E, Ohrbach R, Truelove E, et al. Diagnostic Criteria for Temporomandibular Disorders (DC/TMD) for clinical and research applications. J Oral Facial Pain Headache. 2014;28(1):6–27. https://doi.org/10.11607/jop.1151.
39. Ohrbach R, Dworkin SF. The evolution of TMD diagnosis: past, present, future. J Dent Res. 2022;101(9):1033–40. https://doi.org/10.1177/00220345221104062.

40. Manfredini D, Lobbezoo F. Role of psychosocial factors in the etiology of temporomandibular disorders. J Oral Rehabil. 2021;48(9):1126–35. https://doi.org/10.1111/joor.13186.
41. Slade GD, Ohrbach R, Greenspan JD, et al. Painful temporomandibular disorder: incidence and persistence. J Pain. 2016;17(9):988–1000. https://doi.org/10.1016/j.jpain.2016.06.003.
42. Fillingim RB, Slade GD, Greenspan JD, et al. Long-term changes in pain sensitivity in TMD. Pain. 2020;161(7):1529–37. https://doi.org/10.1097/j.pain.0000000000001867.
43. Greenspan JD, Slade GD, Bair E, et al. Central pain processing in TMD: insights from OPPERA. Pain. 2021;162(1):53–64. https://doi.org/10.1097/j.pain.0000000000002014.
44. Peck CC, Goulet JP, Lobbezoo F, et al. Expanding the taxonomy of the diagnostic criteria for TMD. J Oral Rehabil. 2014;41(1):2–23. https://doi.org/10.1111/joor.12132.
45. Ohrbach R, Durham J. Biopsychosocial aspects of orofacial pain. Dent Clin North Am. 2020;64(2):233–47. https://doi.org/10.1016/j.cden.2019.11.003.
46. Visscher CM, Lobbezoo F, Naeije M. Muscle pain in TMD: clinical and pathophysiologic aspects. J Oral Rehabil. 2021;48(8):935–45. https://doi.org/10.1111/joor.13163.
47. Svensson P, Graven-Nielsen T. Craniofacial muscle pain: mechanisms and clinical implications. J Oral Rehabil. 2020;47(7):904–14. https://doi.org/10.1111/joor.12977.
48. La Touche R, Fernández-de-las-Peñas C. Neck pain and temporomandibular disorders. J Oral Rehabil. 2021;48(7):735–46. https://doi.org/10.1111/joor.13152.
49. Bair E, Brownstein NC, Ohrbach R, et al. Overlapping chronic pain conditions in TMD. Pain. 2022;163(1):e36–46. https://doi.org/10.1097/j.pain.0000000000002312.
50. Manfredini D, Bucci MB, Guarda-Nardini L. Temporomandibular joint sounds. J Oral Rehabil. 2020;47(9):1119–26. https://doi.org/10.1111/joor.13005.
51. Ahmad M, Schiffman EL. Temporomandibular joint disorders and imaging. Dent Clin North Am. 2021;65(1):1–23. https://doi.org/10.1016/j.cden.2020.08.001.
52. Kaur J, Kaur H. Disc displacement without reduction: clinical characteristics. J Oral Maxillofac Surg. 2022;80(4):703–11. https://doi.org/10.1016/j.joms.2021.11.021.
53. Durham J, Newton-John TRO. Functional impairment in TMD. J Oral Rehabil. 2020;47(9):1067–75. https://doi.org/10.1111/joor.13007.
54. Michelotti A, Iodice G. The role of occlusion in TMD. J Oral Rehabil. 2019;46(7):596–605. https://doi.org/10.1111/joor.12792.
55. Türp JC, Greene CS. Occlusion and TMD: evidence-based perspective. J Oral Rehabil. 2022;49(3):331–9. https://doi.org/10.1111/joor.13260.
56. Ohrbach R, Larsson P, List T. The jaw functional limitation scale. J Orofac Pain. 2008;22(3):219–30.
57. Headache Classification Committee of the IHS. ICHD-3. Cephalalgia. 2018;38(1):1–211. https://doi.org/10.1177/0333102417738202.
58. Slade GD, Greenspan JD, Fillingim RB, et al. Pain modulation and risk for TMD. Pain. 2020;161(9):2120–30. https://doi.org/10.1097/j.pain.0000000000001905.
59. Porto De Toledo I, Stefani FM, Porporatti AL, et al. Otologic symptoms and TMD. J Oral Rehabil. 2021;48(5):549–59. https://doi.org/10.1111/joor.13123.
60. Shore SE, Zhou J. Somatosensory influence on auditory perception. J Neurosci. 2021;41(7):1415–25. https://doi.org/10.1523/JNEUROSCI.1800-20.2020.
61. Ward BK, Wenzel A. Multidisciplinary assessment of otologic symptoms. Otol Neurotol. 2022;43(3):e232–39. https://doi.org/10.1097/MAO.0000000000003456.
62. Renton T. Dental pain of non-odontogenic origin. Br Dent J. 2020;228(9):667–75. https://doi.org/10.1038/s41415-020-1470-4.
63. Nixdorf DR, Moana-Filho EJ. Persistent dentoalveolar pain disorders. Pain. 2019;160(3):589–99. https://doi.org/10.1097/j.pain.0000000000001437.
64. Fillingim RB, Ohrbach R. Psychological contributions to chronic TMD pain. Dent Clin North Am. 2021;65(1):145–61. https://doi.org/10.1016/j.cden.2020.08.006.
65. Dworkin SF. Temporomandibular disorders: a biopsychosocial perspective. Clin J Pain. 2020;36(5):355–62. https://doi.org/10.1097/AJP.0000000000000817.

66. Ohrbach R, Slade GD, Bair E, et al. Clinical orofacial characteristics associated with risk of first-onset TMD: the OPPERA prospective cohort study. J Pain. 2013;14(12 Suppl):T33–50. https://doi.org/10.1016/j.jpain.2013.07.018.
67. Schiffman E, Ohrbach R, Truelove E, et al. Diagnostic criteria for temporomandibular disorders (DC/TMD) for clinical and research applications: Recommendations of the International RDC/TMD Consortium Network and Orofacial Pain Special Interest Group. J Oral Facial Pain Headache. 2014;28(1):6–27. https://doi.org/10.11607/jop.1151.
68. Greene CS. Managing the care of patients with temporomandibular disorders: a new guideline for care. J Am Dent Assoc. 2010;141(9):1086–8. https://doi.org/10.14219/jada.archive.2010.0335.
69. Al-Ani Z, Gray RJ, Davies SJ, Sloan P, Glenny AM. Stabilisation splint therapy for temporomandibular pain dysfunction syndrome. Cochrane Database Syst Rev. 2004;1:CD002778. https://doi.org/10.1002/14651858.CD002778.pub2.
70. Manfredini D, Lombardo L, Siciliani G. Temporomandibular disorders and dental occlusion. J Oral Rehabil. 2017;44(6):448–61. https://doi.org/10.1111/joor.12504.
71. Armijo-Olivo S, et al. Effectiveness of manual therapy and therapeutic exercise for temporomandibular disorders. Phys Ther. 2016;96(1):9–25. https://doi.org/10.2522/ptj.20140548.
72. Derry S, et al. Single dose oral NSAIDs for acute postoperative pain. Cochrane Database Syst Rev. 2015;7:CD008659. https://doi.org/10.1002/14651858.CD008659.pub3.
73. Wang YH, et al. Diagnostic efficacy of CBCT, MRI and CBCT–MRI fused images. Dentomaxillofac Radiol. 2022;51(2):20210286 https://doi.org/10.1259/dmfr.20210286.
74. El Shennawy EM, et al. Diagnostic accuracy of MRI–CBCT fused images. Oral Radiol. 2024;40(2):226–33. https://doi.org/10.1007/s11282-023-00727-1.
75. Guo C, Shi Z, Revington P. Arthrocentesis and lavage for TMJ disorders. Cochrane Database Syst Rev. 2015;12:CD004973. https://doi.org/10.1002/14651858.CD004973.pub3.
76. Fillingim RB, et al. Psychosocial influences on TMD pain. Pain. 2011;152(9):2031–7. https://doi.org/10.1016/j.pain.2011.04.028.
77. Al-Moraissi EA, Ellis E. Surgical management of TMJ internal derangement. J Oral Maxillofac Surg. 2020;78(10):1733–48. https://doi.org/10.1016/j.joms.2020.05.028.
78. American Association of Oral and Maxillofacial Surgeons. Clinical guidelines for TMJ disorders. J Oral Maxillofac Surg. 2017;75(4):927–40. https://doi.org/10.1016/j.joms.2017.01.023.
79. Larheim TA, Westesson PL. Maxillofacial imaging: temporomandibular joint. Radiology. 2001;218(3):621–9. https://doi.org/10.1148/radiology.218.3.r01mr17621.
80. Ahmad M, Hollender L, Anderson Q, et al. Research diagnostic criteria for temporomandibular disorders (RDC/TMD): development of image analysis criteria and examiner reliability for image analysis. Oral Surg Oral Med Oral Pathol Oral Radiol Endod. 2009;107(6):844–60. https://doi.org/10.1016/j.tripleo.2009.02.023.
81. Al-Moraissi EA, Ellis E. Arthroscopy versus arthrocentesis for the treatment of internal derangement of the temporomandibular joint: a systematic review and meta-analysis. J Oral Maxillofac Surg. 2015;73(5):934–44. https://doi.org/10.1016/j.joms.2014.12.010.
82. De La Torre Canales G, Câmara-Souza MB, Muñoz Lora VR, et al. Prevalence of psychosocial impairment in temporomandibular disorder patients. J Oral Rehabil. 2020;47(2):152–8. https://doi.org/10.1111/joor.1287.

7 Bell's Palsy

Abstract

Bell's palsy is an acute idiopathic peripheral facial nerve palsy and the most common cause of sudden unilateral lower motor neuron facial weakness. It typically reaches maximal severity within 48–72 h and presents with ipsilateral facial asymmetry, impaired forehead movement, incomplete eye closure, drooping of the mouth corner, and variable non-motor symptoms such as retroauricular pain, taste disturbance, hyperacusis, and lacrimal dysfunction. Because no single diagnostic biomarker exists, diagnosis remains primarily clinical and requires exclusion of secondary causes, including stroke, infection, neoplasm, trauma, otologic disease, and systemic neurological or inflammatory disorders. This chapter examines the epidemiology of Bell's palsy, including incidence, age and sex distribution, pregnancy-related risk, metabolic comorbidity, seasonal variation, and geographic differences. It also presents clinically relevant classification frameworks based on severity, temporal course, and associated conditions. Contemporary concepts of etiopathogenesis are reviewed, emphasizing a multifactorial model in which viral reactivation, immune-mediated inflammation, microvascular susceptibility, and edema of the facial nerve within the fallopian canal converge to produce nerve dysfunction. The chapter further summarizes the characteristic clinical features, prognostic indicators, and evidence-based management of Bell's palsy, highlighting the importance of early corticosteroid therapy, ocular protection, and rehabilitation in optimizing recovery and limiting long-term sequelae such as persistent weakness and synkinesis.

Keywords

Bell's palsy · Peripheral facial nerve palsy · Facial paralysis · Facial nerve · House–Brackmann grading · Retroauricular pain · Synkinesis · Corticosteroid therapy · Facial rehabilitation · Neurostomatology ·

A. Kalanov, *Neurostomatology*, https://doi.org/10.1007/978-981-92-2552-1_7

Definition. Bell's palsy (BP) is an acute disorder of the peripheral facial nerve characterized by the sudden onset of unilateral facial muscle weakness or paralysis of unknown cause. The condition affects the seventh cranial nerve distal to its nucleus, resulting in impaired voluntary and involuntary movements of the muscles of facial expression on the affected side. Clinically, Bell's palsy presents with facial asymmetry, drooping of the mouth corner, flattening of the nasolabial fold, and incomplete eye closure, often accompanied by secondary functional impairments such as difficulty with speech articulation, mastication, and ocular protection. Sensory complaints, including altered taste sensation and hyperacusis, may also be reported. In most cases, symptom onset is rapid, with maximal weakness developing within 48–72 h. The condition is generally self-limiting, and spontaneous recovery occurs in a substantial proportion of patients within weeks to months [1].

Bell's palsy represents the most common cause of acute peripheral facial paralysis, accounting for approximately 60%–75% of unilateral facial nerve palsies encountered in clinical practice [2]. Because no definitive laboratory or imaging marker exists, the diagnosis remains primarily clinical and is established after exclusion of secondary causes such as cerebrovascular disease, neoplasms, trauma, otologic infections, Lyme disease, and systemic inflammatory or demyelinating disorders [3]. The idiopathic nature of Bell's palsy continues to distinguish it from other facial neuropathies and underscores the importance of careful diagnostic evaluation.

7.1 Epidemiology and Incidence

The annual global incidence of Bell's palsy is estimated to range from 15 to 30 cases per 100,000 population, although more recent population-based studies have reported higher rates in certain regions, approaching or exceeding 35 cases per 100,000 [4]. The disorder affects individuals of all ages but demonstrates a clear predilection for young and middle-aged adults. The highest incidence is consistently observed among individuals aged 15–45 years, with a secondary rise in older populations, likely reflecting the increasing prevalence of metabolic and vascular comorbidities [5].

Sex distribution in Bell's palsy is generally balanced, although several contemporary studies have reported a slight female predominance, particularly in association with pregnancy-related cases [6]. Pregnancy, especially during the third trimester and early postpartum period, is a recognized risk factor, with pregnant women demonstrating a two- to fourfold increased risk compared with nonpregnant women of similar age [7]. Metabolic conditions, most notably diabetes mellitus, are also strongly associated with Bell's palsy and are linked to both increased incidence and poorer functional recovery [8]. Hypertension, obesity, and dyslipidemia have likewise been implicated as contributory risk factors in recent epidemiological analyses [9].

7.2 Seasonal Variation

Recent evidence confirms seasonal variation in Bell's palsy incidence. Multiple retrospective cohort studies and national registry analyses have demonstrated a statistically significant increase in new-onset cases during winter and early spring months, with a relative decline during summer periods [10]. This seasonal pattern has been reported across diverse geographic regions, including East Asia, Northern Europe, and North America, suggesting a reproducible epidemiological trend rather than a region-specific phenomenon [11]. Several mechanisms have been proposed to explain this seasonal fluctuation. Increased circulation of viral upper respiratory infections during colder months, seasonal immune modulation, reduced sunlight exposure, and cold-induced vasomotor changes have all been suggested as contributing factors [12]. Environmental cold stress may worsen subclinical facial nerve vulnerability, increasing acute dysfunction risk. Importantly, seasonal variation appears to persist even after adjustment for age, sex, and major comorbidities, reinforcing its potential clinical relevance [13]. Seasonal effects vary by climate, with tropical regions often showing little or no seasonality, implying temperature variability matters more than absolute temperature [14]. These findings highlight the complex interaction between environmental, biological, and population-specific factors in Bell's palsy epidemiology.

7.3 Clinical Classification of Bell's Palsy

Bell's palsy can be classified using several clinically meaningful frameworks that assist in diagnosis, prognostication, and therapeutic decision-making.

Classification by Temporal Course

Based on the temporal pattern of symptom onset, progression, and recovery, Bell's palsy can be classified into several clinical subtypes. Acute Bell's palsy is characterized by a sudden onset of unilateral facial weakness with rapid progression, typically reaching maximal severity within 72 h. In contrast, subacute Bell's palsy demonstrates a more gradual evolution of symptoms over several days, with slower progression to peak facial dysfunction.

A subset of patients experience recurrent Bell's palsy, which is reported in approximately 7%–15% of cases. Recurrences may occur on the same (ipsilateral) or opposite (contralateral) side and can present after variable symptom-free intervals. Finally, persistent Bell's palsy refers to cases in which complete functional recovery is not achieved within 6 months after onset, often accompanied by residual weakness, synkinesis, or facial asymmetry. Both recurrent and persistent forms of Bell's palsy require careful clinical reassessment, as they may indicate alternative or underlying pathologies such as neoplastic, inflammatory, infectious, or systemic neurological disorders [15]. The temporal classification of Bell's palsy based on symptom onset, progression, and recovery is summarized in Table 7.1.

Classification by Associated Conditions

Although idiopathic Bell's palsy accounts for the majority of cases, clinical practice and epidemiological studies often subclassify the condition according to associated

Table 7.1 Classification of Bell's palsy according to temporal course

Associated condition	Typical clinical context	Proposed contributing mechanisms	Clinical considerations
Pregnancy-associated	Third trimester or early postpartum period	Hormonal changes, fluid retention, microvascular ischemia	Often more severe; recovery may be slower
Diabetes-associated	Patients with diabetes mellitus	Microangiopathy, metabolic nerve vulnerability	Higher incidence; delayed recovery possible
Post-infectious	Recent viral illness (e.g., HSV-1)	Inflammatory edema, viral-induced nerve injury	Responds well to early corticosteroids
Post-vaccination	Temporal proximity to vaccination	Likely coincidental; background incidence	No proven causal relationship; reassurance essential

physiological or systemic factors. This approach is useful for risk stratification, prognostic assessment, and clinical counseling.

Pregnancy-associated Bell's palsy most frequently occurs during the third trimester or early postpartum period and has been linked to fluid retention, hormonal changes, and microvascular ischemia of the facial nerve. Outcomes may be less favorable compared with nonpregnant populations, underscoring the need for early recognition and management.

Diabetes-associated Bell's palsy reflects the role of metabolic dysfunction and microangiopathy in facial nerve vulnerability. Patients with diabetes mellitus have a higher incidence of facial nerve palsy and may experience delayed recovery, particularly in the presence of poor glycemic control.

Post-infectious Bell's palsy is commonly associated with viral infections, particularly herpes simplex virus type 1, and is supported by clinical, serological, and imaging findings suggesting inflammatory edema and nerve compression within the facial canal. Post-vaccination Bell's palsy has been the subject of increased scrutiny in recent years. Pharmacovigilance analyses have investigated temporal associations between vaccination—most notably COVID-19 vaccines—and the onset of facial paralysis.

However, large-scale surveillance data indicate that reported incidence rates do not exceed expected background levels in the general population, supporting the conclusion that vaccination does not confer a causal increased risk of Bell's palsy [16]. These associated conditions and their key clinical features are summarized in Table 7.2.

Classification by Severity

The **House–Brackmann Facial Nerve Grading System** remains the most widely used method for assessing facial nerve dysfunction (see Table 7.3). This system grades facial function on a six-point scale, ranging from Grade I (normal function) to Grade VI (complete paralysis). Severity at presentation is a key prognostic indicator, with higher grades associated with delayed or incomplete recovery [17].

Table 7.2 Classification of Bell's palsy based on associated conditions

Associated condition	Typical clinical context	Proposed contributing mechanisms	Clinical considerations
Pregnancy-associated	Third trimester or early postpartum period	Hormonal changes, fluid retention, microvascular ischemia	Often more severe; recovery may be slower
Diabetes-associated	Patients with diabetes mellitus	Microangiopathy, metabolic nerve vulnerability	Higher incidence; delayed recovery possible
Post-infectious	Recent viral illness (e.g., HSV-1)	Inflammatory edema, viral-induced nerve injury	Responds well to early corticosteroids
Post-vaccination	Temporal proximity to vaccination	Likely coincidental; background incidence	No proven causal relationship; reassurance essential

Table 7.3 Severity Grading of Bell's Palsy Using the Modified House–Brackmann System

Grade	Description	Key Characteristics	Facial Motion
🟢**I (Normal)**	Full facial function	No weakness, normal symmetry	All movements normal and strong
🟡**II (Mild Dysfunction)**	Slight weakness, subtle asymmetry	Normal at rest. may have mild synkinesis	**Forehead:** Good function **Eye:** Complete closure with minimal effort **Mouth:** Slight asymmetry
🟠**III (Moderate Dysfunction)**	Noticeable weakness, but not disfiguring	Normal at rest. may have mild synkinesis, contracture, or hemifacial spasm	**Forehead:** Slight to moderate movement **Eye:** Complete closure with effort **Mouth:** Slight weakness with maximal effort
🟠🔴**IV (Moderate to Severe Dysfunction)**	Obvious weakness, possible disfiguring asymmetry	Normal at rest, clear asymmetry during motion	**Forehead:** No movement **Eye:** Incomplete closure **Mouth:** Asymmetrical with effort
🔴**V (Severe Dysfunction)**	Barely perceptible motion, major asymmetry	Noticeable asymmetry at rest. severe dysfunction	**Forehead:** No movement **Eye:** Incomplete closure **Mouth:** Slight movement
🔴**VI (Total Paralysis)**	No facial movement	Complete loss of function	No movement in forehead, eye, or mouth

Grade I (Normal), shown with a green circle, represents full facial function in its entirety. Key characteristics include no weakness of any kind and facial symmetry that remains completely normal in every respect. Facial motion shows all movements across the face are entirely normal and strong, with full power and no limitations observed anywhere.

Grade II (Mild Dysfunction), shown with a yellow circle, involves slight weakness accompanied by subtle asymmetry. Key characteristics are that the face appears normal when at rest, although mild synkinesis may sometimes be present. In terms of facial motion, the forehead maintains good function, the eye achieves complete closure with only minimal effort required, and the mouth shows only slight asymmetry noticeable.

Grade III (Moderate Dysfunction), shown with an orange circle, features noticeable weakness, yet it is still not disfiguring overall. Key characteristics include that the face looks normal at rest, but mild synkinesis, contracture, or hemifacial spasm may occasionally appear. Regarding facial motion, the forehead has slight to moderate movement possible, the eye achieves complete closure, but it requires noticeable effort, and the mouth shows slight weakness even when the patient makes maximal effort.

Grade IV (Moderate to Severe Dysfunction), shown with an orange circle plus a red circle, is marked by obvious weakness that can possibly result in disfiguring asymmetry. Key characteristics are that the face remains normal when at rest; however, clear asymmetry becomes evident during any attempted motion. In facial motion, the forehead has no movement at all possible, the eye shows incomplete closure, and the mouth has movement that is asymmetrical even with effort.

Grade V (Severe Dysfunction), shown with a red circle, shows barely perceptible motion along with major asymmetry. Key characteristics include noticeable asymmetry visible even at rest, combined with severe overall dysfunction. For facial motion, the forehead has no movement possible, the eye shows incomplete closure, and the mouth has only slight movement that can be detected.

Grade VI (Total Paralysis), shown with a red circle, indicates no facial movement whatsoever. Key characteristics are a complete loss of function throughout the face. In terms of facial motion, no movement whatsoever is present in the forehead, eye, or mouth regions.

7.4 Geographic Distribution

Geographic variability in the incidence of Bell's palsy has been increasingly recognized. Higher incidence rates are generally reported in higher-latitude and colder-climate regions, particularly in areas with marked seasonal temperature changes [18]. Population-based studies from Scandinavia, Canada, and northern parts of East Asia consistently demonstrate higher incidence figures compared with studies from equatorial regions [19]. In addition to climate, population health characteristics appear to influence geographic differences. Regions with a higher prevalence of diabetes mellitus, metabolic syndrome, and cardiovascular risk factors tend to exhibit increased rates of Bell's palsy and worse functional outcomes [20]. Socioeconomic factors, access to early corticosteroid treatment, and differences in healthcare utilization may further contribute to observed regional disparities [21]. Urban–rural differences have also been described, with some studies reporting higher incidence in urban populations, potentially reflecting greater exposure to environmental stressors or improved case detection [22].

Etiopathogenesis of Bell's palsy. Numerous lines of evidence support immunological, infectious, and ischemic–edematous pathways in Bell's palsy (BP); however, "classical" BP remains a syndromic diagnosis defined as idiopathic acute peripheral facial palsy after exclusion of structural, infectious, neoplastic, and systemic causes. Contemporary models increasingly conceptualize BP not as a single disease entity, but as a final common pathway of facial nerve inflammation and edema occurring within the anatomically constrained fallopian canal. This process results in transient conduction block and, in more severe cases, axonal injury and delayed recovery. This conceptual shift is reinforced by modern clinical reviews and mechanistic summaries emphasizing acute inflammatory swelling, microvascular vulnerability, and frequent viral associations, even when a single causative pathogen cannot be identified in all patients [23].

Importantly, this framework accommodates the marked clinical heterogeneity observed in BP, including variability in severity at onset, response to corticosteroid therapy, and recovery trajectory. Rather than implying diagnostic uncertainty, the syndromic definition of BP reflects convergence of diverse upstream triggers on a shared downstream mechanism of facial nerve dysfunction.

7.5 Viral Reactivation as a Trigger: Host Response Rather than Virus Alone

Within this mechanistic framework, viral reactivation is best understood as a frequent initiating trigger rather than a universal cause, with host immune susceptibility and inflammatory response determining clinical expression and severity. The herpes simplex virus type 1 (HSV-1) reactivation hypothesis centered on the geniculate ganglion remains biologically plausible. Latent HSV-1 within sensory ganglia may reactivate under conditions of systemic stress or immune modulation, provoking localized neuritis, endothelial activation, and immune-cell recruitment around the facial nerve. Increasingly, HSV-1 is viewed not solely as a direct cytopathic agent, but as a trigger that initiates a disproportionate local immune response, culminating in endoneurial edema and entrapment neuropathy.

Contemporary discussions place HSV-1 alongside varicella-zoster virus and other neurotropic viruses within a broader herpes-family reactivation spectrum, in which the magnitude and regulation of host inflammatory responses shape the clinical phenotype [24].

Beyond HSV-1, the absence of definitive viral isolation in many patients likely reflects early viral clearance, "hit-and-run" immune activation, compartmentalization within the nerve or ganglia, or etiologic heterogeneity, with some cases being virally triggered and others driven by noninfectious inflammatory mechanisms. Recent clinical syntheses of peripheral facial palsy consistently list viral triggers among the most biologically plausible initiating factors, while emphasizing that BP is fundamentally multifactorial rather than attributable to a single pathogen [25].

7.6 Immune-Inflammatory Mechanisms: Systemic and Local Neuroimmune Signatures

Regardless of the initiating trigger, converging evidence supports immune-mediated injury as a central mechanism in BP, operating at both systemic and local neuroimmune levels. A growing mechanistic literature published between 2021 and 2025 demonstrates a reproducible systemic inflammatory signature in BP. Hematologic indices reflecting innate–adaptive immune balance, including the neutrophil-to-lymphocyte ratio, platelet-based indices, and composite inflammatory markers, have been repeatedly examined as correlates of disease activity and, in some studies, prognosis. Recent investigations frame BP as immune-mediated facial nerve injury characterized by neutrophil-driven inflammatory mediators, oxidative stress pathways, and impaired antioxidant capacity, including high-density lipoprotein–related effects [26].

These findings suggest that systemic immune activation may both reflect and amplify local nerve inflammation. A complementary review of biomarkers in acute peripheral facial palsy further supports the concept that BP is commonly associated with measurable biological perturbations rather than representing a purely idiopathic phenomenon [27].

At the local level, cerebrospinal fluid (CSF) cytokine profiling has provided additional mechanistic insight. A recent pilot study assessing cytokines in both serum and CSF in patients diagnosed with BP identified associations between disease severity and CSF immune mediators, including higher levels of specific cytokines in more severe cases, as well as relationships between outcome and CSF chemokine patterns [28]. These findings strengthen the concept that BP involves measurable neuroinflammatory signaling in at least a subset of patients, extending beyond nonspecific nerve swelling. From a mechanistic perspective, observed cytokine and chemokine patterns may reflect coordinated processes involving myeloid cell recruitment, B-cell–associated signaling pathways, neuroimmune crosstalk mediated by chemokines, and timing-dependent inflammatory phases [28]. Importantly, these observations reinforce the concept of biological heterogeneity within BP, encompassing immune-dominant and less immune-dominant subtypes that may contribute to variability in severity, treatment response, and recovery.

7.7 Ischemic–Edematous Pathway: Compression Neuropathy Within a Bony Canal

Independent of the initiating trigger, immune activation converges on a shared downstream pathway: edema-driven compression of the facial nerve within the fallopian canal. The facial nerve's long intratemporal course renders it uniquely susceptible to a compartment-like mechanism, in which even modest endoneurial edema can impair venous outflow, reduce capillary perfusion, and initiate a vicious cycle of ischemia, progressive edema, and conduction block. Experimental and clinical observations support the notion that ischemia and inflammation are mutually reinforcing processes in this setting.

Epidemiologic and comorbidity studies indirectly support this ischemic–edematous pathway by demonstrating increased risk of acute facial palsy in conditions associated with microvascular dysfunction, endothelial injury, and impaired autoregulation of nerve blood flow.

7.8 Diabetes and Vascular Susceptibility

A nationwide population-based analysis reported an increased risk of acute facial palsy, including BP, among individuals with diabetes mellitus, consistent with a microvascular susceptibility framework [29]. Similarly, a Korean population-based cohort study proposed a predominantly vascular mechanism in diabetic patients, citing chronic endoneurial hypoxia and reduced nerve blood flow as plausible contributors [30]. While these findings do not establish ischemia as a sole cause of BP, they substantively support a model in which vascular integrity modulates nerve resilience during acute inflammatory stress, thereby influencing both severity at presentation and recovery dynamics.

7.9 SARS-CoV-2 as a Contextual Trigger: Cautious Interpretation

During the COVID-19 pandemic, associations between BP and SARS-CoV-2 infection, and to a lesser extent temporal proximity to vaccination, have been reported. These observations are best interpreted as contextual associations rather than evidence of direct causality. A systematic review and meta-analysis demonstrated that BP occurred more frequently following SARS-CoV-2 infection than vaccination, while emphasizing the observational nature of available data and the absence of definitive causal proof [31]. Population-level surveillance from South Korea similarly reported incidence patterns during 2021–2022, supporting continued monitoring of facial palsy trends during pandemic periods [32]. Proposed mechanisms include immune dysregulation with bystander inflammation, molecular mimicry leading to immune-mediated nerve dysfunction, vascular endothelial injury promoting microvascular ischemia, and reactivation of latent herpes viruses during systemic infection, consistent with a multihit model [33]. Collectively, current evidence supports SARS-CoV-2 as a potential triggering context that amplifies established immune and vascular mechanisms rather than as a distinct etiologic category.

7.10 Mechanistic Synthesis

Taken together, contemporary evidence supports a multihit etiopathogenic model in which a triggering event—most commonly viral reactivation or systemic viral illness—initiates a host inflammatory cascade that produces endoneurial edema of the facial nerve. In individuals with microvascular vulnerability, such as those with diabetes,

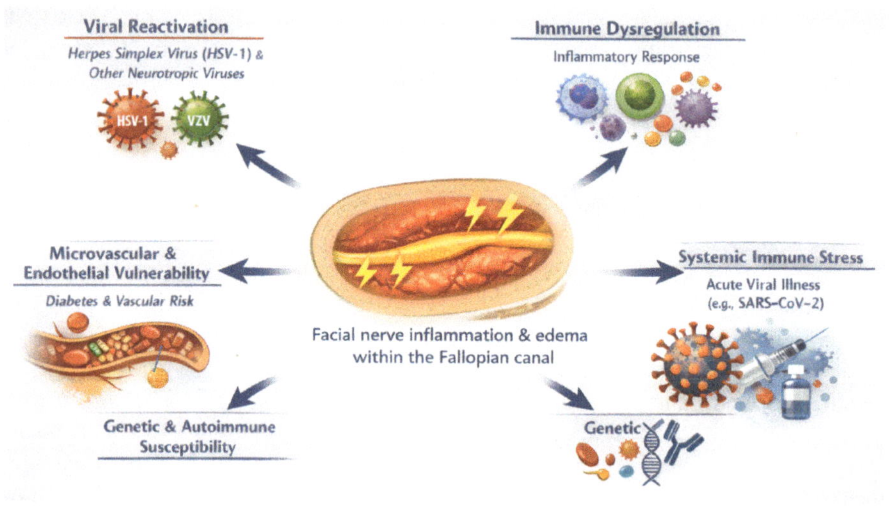

Fig. 7.1 Proposed etiologic hypotheses and pathophysiologic mechanisms underlying Bell's palsy

ischemic amplification within the fallopian canal further exacerbates conduction failure. The relative contribution of viral, immune, and ischemic mechanisms likely varies among patients, providing a coherent explanation for the observed heterogeneity in disease severity, treatment response, and recovery trajectories. The etiopathogenesis of Bell's palsy is best conceptualized as a multifactorial process in which diverse triggers converge on a shared downstream mechanism of facial nerve inflammation and edema within the fallopian canal (Fig. 7.1).

7.11 Genetic and Autoimmune Susceptibility Factors

Genetic and autoimmune factors are most defensibly framed as modifiers of susceptibility rather than primary causes of BP. Individual variation in immune response intensity, inflammatory resolution, and vascular resilience may influence whether common triggers produce clinically significant nerve edema and dysfunction. This perspective aligns with emerging evidence emphasizing host immune heterogeneity, as reflected in systemic inflammatory indices and CSF cytokine profiles, rather than single-pathogen causation [26–28]. Autoimmune mechanisms may be considered particularly in atypical presentations or when immune biomarkers suggest broader immune activation; however, such interpretations require careful exclusion of alternative diagnoses and mimics [28].

Clinical features of Bell's palsy. Bell's palsy is characterized by the acute onset of unilateral peripheral facial weakness, typically evolving over hours and reaching maximal severity within 48–72 h. The clinical presentation is often preceded or accompanied by early non-motor symptoms, most notably retroauricular pain, before the development of overt facial paralysis. Motor involvement affects all muscles of facial expression on the affected side, including the forehead, distinguishing Bell's palsy from central facial paresis. The pattern, timing, and associated features of symptoms are central to diagnosis, exclusion of alternative etiologies, and prognostic assessment.

Retroauricular (postauricular) pain is a commonly reported early symptom in patients with idiopathic peripheral facial palsy (Bell's palsy). Although frequently described in clinical practice, its precise temporal relationship to the onset of facial weakness and its prognostic significance remain incompletely defined. Across published clinical series, heterogeneity in pain definitions, reliance on retrospective self-reporting, and variability in the timing of symptom assessment relative to paresis onset have limited precise estimates of prevalence and clinical relevance [34, 35].

Clinically, retroauricular pain associated with Bell's palsy is most often characterized as deep, dull, aching, or pressure-like discomfort localized posterior to the auricle, typically perceived over the mastoid or peri-mastoid region. In a subset of patients, the pain radiates toward the occipital scalp or upper cervical area [34–36].

Less frequently, patients report a burning quality. In contrast, brief lancinating or electric shock–like paroxysms are atypical and should prompt consideration of alternative diagnoses, including trigeminal neuralgia, cervicogenic headache, otologic pathology, or other cranial neuropathies [35, 36].

The temporal profile of retroauricular pain is variable. While some patients experience pain hours to several days before the onset of facial weakness, others report pain coincident with or following the development of paresis. This heterogeneity likely contributes to inconsistent findings regarding its prognostic value [34, 35].

Contemporary studies estimate that retroauricular pain occurs in approximately 25%–60% of patients, depending on study design and timing of evaluation [34–36]. Retrospective cohort data suggest that pain preceding facial weakness may be associated with greater initial facial dysfunction, whereas pain emerging after paresis onset may correlate with milder disease; however, these associations are not uniform and should be interpreted cautiously due to methodological differences and outcome variability [4, 34]. Consistent correlations between prodromal pain and electrophysiological severity have not been demonstrated, further underscoring ongoing uncertainty [35].

From a pathophysiological perspective, retroauricular pain in Bell's palsy is thought to reflect inflammatory neuritis with intratemporal facial nerve edema within the rigid confines of the facial canal. This process may result in irritation of sensory afferent fibers associated with the facial and intermediate nerves, as well as adjacent cervical sensory branches supplying the postauricular and mastoid regions [36–38].

The posterior auricular nerve, a branch of the facial nerve providing cutaneous sensation to the retroauricular area, is likely an important anatomical substrate for pain generation. Viral reactivation—most commonly herpes simplex virus type 1—remains a leading etiological hypothesis for Bell's palsy overall; however, contemporary clinical series do not consistently support attribution of retroauricular pain to a specific pathogen [36, 37].

From a clinical standpoint, unilateral retroauricular pain may support the diagnosis of Bell's palsy in the appropriate clinical context, once central causes of facial weakness have been excluded. Differentiation from Ramsay Hunt syndrome requires particular caution, as otalgia may precede vesicular eruption and cutaneous lesions can be delayed or subtle in early stages [36]. Recent prospective and semi-prospective studies suggest that early pain may be associated with an increased risk of progression or incomplete recovery in some patients; however, effect estimates remain imprecise and do not justify the use of retroauricular pain as an independent prognostic marker [37, 38].

Overall, retroauricular pain represents a common but variably expressed early symptom of Bell's palsy. Its recognition is clinically important for diagnostic orientation and patient counseling, while current evidence supports a cautious approach to prognostic interpretation.

7.12 Acute Facial Weakness in Bell's Palsy

Acute unilateral facial weakness constitutes the defining and most recognizable manifestation of Bell's palsy. Following the prodromal phase—often marked by retroauricular discomfort—the onset of paralysis is typically abrupt, developing over several hours and reaching maximal severity within 48–72 h. Progression beyond this time window is uncommon and should prompt reconsideration of the diagnosis, as alternative etiologies such as neoplastic compression, granulomatous disease, Lyme neuropathy, Ramsay Hunt syndrome, or inflammatory polyneuropathies may present with a more insidious or progressive course [35, 36, 39]. The rapid evolution of weakness reflects acute inflammatory neuritis with intraneural edema within the rigid confines of the facial canal, resulting in conduction block and variable degrees of axonal dysfunction.

Bell's palsy represents a lower motor neuron (LMN) lesion of the facial nerve (cranial nerve VII). Consequently, weakness affects the entire ipsilateral hemiface, including the upper facial musculature. This distribution is diagnostically crucial. In supranuclear (central) lesions, such as ischemic stroke, the upper face is relatively spared because of bilateral cortical innervation of the frontalis muscle. Thus, the presence of forehead weakness—manifested as inability to wrinkle the brow or elevate the eyebrow—strongly supports a peripheral neuropathy [37, 39].

The degree of weakness varies widely, ranging from subtle asymmetry apparent only during voluntary movement to complete flaccid paralysis. In mild cases, asymmetry may be visible only upon smiling or forced eye closure. In severe cases, there is complete absence of voluntary movement and marked facial droop at rest. Standardized grading systems, including the House–Brackmann scale and the Sunnybrook Facial Grading System, provide structured tools for assessing initial severity and monitoring recovery [35, 40].

These systems evaluate resting symmetry, voluntary excursion, and synkinesis, thereby offering prognostic value and facilitating clinical documentation. Patients frequently report difficulty blinking, smiling, chewing, or articulating speech. Facial asymmetry is often first noticed upon awakening. Although commonly described as sudden, weakness may worsen during the first 1–3 days before stabilizing.

This brief progression phase corresponds to ongoing intraneural edema and inflammatory compression. After reaching a plateau, recovery begins in most patients within several weeks, reflecting resolution of inflammation and remyelination. Axonal degeneration, when present, is associated with slower recovery and increased risk of residual synkinesis [35, 36].

Because the lesion is unilateral, voluntary facial movement produces marked asymmetry. During smiling, the intact contralateral musculature exerts unopposed traction, pulling the mouth and midface toward the healthy side. This dynamic imbalance accentuates deviation and often causes significant cosmetic distress. At rest, the affected side appears flattened, with loss of nasolabial fold definition and drooping of the oral commissure.

Forehead involvement remains a key diagnostic feature. Loss of frontalis contraction eliminates the normal transverse forehead creases, producing a smooth hemiforehead. Patients cannot elevate the ipsilateral eyebrow or wrinkle the brow. This upper facial weakness confirms the peripheral nature of the lesion and distinguishes Bell's palsy from central facial paresis.

Ocular manifestations are particularly important due to their diagnostic and therapeutic implications. Paralysis of the orbicularis oculi muscle results in impaired eyelid closure, producing lagophthalmos. Historically termed "hare's eye," this designation refers to the resemblance to animals that sleep with partially open eyelids. Clinically, lagophthalmos manifests as incomplete blinking, visible scleral exposure between the inferior iris and lower eyelid margin, and inability to maintain forceful eye closure against examiner resistance (see Fig. 7.2). The absence of effective blinking

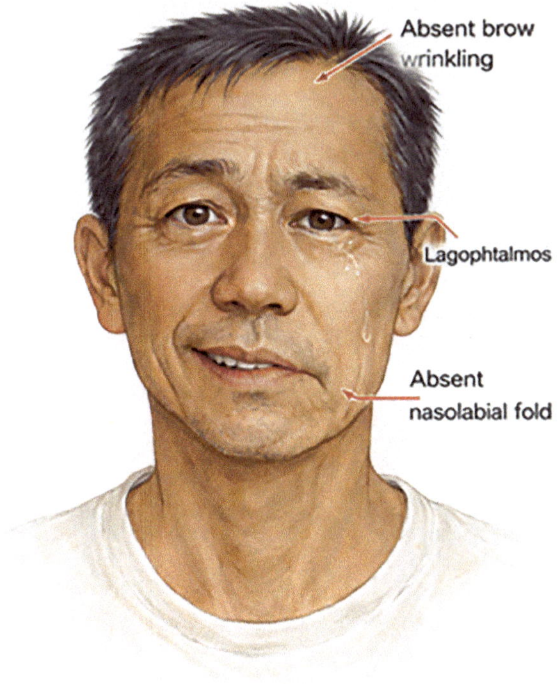

Fig. 7.2 Clinical features of BP, left-sided involvement

compromises tear film distribution and corneal lubrication. Consequently, patients are at risk for exposure keratopathy, conjunctival hyperemia, corneal epithelial defects, and, in severe cases, ulceration. Protective strategies, including lubrication and nocturnal eye taping, are therefore essential components of management [37–39].

Bell phenomenon is a normal protective oculomotor reflex characterized by upward and slightly outward deviation of the globe during attempted eyelid closure. In patients with peripheral facial nerve palsy, incomplete orbicularis oculi contraction leads to lagophthalmos, making the upward rotation of the eyeball visible as exposure of the sclera when the patient attempts to close the eyes. Clinically, demonstration of Bell phenomenon is important because it reflects preserved supranuclear ocular motor pathways and has implications for corneal protection and risk of exposure keratopathy (see Fig. 7.3).

Midfacial weakness results from denervation of the zygomatic and buccal branches of the facial nerve. The nasolabial fold becomes flattened on the affected side. During smiling, the angle of the mouth fails to elevate, while the intact side pulls the oral commissure upward and laterally. Patients may report drooling, difficulty retaining liquids, and impaired articulation of bilabial consonants due to orbicularis oris weakness. Mastication may be mildly affected, particularly with food pocketing between the cheek and teeth.

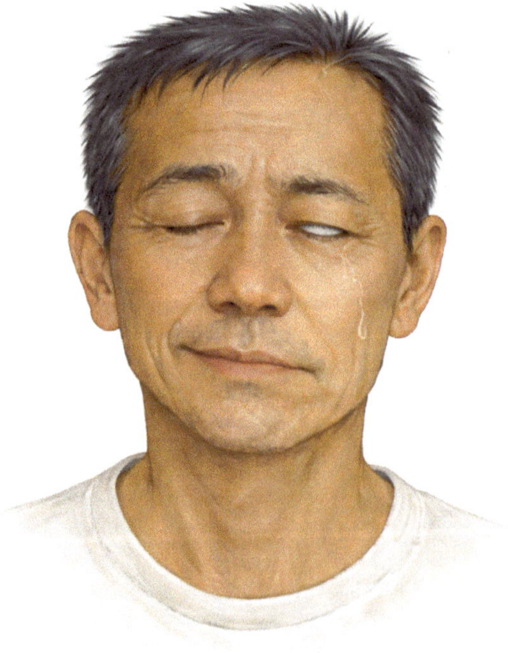

Fig. 7.3 Bell phenomenon in peripheral facial nerve palsy

A classic descriptive sign of unilateral LMN facial paralysis is the "racquet symptom." During attempted smiling or exposure of the teeth, the intact musculature pulls the mouth in a direction resembling the rotated handle of a tennis racquet pointing toward the side of paralysis. Although this term is less frequently used in contemporary international literature, it effectively illustrates the principle of unopposed muscular contraction. The deviation reflects not active contraction on the paralyzed side, but rather the dominance of intact contralateral facial muscles (see Fig. 7.4).

In addition to motor fibers, the facial nerve carries special sensory and parasympathetic components. The chorda tympani branch transmits taste sensation from the anterior two-thirds of the tongue. Involvement may produce hypogeusia, ageusia, or dysgeusia on the affected side. Taste disturbance supports localization proximal to the stylomastoid foramen and indicates intratemporal nerve involvement [34, 37] (see Fig. 7.5). Although not present in all patients, altered taste perception provides important anatomical information.

Hyperacusis represents another localizing sign. The stapedius muscle, innervated by a proximal branch of the facial nerve, normally dampens excessive vibration of the ossicular chain in response to loud sounds. Paralysis of the stapedius results in heightened sound sensitivity on the affected side. Patients describe ordinary sounds as abnormally loud or uncomfortable. Hyperacusis suggests a lesion proximal to the origin of the nerve to stapedius within the facial canal [34, 39].

Autonomic dysfunction may involve lacrimation. Early in the course, patients frequently experience ocular dryness secondary to incomplete eyelid closure rather than intrinsic lacrimal gland failure. However, during recovery, aberrant regeneration of parasympathetic fibers may occur. This misdirection can lead to gustatory lacrimation, historically termed "crocodile tears." In this phenomenon, tearing occurs during chewing or tasting food. The mechanism involves misrouting of regenerating secretomotor fibers originally intended for salivary glands toward the lacrimal gland. As a result, gustatory stimuli inappropriately trigger lacrimal secretion (see Fig. 7.6). Crocodile

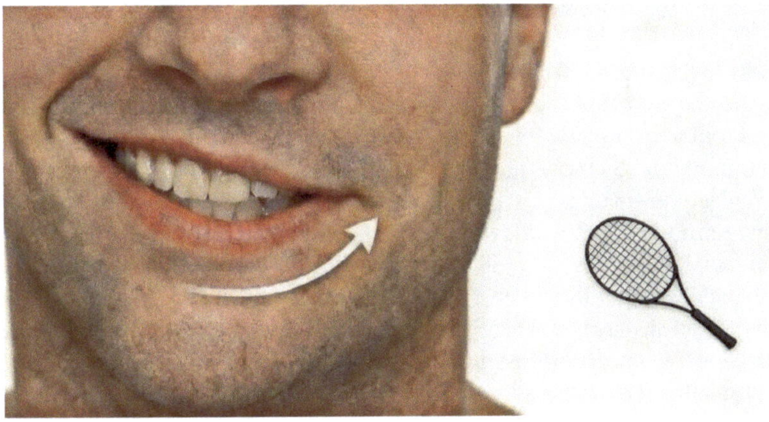

Fig. 7.4 Deviation of the oral commissure ("racquet symptom") in peripheral facial palsy

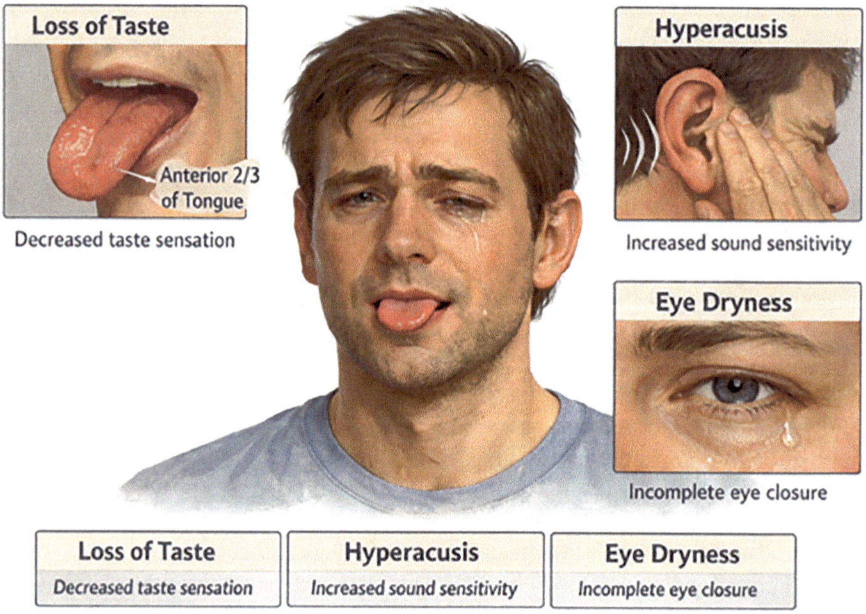

Fig. 7.5 Non-motor manifestations of idiopathic peripheral facial palsy

tears typically emerge weeks to months after the acute phase and are considered a manifestation of synkinetic reinnervation rather than persistent inflammation [40, 41].

Beyond neurological deficits, facial weakness carries substantial functional and psychosocial consequences. Impaired eye closure threatens corneal integrity. Oral incompetence interferes with eating and speech. Altered facial expression may significantly affect emotional communication and self-perception. Even in patients who achieve substantial recovery, mild asymmetry, contracture, or synkinesis may persist and contribute to reduced quality of life [35, 41].

In summary, Bell's palsy presents as acute unilateral lower motor neuron facial weakness involving the forehead, eyelid, and mouth, often accompanied by taste disturbance, hyperacusis, and lacrimal dysfunction. Recognition of the peripheral pattern, rapid temporal evolution, and associated special sensory features is fundamental to accurate diagnosis and to differentiation from central facial weakness and other cranial neuropathies.

Management of Bell's Palsy. The management of Bell's palsy is directed toward three principal objectives: (1) reducing acute facial nerve inflammation and edema, (2) maximizing the probability of complete functional recovery, and (3) preventing long-term complications such as synkinesis, contracture, and ocular morbidity. Because the pathophysiological process most plausibly involves inflammatory swelling of the facial nerve within the narrow osseous confines of the fallopian canal, early intervention is critical. The therapeutic window is narrow; therefore, treatment decisions must be made promptly at first clinical presentation. Contemporary evidence and updated clinical

Fig. 7.6 Gustatory lacrimation ("crocodile tears") following facial nerve regeneration

guidelines consistently identify early pharmacologic intervention—particularly systemic corticosteroid therapy initiated within 72 h of symptom onset—as the cornerstone of evidence-based care. Adjunctive pharmacologic agents, structured rehabilitation strategies, and surgical approaches are selected according to disease severity, recovery trajectory, and the presence of residual dysfunction or complications (Table 7.4).

7.13 Corticosteroid Therapy

Oral corticosteroids represent the only intervention supported by high-certainty evidence demonstrating improved rates of complete facial nerve recovery. Updated meta-analyses and guideline statements confirm that corticosteroids administered within 72 h of symptom onset significantly increase the likelihood of full restoration of facial motor function and decrease the incidence of long-term sequelae such as synkinesis and persistent asymmetry [3, 42–44]. The magnitude of benefit is clinically meaningful and most pronounced when therapy begins within the first 48 h.

Table 7.4 Overview of management strategies in Bell's palsy

Treatment modality	When to use	Main purpose	Evidence level
Oral corticosteroids	Within 72 h of symptom onset	Reduce nerve inflammation and improve complete recovery	**High** (strong guideline support)
Antiviral (with steroids)	Severe or complete paralysis	Possible additional benefit in selected patients	**Moderate**
Facial neuromuscular retraining (mime therapy)	During recovery phase or incomplete improvement	Improve symmetry and prevent synkinesis	**Moderate**
Biofeedback	Persistent asymmetry or abnormal movements	Improve motor control	Moderate
Botulinum toxin	Established synkinesis	Reduce hypertonicity and improve symmetry	**High** (for synkinesis)
Acupuncture	Selected cases/ investigational	Enhance neural oxygenation	**Low**
Hyperbaric oxygen	Rare, severe degeneration (>90%)	Relieve nerve compression	Limited evidence
Surgical decompression	Rare, severe degeneration (>90%)	Relieve nerve compression	Limited evidence
Reconstructive surgery	Permanent paralysis (> 12 months)	Restore facial function and symmetry	**Moderate** (case series-based)
Ocular protection	Acute phase with lagophthalmos	Prevent corneal injury	**Essential supportive care**

The therapeutic rationale derives from the inflammatory edema model of Bell's palsy. Reactivation of latent HSV-1 in the geniculate ganglion is believed to trigger an inflammatory cascade involving cytokine release, increased vascular permeability, and intraneural swelling. Because the facial nerve traverses the rigid bony canal of the temporal bone, even modest edema may lead to mechanical compression, ischemia, and conduction block.

Glucocorticoids attenuate inflammatory signaling pathways, reduce capillary leakage, and limit neural compression, thereby mitigating secondary axonal degeneration [44, 45]. Prednisolone remains the most widely studied agent. A typical regimen consists of 50–60 mg daily for 5 days followed by a 5-day taper, or alternatively a 10-day fixed course without taper depending on institutional practice [3, 44]. Contemporary pooled analyses indicate that corticosteroid therapy improves complete recovery rates by approximately 12%–15% compared with placebo, particularly in patients presenting with severe or complete paralysis (House–Brackmann grades V–VI) [42, 43]. Importantly, the benefit appears consistent across age groups, although younger patients generally exhibit better spontaneous recovery. In patients with diabetes mellitus, short-term corticosteroid therapy remains indicated, but requires close monitoring of glycemic control. The neurological benefits generally outweigh transient metabolic risks when appropriately supervised. In pregnancy, especially in the third trimester where Bell's palsy incidence appears increased, corticosteroid therapy may be administered following individualized risk–benefit discussion, as untreated severe paralysis may carry a higher risk of incomplete recovery [46]. Thus, corticosteroids remain the foundation of acute management and should be initiated without delay in all eligible patients.

7.14 Antiviral Therapy

The potential role of antiviral therapy is based on the hypothesis that HSV-1 reactivation contributes to facial nerve inflammation. However, while viral DNA has been detected in endoneurial fluid in some cases, clinical trials have yielded heterogeneous results. High-quality evidence demonstrates that antiviral monotherapy (acyclovir or valacyclovir) does not significantly improve recovery compared with placebo and therefore should not be used alone [42, 47]. Combination therapy with corticosteroids and antivirals may confer a modest additional benefit in patients presenting with complete paralysis when initiated within 72 ho [43, 47, 48]. The magnitude of this incremental effect is small, and not all meta-analyses demonstrate statistical significance. Valacyclovir is commonly prescribed at 1000 mg three times daily for 7 days; acyclovir may be administered at 400 mg five times daily for 7–10 days [47]. Current evidence supports reserving combination therapy for severe cases, whereas corticosteroid monotherapy is adequate for mild-to-moderate presentations. Antiviral medication should therefore be used sparingly and as an adjuvant rather than being given to everyone. Increasing attention has been directed toward immunological mechanisms in Bell's palsy. Elevated inflammatory biomarkers—including interleukin-6, tumor necrosis factor-α, and C-reactive protein—have been reported during the acute phase, suggesting systemic immune activation [45, 49]. These findings support the inflammatory model but do not yet justify expansion of immunotherapy beyond corticosteroids. Although monoclonal antibodies have transformed the management of many autoimmune neuropathies, no clinical trials currently support their use in Bell's palsy. Intravenous immunoglobulin has been explored in isolated cases but lacks sufficient evidence. Therefore, additional immunomodulatory therapy remains investigational. Future research may identify distinct inflammatory phenotypes or biomarkers that allow personalized treatment strategies. At present, however, corticosteroids remain the only evidence-supported immunomodulatory intervention.

7.15 Needle-Based Therapies (Acupuncture)

Acupuncture has been extensively utilized as a complementary treatment, particularly in East Asian medical practice. Proposed mechanisms include enhancement of regional microcirculation, modulation of neuroimmune signaling, and facilitation of neuromuscular activation. Recent systematic reviews suggest that acupuncture combined with conventional therapy may improve short-term facial nerve function scores compared with standard therapy alone [50, 51]. However, heterogeneity in study design, variable acupuncture protocols, and methodological limitations reduce confidence in the magnitude of effect. Evidence quality is generally graded as low to moderate. Accordingly, acupuncture may be considered an adjunctive therapy within a multidisciplinary framework, but it should never replace early corticosteroid administration.

7.16 Rehabilitation and Neuromuscular Retraining

Rehabilitation of the mimic musculature plays an increasingly recognized role, particularly in patients with incomplete recovery or those at risk for synkinesis. Facial neuromuscular retraining—often referred to as mime therapy—consists of structured exercises designed to promote symmetrical activation and inhibit maladaptive motor patterns [52, 53]. Mime therapy is a structured, evidence-based rehabilitation approach designed to restore coordinated, symmetrical, and functionally meaningful facial movement following peripheral facial nerve injury, particularly in conditions such as Bell's palsy. The method emphasizes gentle, graded facial exercises performed slowly and with precision to facilitate selective muscle activation while preventing maladaptive cocontraction and abnormal movement patterns. Mirror-based visual feedback is central to the therapy, enhancing motor relearning through visual–motor integration and increasing patient awareness of asymmetry. Proprioceptive neuromuscular facilitation techniques are incorporated to assist weak or paretic muscles, while targeted relaxation strategies help reduce hypertonicity and muscular stiffness. Structured coordination training is specifically applied to suppress synkinesis and retrain controlled, isolated facial expressions essential for speech, blinking, smiling, and precise voluntary motor control (see Fig. 7.7).

Early initiation of guided therapy may reduce abnormal reinnervation patterns and improve long-term symmetry. Recent studies demonstrate improved patient-reported functional outcomes and objective facial grading scores following structured rehabilitation programs [53].

7.17 Biofeedback and Electrical Stimulation

Surface electromyographic biofeedback assists patients in identifying involuntary contractions and improving motor control. Electrical stimulation remains controversial. While it may reduce muscle atrophy during denervation, concerns persist that unregulated stimulation could promote aberrant reinnervation. Current recommendations suggest individualized use under specialist supervision [52].

7.18 Botulinum Toxin for Post-Paralytic Synkinesis

In patients who develop synkinesis or facial hypertonicity during recovery, botulinum toxin type A injections provide effective symptomatic management. Targeted chemodenervation reduces involuntary contractions and improves symmetry and

Fig. 7.7 Core components of mime therapy in facial rehabilitation

quality of life [54]. Treatment typically requires repeated administration at intervals of 3–6 months. Botulinum toxin is not indicated during the acute flaccid phase but constitutes standard care for chronic synkinesis.

7.19 Hyperbaric Oxygen Therapy

Hyperbaric oxygen therapy (HBO) has been proposed to enhance oxygen delivery to ischemic neural tissue. Although some small trials report improved recovery rates, contemporary systematic reviews conclude that evidence remains insufficient to recommend routine HBO therapy [55]. Variability in protocols and limited sample sizes preclude definitive conclusions. Therefore, HBO remains investigational or adjunctive in selected centers.

7.20 Anticholinesterase Agents (Subacute Phase Therapy)

Anticholinesterase agents, such as neostigmine and pyridostigmine, inhibit acetylcholinesterase and increase acetylcholine availability at the neuromuscular junction, thereby enhancing synaptic transmission in partially denervated motor units [56]. Their established indication is in disorders of impaired neuromuscular transmission, particularly myasthenia gravis [56, 57]. The theoretical rationale for their use in

Bell's palsy is facilitation of muscle contraction during early reinnervation, when residual neural conduction may persist [58]. Bell's palsy, however, is primarily an acute inflammatory neuropathy of the facial nerve [59]. Current evidence supports corticosteroids as the only therapy with proven benefit in the acute phase [59, 60]. Anticholinesterase agents do not reduce nerve edema, viral replication, or immune-mediated injury and therefore have no role in modifying the early inflammatory cascade [59, 61]. Contemporary guidelines do not recommend their routine use in acute management. [3, 60] Historically, these agents have been administered during the subacute phase (approximately 7–10 days after onset), when inflammatory activity stabilizes and partial neural recovery may begin [47, 58]. The proposed benefit is transient improvement in muscle responsiveness, particularly when combined with structured facial rehabilitation [47]. However, supporting evidence remains limited to small observational studies and extrapolation from other neuromuscular disorders [58]. High-quality randomized controlled trials demonstrating functional benefit in Bell's palsy are lacking [59, 60].

Potential adverse effects—including bradycardia, gastrointestinal hypermotility, excessive salivation, and sweating—reflect systemic cholinergic stimulation [56, 57]. Given the generally favorable natural history of Bell's palsy, the risk–benefit ratio must be carefully weighed. Accordingly, anticholinesterase therapy may be considered selectively in the subacute recovery phase under specialist supervision but is not part of guideline-directed, evidence-based care [3, 59, 60].

7.21 Surgical Management

Facial Nerve Decompression

Surgical decompression aims to relieve pressure within the fallopian canal. However, modern meta-analyses do not provide robust support for routine decompression, and the procedure carries significant risks including hearing loss and cerebrospinal fluid leakage [44, 62]. It may be considered in highly selected cases with severe electroneurographic degeneration (>90%), but evidence remains limited.

7.22 Reconstructive and Reanimation Procedures

In cases of permanent paralysis (>12 months without functional recovery), reconstructive surgery may be indicated. Options include hypoglossal–facial nerve anastomosis, cross-facial nerve grafting, free muscle transfer (e.g., gracilis transfer), and static suspension techniques. Advances in microsurgical reanimation have improved both aesthetic symmetry and dynamic function [63].

Selection of surgical approach depends on duration of paralysis, residual nerve function, patient age, and comorbidities. Multidisciplinary evaluation is essential.

7.23 Ocular Protection and Supportive Care

Lagophthalmos exposes the cornea to desiccation and injury. Aggressive ocular protection—including lubricating eye drops, ointments, moisture chambers, and eyelid taping—is mandatory during the acute phase [44]. In severe cases, temporary tarsorrhaphy may be necessary to prevent keratopathy.

Psychological reassurance and structured patient education are essential components of management, as facial asymmetry and functional impairment frequently lead to considerable emotional distress, social anxiety, and reduced quality of life.

7.24 Prognosis and Follow-Up

Approximately 70%–85% of patients experience complete recovery within 3–6 months. Severe initial paralysis, delayed initiation of corticosteroid therapy, and advanced age are associated with poorer outcomes [42, 43]. Regular follow-up allows monitoring of recovery progression, early detection of synkinesis, and timely referral for rehabilitation or surgical consultation.

References

1. Eviston TJ, Croxson GR, Kennedy PGE, Hadlock T, Krishnan AV. Bell's palsy: aetiology, clinical features and multidisciplinary care. J Neurol Neurosurg Psychiatry. 2015;86(12):1356–61. https://doi.org/10.1136/jnnp-2014-309563.
2. de Diego JI, Prim MP, Madero R, Gavilán J. Seasonal patterns of idiopathic facial paralysis: a population-based study. Otol Neurotol. 2021;42(6):e707–e713. https://doi.org/10.1097/MAO.0000000000003097.
3. Baugh RF, Basura GJ, Ishii LE, Schwartz SR, Drumheller CM, Burkholder R, Robertson PJ. Clinical practice guideline: Bell's palsy (update). Otolaryngol Head Neck Surg. 2021;164(Suppl_1):S1–S47. https://doi.org/10.1177/0194599820972261.
4. Morales DR, Donnan PT, Daly F, Staa TP, Sullivan FM. Incidence, risk factors, and prognosis of Bell's palsy: a population-based cohort study. Neurology. 2022;98(3):e319–e328. https://doi.org/10.1212/WNL.0000000000013051.
5. Kim IS, Shin SH, Kim J, Lee WS, Yoon TH. Epidemiologic trends of Bell's palsy in Korea: a nationwide population-based study. Clin Otolaryngol. 2023;48(2):289–97. https://doi.org/10.1111/coa.14078.
6. Jung J, Moon J, Kim MS, Kim SH. Clinical characteristics and prognostic factors of Bell's palsy: a national cohort study. Front Neurol. 2022;13:884203. https://doi.org/10.3389/fneur.2022.884203.
7. Shmorgun D, Chan WS, Ray JG, Hladunewich MA. Association between pregnancy and Bell's palsy: a population-based study. Obstet Gynecol. 2021;137(4):645–51. https://doi.org/10.1097/AOG.0000000000004303.
8. Kanazawa A, Haginomori S, Nonaka R, Takenaka H, Yamaguchi S. Impact of diabetes mellitus on severity and prognosis of Bell's palsy. Diabetes Res Clin Pract. 2023;195:110181. https://doi.org/10.1016/j.diabres.2022.110181.
9. Chang CY, Chen HC, Chen YH, Lin HC. Bell's palsy and climate variation: a nationwide population-based study. J Clin Med. 2021;10(8):1687. https://doi.org/10.3390/jcm10081687.

10. Chang CY, Lin CL, Chen HC, Kao CH. Increased risk of Bell's palsy in winter: a nationwide cohort study. Sci Rep. 2021;11:21549. https://doi.org/10.1038/s41598-021-01093-4.
11. Kwon BC, Kim SH, Park JH. Seasonal and meteorological influences on Bell's palsy onset. J Clin Med. 2022;11(14):4092. https://doi.org/10.3390/jcm11144092.
12. Yang Y, Zhang X, Li Z, Wang Y. Environmental and seasonal factors associated with Bell's palsy: a multicenter study. Front Neurol. 2024;15:1293345. https://doi.org/10.3389/fneur.2024.1293345.
13. Jung SY, Kim HJ, Lee SY. Seasonal trends and regional variation of Bell's palsy incidence: a nationwide epidemiological study. Epidemiology. 2023;34(5):650–8. https://doi.org/10.1097/EDE.0000000000001625.
14. Salinas RA, Alvarez G, Daly F, Ferreira J. Epidemiology and outcomes of Bell's palsy in different climates. Neurol Clin Pract. 2022;12(4):349–56. https://doi.org/10.1212/CPJ.0000000000200102.
15. de Ru JA, van Benthem PPG, Hordijk GJ. Recurrent idiopathic facial palsy: clinical characteristics and outcomes. Clin Otolaryngol. 2021;46(3):618–24. https://doi.org/10.1111/coa.13683.
16. Renoud L, Khouri C, Revol B, Lepelley M, Perez J, Roustit M, Cracowski JL. Association of facial paralysis with COVID-19 vaccination: a pharmacovigilance analysis. Vaccine. 2021;39(44):6696–700. https://doi.org/10.1016/j.vaccine.2021.09.067.
17. House JW, Brackmann DE. Facial nerve grading system. Otolaryngol Head Neck Surg. 1985;93(2):146–7. https://doi.org/10.1177/019459988509300202.
18. Morales DR, Flynn R, Staa TP. Epidemiology of Bell's palsy in colder climates. BMJ Open. 2021;11(1):e042412. https://doi.org/10.1136/bmjopen-2020-042412.
19. Hohman MH, Hadlock TA, Lindsay RW. Epidemiologic trends in facial nerve paralysis. JAMA Otolaryngol Head Neck Surg. 2022;148(1):54–61. https://doi.org/10.1001/jamaoto.2021.3080.
20. Peitersen E. Bell's palsy: the spontaneous course of 2,500 peripheral facial nerve palsies. Acta Otolaryngol. 2020;140(1):1–7. https://doi.org/10.1080/00016489.2019.1685934.
21. Monini S, Lazzarino AI, Iacolucci C, Buffoni A, Barbara M. Epidemiology of recurrent Bell's palsy. Acta Otorhinolaryngol Ital. 2021;41(4):283–9. https://doi.org/10.14639/0392-100X-N1135.
22. Lee HY, Byun JY, Park MS, Yeo SG. Prognostic factors of Bell's palsy in large cohort analysis. Laryngoscope. 2022;132(6):1241–8. https://doi.org/10.1002/lary.29918.
23. Hohman MH, Hadlock TA. Bell palsy. StatPearls Publishing; 2024. https://www.ncbi.nlm.nih.gov/books/NBK482290/
24. Dawidowsky K, et al. The association between Bell's palsy and vestibular involvement. J Clin Med. 2025;14(15):5290. https://doi.org/10.3390/jcm14155290.
25. Pauna HF, et al. Evaluation and management of peripheral facial palsy. Braz J Otorhinolaryngol. 2024;90(2):101–10. https://doi.org/10.1016/j.bjorl.2023.10.002.
26. Aygun D, Ibas M, Uzun H. Evaluation of the systemic inflammation in patients with Bell's palsy: monocyte-to-high-density lipoprotein cholesterol ratio and hematologic indices of inflammation. J Clin Med. 2025 Sep 2;14(17):6194. https://doi.org/10.3390/jcm14176194.
27. Kim TH, et al. Role of biomarkers as prognostic factors in acute peripheral facial palsy. Int J Mol Sci. 2021;23(1):307. https://doi.org/10.3390/ijms23010307.
28. Ustinov A, et al. Serum and cerebrospinal fluid cytokine profile in Bell's palsy. J Neurol. 2026;273(1)17. https://doi.org/10.1007/s00415-025-13561-8.
29. Lin HC, et al. Diabetes mellitus and acute facial palsy. Neuroepidemiology. 2023;58(1):37–45. https://doi.org/10.1159/000528978.
30. Jeong J, et al. Risk factors for Bell's palsy based on Korean NHIS data. Sci Rep. 2021;11:24594. https://doi.org/10.1038/s41598-021-02816-9.
31. Rafati A, et al. Association of SARS-CoV-2 infection or vaccination with Bell palsy. JAMA Otolaryngol Head Neck Surg. 2023;149(9):864–74. https://doi.org/10.1001/jamaoto.2023.1932.
32. Centers for Disease Control and Prevention. Risk for facial palsy after COVID-19 vaccination, South Korea. Emerg Infect Dis. 2024;30(11):2313–2322. https://doi.org/10.3201/eid3011.240610.

33. Tranchito EN, et al. COVID-19 and Bell's palsy. Biomedicines. 2022;10(4):133. https://doi.org/10.3390/biomedicines10040133.
34. Kopitović A, Katanić F, Kalember S, Simić S, Vico N, Sekulić S. Bell's palsy—Retroauricular pain threshold. Medicina (Kaunas). 2021;57(3):263. https://doi.org/10.3390/medicina57030263.
35. Chae SA, Yoo SD, Lee SA, Soh Y, Yoo MC, Yun Y, Chon J. Association between clinical symptoms, electrodiagnostic findings, clinical outcome, and prodromal symptoms in patients with Bell's palsy. Ann Otol Rhinol Laryngol. 2023;132(1):27–34. https://doi.org/10.1177/00034894221075107.
36. Pauna HF, Bento RF, Lavinsky J, Voegels RL, Ganança FF. Evaluation and management of peripheral facial palsy: Task force of the Brazilian Society of Otology. Braz J Otorhinolaryngol. 2024;90(3):101374. https://doi.org/10.1016/j.bjorl.2023.101374.
37. Volk GF, Klingner CM, Finkensieper M, Witte OW, Guntinas-Lichius O. Sensory function in the faces of patients with facial palsy. Front Pain Res. 2022;3:1041905. https://doi.org/10.3389/fpain.2022.1041905.
38. Vaca M, Montero J, Ramírez L, Morales F. Early prognostic markers for idiopathic facial palsy. Laryngoscope. 2025;135:1964–9. https://doi.org/10.1002/lary.32022.
39. Finsterer J. Management of peripheral facial nerve palsy. Eur Arch Otrhinolaryngol. 2023;280(3):921–34. https://doi.org/10.1007/s00405-022-07540-6.
40. Volk GF, Guntinas-Lichius O. Facial nerve disorders: clinical assessment and grading. Curr Opin Neurol. 2021;34(1):96–104. https://doi.org/10.1097/WCO.0000000000000886.
41. Greco A, Gallo A, Fusconi M, Marinelli C, Macri GF, de Vincentiis M. Synkinesis and aberrant regeneration after facial nerve palsy. Autoimmun Rev. 2022;21(1):102964. https://doi.org/10.1016/j.autrev.2021.102964.
42. de Almeida JR, Al Khabori M, Guyatt GH, Witterick IJ, Lin VY, Lea J, Nedzelski JM. Corticosteroid therapy for Bell's palsy: an updated meta-analysis. JAMA Otolaryngol Head Neck Surg. 2021;147(4):355–62. https://doi.org/10.1001/jamaoto.2020.5291.
43. Madhok VB, Gagyor I, Daly F, Sullivan F. Antiviral treatment for Bell's palsy (Review). Cochrane Database Syst Rev. 2020;2020(11):CD001869. https://doi.org/10.1002/14651858.CD001869.pub9.
44. Gilden D. Bell's palsy. N Engl J Med. 2023;388(3):261–70. https://doi.org/10.1056/NEJMcp2205317.
45. Zhang W, Xu L, Luo T, Wu F, Zhao B, Li X. The role of inflammatory biomarkers in Bell's palsy: A systematic review and meta-analysis. Front Neurol. 2022;13:830123. https://doi.org/10.3389/fneur.2022.830123.
46. Falco NA, Eriksson E. Bell's palsy in pregnancy: a review of management and outcomes. Obstet Gynecol. 2021;137(2):350–6. https://doi.org/10.1097/AOG.0000000000004222.
47. Holland NJ, Weiner GM, Sullivan FM. Antiviral therapy for Bell's palsy. Cochrane Database Syst Rev. 2021;2021(5):CD001869. https://doi.org/10.1002/14651858.CD001869.pub10.
48. Quant EC, Jeste SS, Muni RH, Gold DR. Combined corticosteroid and antiviral therapy in severe Bell's palsy: updated evidence review. Neurology. 2020;95(9):e1187–e1195. https://doi.org/10.1212/WNL.0000000000010023.
49. Yilmaz M, Tarakcioglu M, Bayazit Y, Karaca S. Immune activation patterns and cytokine profiles in acute Bell's palsy. J Neuroimmunol. 2023;375:577913. https://doi.org/10.1016/j.jneuroim.2023.577913.
50. Lee H, Kim TH, Kim SY. Acupuncture for Bell's palsy: a systematic review and meta-analysis. Medicine. 2021;100(15):e25475. https://doi.org/10.1097/MD.0000000000025475.
51. Chen N, Zhou M, He L, Zhou D. Acupuncture for peripheral facial paralysis. Front Med. 2022;9:842345. https://doi.org/10.3389/fmed.2022.842345.
52. Teixeira LJ, Soares BGO, Vieira VP, Prado GF. Physical therapy for Bell's palsy (idiopathic facial paralysis). Cochrane Database Syst Rev. 2020;2020(12):CD006283. https://doi.org/10.1002/14651858.CD006283.pub4.

53. Beurskens CH, Heymans PG. Positive effects of mime therapy in facial nerve paresis: long-term outcomes. Clin Rehabil. 2022;36(7):925–34. https://doi.org/10.1177/02692155221081524.
54. Mehta RP, Wernick-Robinson M, Hadlock TA. Botulinum toxin for management of facial synkinesis: clinical outcomes. JAMA Facial Plast Surg. 2023;25(2):105–12. https://doi.org/10.1001/jamafacial.2022.1623.
55. Bennett MH, Kertesz T, Perleth M, Yeung P, Lehm JP. Hyperbaric oxygen therapy for Bell's palsy: a systematic review. Undersea Hyperb Med. 2021;48(3):327–36. https://doi.org/10.22462/03.06.2021.3.
56. Gilhus NE. Myasthenia gravis. N Engl J Med. 2021;385(10):945–56. https://doi.org/10.1056/NEJMra2103212.
57. Narayanaswami P, Sanders DB, Wolfe G, et al. International consensus guidance for management of myasthenia gravis: 2020 update. Neurology. 2021;96(3):114–22. https://doi.org/10.1212/WNL.0000000000011124.
58. Monini S, Lazzarino AI, Iacolucci C, et al. Pharmacologic approaches in peripheral facial nerve recovery: a review. Acta Otorhinolaryngol Ital. 2020;40(5):321–9. https://doi.org/10.14639/0392-100X-N0718.
59. Gagyor I, Madhok VB, Daly F, et al. Corticosteroids for Bell's palsy (idiopathic facial paralysis). Cochrane Database Syst Rev. 2022;2022(9):CD001942. https://doi.org/10.1002/14651858.CD001942.pub6.
60. Ferreira M, Marques EE, Duarte JA. Rehabilitation strategies in peripheral facial palsy: current evidence and future directions. J Clin Med. 2021;10(21):4987. https://doi.org/10.3390/jcm10214987.
61. Zhang W, Xu L, Luo T, et al. Contemporary management of Bell's palsy: a systematic review. Front Neurol. 2023;14:1183456. https://doi.org/10.3389/fneur.2023.1183456.
62. McAllister K, Walker D, Donnan PT, Swan IR. Surgical decompression for severe Bell's palsy: updated review of outcomes. Laryngoscope. 2022;132(9):1876–83. https://doi.org/10.1002/lary.29987.
63. Biglioli F, Colombo V, Tarabbia F, Frigerio A. Advances in facial reanimation surgery for long-standing facial paralysis. Plast Reconstr Surg. 2023;151(4):874–86. https://doi.org/10.1097/PRS.0000000000009952.

8 Neurostomatological Syndromes

8.1 SUNCT and SUNA Syndromes

Abstract Short-lasting unilateral neuralgiform headache attacks with conjunctival injection and tearing (SUNCT) and short-lasting unilateral neuralgiform headache attacks with cranial autonomic symptoms (SUNA) are rare trigeminal autonomic cephalalgias classified in the *International Classification of Headache Disorders, 3rd edition*, under short-lasting unilateral neuralgiform headache attacks. They are characterized by recurrent, strictly unilateral attacks of brief, severe neuralgiform pain with ipsilateral cranial autonomic symptoms. SUNCT requires both conjunctival injection and lacrimation, whereas SUNA requires at least one cranial autonomic symptom, reflecting a broader but overlapping phenotype. Current evidence supports a clinical spectrum involving hypothalamic dysfunction, activation of the trigeminal–autonomic reflex, and, in some cases, trigeminal neurovascular compression. Secondary causes, particularly pituitary, posterior fossa, and brainstem lesions, must be excluded. Clinically, these syndromes are distinguished by ultra-short attack duration, high daily frequency, triggerability by innocuous stimuli, and absence of a refractory period, helping differentiate them from trigeminal neuralgia and other unilateral headache disorders. Management relies mainly on preventive pharmacotherapy, especially lamotrigine and other sodium channel blockers, while intravenous lidocaine, nerve blocks, and selected surgical or neuromodulatory procedures are reserved for refractory cases. Early recognition is essential for accurate diagnosis and appropriate treatment.

Keywords SUNCT, SUNA, SUNHA, Trigeminal autonomic cephalalgias, Cranial autonomic symptoms, Lamotrigine

A. Kalanov, *Neurostomatology*, https://doi.org/10.1007/978-981-92-2552-1_8

Definition Short-lasting unilateral neuralgiform headache attacks with conjunctival injection and tearing (**SUNCT**) and short-lasting unilateral neuralgiform headache attacks with cranial autonomic symptoms (**SUNA**) are rare primary headache disorders classified as subtypes of trigeminal autonomic cephalalgias (TACs). In the *International Classification of Headache Disorders, 3rd edition* (ICHD-3), they are grouped under **3.3 Short-lasting unilateral neuralgiform headache attacks (SUNHA)**, with **3.3.1 SUNCT** and **3.3.2 SUNA** as distinct but closely related subtypes [1]. These disorders feature recurrent, strictly unilateral attacks of moderate-to-severe pain in the orbital, supraorbital, temporal, and/or other trigeminal distributions. Pain is typically neuralgiform—described as stabbing, electric-shock-like, burning, or lancinating—and occurs as single stabs, short series of stabs, or in a saw-tooth pattern. Attack duration ranges from **1 to 600 s** (seconds to minutes), with a frequency of at least one per day, though often much higher (several to over 100 daily in active phases) [1]. Attacks are usually side-locked and may be episodic (with remissions) or chronic (persistent without significant remission).

For **SUNCT** (3.3.1), both ipsilateral conjunctival injection **and** lacrimation (tearing) must occur during attacks. Other cranial autonomic symptoms (e.g., nasal congestion, rhinorrhea, eyelid edema, forehead/facial sweating, miosis, ptosis) may accompany but are not required [1].

SUNA (3.3.2) requires at least one ipsilateral cranial autonomic symptom (which may include conjunctival injection and/or lacrimation, but not both mandatorily). This broader autonomic criterion distinguishes SUNA, though recent meta-analyses and reviews suggest SUNCT and SUNA represent a clinical spectrum rather than entirely separate entities, with overlapping features in many patients [6,7]. A key diagnostic hallmark is the presence of triggers: cutaneous or intraoral innocuous stimuli (e.g., touching the face, chewing, talking, wind on the face) often precipitate attacks, similar to trigeminal neuralgia but without a refractory period after triggering [1]. Migrainous features (photophobia, phonophobia, nausea) can co-occur in up to 40%–50% of cases, complicating differential diagnosis with migraine or other TACs [1, 8].

Epidemiologically, SUNHA is extremely rare, with prevalence estimates <1 per 100,000. It affects both sexes (slight male predominance in some series) across a wide age range, though onset is typically in adulthood (mean ~40–50 years) [9, 10]. SUNCT may present more prominently with autonomic features, while SUNA shows greater variability [11]. Differential diagnosis includes trigeminal neuralgia (shorter attacks, refractory period, no autonomic symptoms), cluster headache (longer duration), paroxysmal hemicrania (responsive to indomethacin), and secondary causes requiring neuroimaging [1, 12].

Etiopathogenesis of SUNCT and SUNA Syndromes The pathophysiology of SUNCT and SUNA involves a complex interplay of central and peripheral mechanisms within the trigeminal-autonomic system. Central elements dominate, with strong evidence implicating hypothalamic dysfunction as a key driver, similar to other TACs like cluster headache. Functional neuroimaging (PET and fMRI) consistently shows activation in the posterior hypothalamus during spontaneous or triggered attacks in SUNCT and SUNA patients [2]. This activation correlates with pain intensity and autonomic features, often ipsilateral or bilateral [13, 14]. Earlier BOLD-fMRI studies in small cohorts (e.g., nine SUNCT and two SUNA cases) demonstrated hypothalamic BOLD signal changes matching attack timing, supporting a permissive or triggering role for the hypothalamus in generating attacks [15]. Recent reviews confirm this, with hypothalamic overactivation leading to disinhibition of the trigeminal-autonomic reflex arc [16].

The trigeminal-autonomic reflex involves afferent trigeminal input synapsing in the trigeminocervical complex (trigeminal nucleus caudalis and upper cervical horns), with efferent parasympathetic outflow via the superior salivatory nucleus, greater superficial petrosal nerve, and sphenopalatine ganglion. This pathway releases vasoactive intestinal peptide (VIP) and other mediators, causing ipsilateral cranial autonomic symptoms (lacrimation, conjunctival injection, rhinorrhea). Hypothalamic modulation—via trigemino-hypothalamic tracts—heightens sensitivity to noxious stimuli and triggers autonomic outflow [17]. In TACs, including SUNHA, posterior hypothalamic activation is thought to act as an "internal trigger," explaining the circadian/autonomic elements and lack of consistent external precipitants in some cases [3]. Deep brain stimulation (DBS) of the posterior hypothalamus has relieved drug-resistant SUNCT in select cases, reinforcing this central role [19].

Peripheral mechanisms contribute significantly, particularly neurovascular compression at the trigeminal root entry zone. High-resolution MRI studies show ipsilateral vascular contact (often superior cerebellar artery) with the symptomatic trigeminal nerve in a substantial proportion of SUNCT/SUNA patients—more than on the asymptomatic side—suggesting focal demyelination or ectopic firing akin to trigeminal neuralgia [20]. This peripheral irritation may generate short, high-frequency neuralgiform pain, while central disinhibition (hypothalamic-driven) amplifies autonomic features absent in pure trigeminal neuralgia. Therapeutic overlap with sodium channel blockers (e.g., lamotrigine, carbamazepine) supports sodium channel dysfunction as a pathophysiological hallmark, potentially at peripheral or central levels [21]. Secondary (symptomatic) SUNCT/SUNA occurs in ~10%–20% of cases, most commonly due to pituitary lesions (micro- or macroadenomas, often prolactin- or growth hormone-secreting) or posterior fossa abnormalities (vascular malformations, infarcts, tumors). Pituitary tumors are ipsilateral to pain in most reported cases, with mechanical compression (e.g., cavernous sinus involvement) or neurohormonal effects postulated [4]. Symptoms often resolve post-surgical removal, confirming causality in many instances [23, 24]. Other secondary causes include brainstem demyelination (e.g., neuromyelitis optica), vascular events, or Rathke's cleft cysts [25]. Neuroimaging (high-resolution MRI with pituitary protocol) is essential to exclude these, as secondary forms may precede endocrine symptoms by years [1, 26].

Overall, SUNCT and SUNA likely represent a clinical spectrum within the broader category of short-lasting unilateral neuralgiform headache attacks (SUNHA), rather than entirely distinct entities. This continuum blends the peripheral, trigeminal neuralgia-like features—such as innocuous cutaneous or intraoral triggers eliciting brief, high-frequency stabbing pain—with the central TAC-like autonomic dysregulation characteristic of cluster headache and related disorders. The strict diagnostic separation in ICHD-3, where SUNCT requires both conjunctival injection and lacrimation while SUNA allows greater variability in cranial autonomic symptoms, is increasingly challenged by overlapping clinical presentations observed in many patients. Emerging evidence from recent meta-analyses and systematic reviews questions the rigid subtype distinction, instead supporting a unified pathophysiological model in which hypothalamic overactivation drives central sensitization and autonomic outflow, while peripheral neurovascular compression or sodium channel dysfunction contributes to the neuralgiform pain quality [6, 7].

This integrated perspective better accounts for the shared treatment responses, imaging findings, and transitional cases reported in the literature. Further research is essential to clarify genetic predispositions, refine advanced neuroimaging techniques (e.g., high-resolution fMRI or PET), and develop more targeted therapies for this rare and often refractory condition.

Clinical Features of SUNCT and SUNA Syndromes Short-lasting unilateral neuralgiform headache attacks with conjunctival injection and tearing (SUNCT) and short-lasting unilateral neuralgiform headache attacks with cranial autonomic symptoms (SUNA) are rare primary headache disorders classified under trigeminal autonomic cephalalgias (TACs) in the *International Classification of Headache Disorders, 3rd edition* (ICHD-3) [1]. These conditions are characterized by recurrent, strictly unilateral episodes of moderate-to-severe head pain accompanied by prominent ipsilateral cranial autonomic symptoms. They can be distinguished from other TACs, such as paroxysmal hemicrania and cluster headache, by their ultra-short attack duration (1–600 s), high frequency (often multiple to hundreds per day), and distinctive neuralgiform pain quality—sharp, stabbing, or electric-shock-like—primarily involving the first division of the trigeminal nerve (V1) [11, 29].

Patients typically describe the pain as scorching, stabbing, electrical, pulsing, electrifying, or searing [29]. Recent clinical studies and meta-analyses confirm that the intensity is most often moderate to severe, with many individuals experiencing excruciating levels that profoundly affect daily functioning and quality of life [26, 31]. Attacks manifest in three main patterns based on duration and grouping: single stabs (brief, isolated jolts lasting seconds), clusters of stabs (rapid sequences of multiple stabs over a short interval), and saw-tooth attacks (prolonged episodes with fluctuating, jagged intensity due to superimposed stabs) [9, 15].

Longer variants, especially saw-tooth patterns, may escalate in severity during the episode, potentially causing heightened distress; some patients display restlessness or agitation, though this is generally milder than in cluster headache and likely secondary to intense pain rather than primary psychological mechanisms [9, 22, 26].

The pain remains strictly unilateral and side-locked in the vast majority of cases, predominantly localized to the orbital, supraorbital, or temporal regions. However, it may extend to the side, top, or back of the head, as well as the second (V2) and third (V3) trigeminal divisions, teeth, neck, or ear [9]. Migrainous features, including photophobia or phonophobia, co-occur in up to 50% of patients, which can complicate differential diagnosis with migraine variants [10].

A defining feature of SUNCT is the obligatory presence of both ipsilateral conjunctival injection (red eye) and lacrimation (tearing) during attacks, serving as key diagnostic markers [10, 18, 23, 27]. Additional autonomic symptoms—such as nasal congestion, rhinorrhea, eyelid edema, facial sweating, miosis, or ptosis—may be present but are not required. In contrast, SUNA requires at least one ipsilateral cranial autonomic symptom, without mandating the combination of conjunctival injection and lacrimation [1, 5]. (see Fig. 8.1).

Recent cross-sectional and meta-analytic studies reveal considerable clinical overlap between SUNCT and SUNA, with shared core phenotypes including attack characteristics, triggers, and autonomic involvement in many patients, supporting the concept of a clinical spectrum rather than fully separate disorders [10, 20, 26]. For instance, multicenter cohorts have shown similarities in qualitative pain descriptions (e.g., stabbing 44.7%, electric-shock-like 36.8%) and common autonomic features like lacrimation as the most frequent symptom, while differences may appear in attack site predominance (e.g., temporal in SUNCT, parietal in some SUNA cases) [26, 28].

The principal diagnostic similarities and distinctions between SUNCT and SUNA are systematically synthesized in Table 8.1 to facilitate structured comparison, enhance diagnostic precision, and support accurate clinical differentiation within the spectrum of trigeminal autonomic cephalalgias.

Triggers represent a prominent aspect of both conditions, with attacks frequently precipitated by otherwise innocuous cutaneous or intraoral stimuli, including light touch to the face, chewing, talking, toothbrushing, or exposure to wind and cold air,

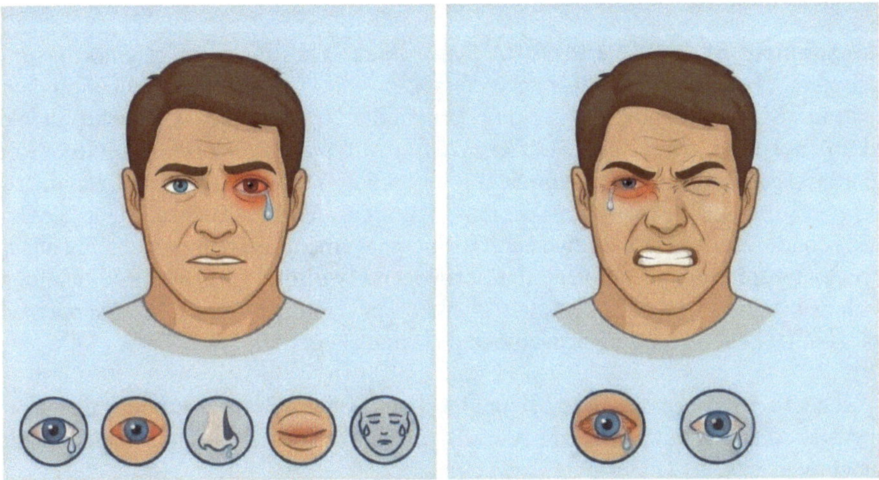

Fig. 8.1 Illustration of cranial autonomic symptoms in SUNCT and SUNA. Left panel: Patient exhibiting unilateral conjunctival injection and lacrimation. Right panel: Intensified symptoms with facial grimacing, additional tearing, and distress. Bottom icons depict associated features including normal versus injected eye, rhinorrhea, oral involvement or flushing, and emotional distress

Table 8.1 Diagnostic comparison of SUNCT and SUNA

Diagnostic criterion	SUNCT	SUNA
Nosological group	SUNHA (TAC subtype)	SUNHA (TAC subtype)
Pain laterality	Strictly unilateral	Strictly unilateral
Pain location	Orbital, supraorbital, temporal, and trigeminal distribution	Orbital, supraorbital, temporal, and trigeminal distribution
Pain quality	Stabbing, electric, neuralgiform	Same as SUNCT
Attack duration	1–600 s	1–600 s
Attack frequency	≥1/day (often 3–200/day)	≥1/day
Cranial autonomic symptoms	Both conjunctival injection AND lacrimation are required	≥1 cranial autonomic symptom required
Other possible autonomic features	Nasal congestion, rhinorrhea, eyelid edema, sweating, miosis, ptosis	Same spectrum
Refractory period	Typically absent	Typically absent

all of which can reliably provoke painful paroxysms in susceptible individuals during daily activities and routines.

Significant pathophysiological overlap with TN is supported by shared clinical traits: adult onset, neuralgiform pain quality, mechanical triggers, and neuroimaging evidence of trigeminal neurovascular compression in approximately 50% of cases [20, 29, 30, 32]. Both respond to similar treatments, particularly sodium channel blockers like lamotrigine. This convergence suggests a unified trigeminal-autonomic spectrum, where hypothalamic dysregulation in SUNCT/SUNA amplifies the autonomic outflow absent in pure TN [10, 32].

Management of SUNCT and SUNA Syndromes The treatment of Short-lasting Unilateral Neuralgiform headache attacks with Conjunctival injection and Tearing (SUNCT) and Short-lasting Unilateral Neuralgiform headache attacks with Autonomic symptoms (SUNA) syndromes typically involves a multifaceted approach combining pharmacological interventions, transitional therapies, and, in refractory cases, surgical options. As rare trigeminal autonomic cephalalgias (TACs), these conditions lack a definitive cure, with management focused on reducing attack frequency, duration, and severity to improve quality of life. Lifestyle modifications, such as avoiding triggers like touch, chewing, or wind exposure, play a supportive role, but evidence-based pharmacotherapy remains central.

Recent advances emphasize personalized algorithms based on efficacy data from open-label studies and meta-analyses, highlighting sodium channel blockers and anticonvulsants.

Preventive medications form the cornerstone of long-term management. Lamotrigine is widely regarded as the first-line preventive agent, demonstrating efficacy in approximately 56%–68% of patients across SUNCT and SUNA, with higher response rates in episodic forms. It often requires titration up to 200–700 mg/day for optimal effect, reducing attack frequency rather than eliminating them entirely. A 2021 prospective study of 161 patients reported 56% responders (≥50% improvement), with similar outcomes in SUNCT (54%) and SUNA (58%). Second-line options include oxcarbazepine (46% response rate), topiramate (25%–50%), carbamazepine (26%–33%), duloxetine (30%), and gabapentin (10-48%).

Oxcarbazepine, a sodium channel blocker, shows promise at doses up to 2400 mg/day, with 59%–71% improvement in recent reviews. Combination therapy may enhance outcomes; for instance, oxcarbazepine with gabapentin has been effective in refractory cases, offering rapid relief with minimal adverse effects compared to gradual titration drugs like topiramate or lamotrigine [33]. A 2024 update reinforces lamotrigine as the primary choice, with topiramate, oxcarbazepine, and gabapentin as alternatives for non-responders. Emerging data from 2023 suggest categorizing treatments into acute, preventive, and bridge therapies, with preventive agents like these reducing daily attacks that can number in the hundreds.

For episodic SUNCT/SUNA, short-term interventions can provide rapid resolution. Oral methylprednisolone has shown complete headache cessation in affected individuals, with no recurrence upon discontinuation at the cycle's end [34]. This corticosteroid approach is particularly useful for predictable bouts, offering quick symptom control.

Transitional or bridge therapies are vital during acute exacerbations or while titrating oral preventives. Intravenous lidocaine is highly effective, achieving pain freedom in 80%–94% of cases and serving as a diagnostic tool if attacks subside during infusion. A 2010 case reported sustained response in SUNCT [35]. Greater occipital nerve (GON) blocks with bupivacaine provide safe, temporary relief, with variable duration (1 week to 6 months) and 28% good response rate [36]. These minimally invasive options are recommended for bridging to stable preventive regimens.

Adjuvant therapies offer additional avenues for incomplete responders. OnabotulinumtoxinA injections have demonstrated sustained efficacy in SUNCT, proposed as a second-line or add-on for antiepileptic non-responders [37]. Recent case reports from 2021 highlight galcanezumab, an anti-CGRP monoclonal antibody, as beneficial in refractory SUNCT, suggesting a role for CGRP pathways in pathophysiology.

Surgical interventions are reserved for medically intractable cases, categorized into trigeminal nerve ablative procedures, microvascular decompression (MVD), and neurostimulation. Outcomes are promising, with significant attack frequency reductions [32]. MVD yields 75% efficacy when neurovascular conflict is identified on MRI (prevalent in 17% of cases). Occipital nerve stimulation (ONS) achieves 77% response (≥50% improvement), while deep brain stimulation (DBS) of the posterior hypothalamus or ventral tegmentum offers 82% response, with up to 78% attack reduction. Sphenopalatine ganglion pulsed radiofrequency provides 78% response at 6 months.

These advances, particularly in neuromodulation, represent key developments for chronic, refractory patients. The prominence of sodium channel blockers (e.g., oxcarbazepine, carbamazepine) in treatment ladders supports an underlying sodium channelopathy, potentially overlapping with conditions like hemiplegic migraine [38]. Molecular analyses could pave the way for targeted therapies. Pediatric considerations mirror adults, with lamotrigine favored, though data are limited. A tiered algorithm—lamotrigine first, followed by oxcarbazepine/duloxetine/topiramate, then gabapentin/pregabalin—guides practice, informed by 2021 meta-analyses. Ongoing research into CGRP antagonists and refined neuromodulation promises further progress.

8.2 Trigeminal Postherpetic Neuralgia: Involvement of the Trigeminal (Semilunar/Gasserian) Ganglion

Abstract Trigeminal postherpetic neuralgia (TG-PHN) is a chronic neuropathic facial pain disorder caused by reactivation of varicella-zoster virus within the trigeminal (Gasserian/semilunar) ganglion. It most often follows herpes zoster affecting the ophthalmic division, particularly in older adults and immunocompromised patients. Persistent unilateral facial pain lasting more than 3 months after herpes zoster, accompanied by burning, dysesthesia, allodynia, hyperalgesia, paresthesia, or sensory loss, is characteristic. Ophthalmic involvement is clinically significant because of the risk of keratitis, uveitis, corneal hypoesthesia, and neurotrophic keratopathy. Diagnosis is primarily clinical, supported by a history of trigeminal zoster and sensory abnormalities on examination, with imaging used to exclude secondary causes. Management includes recombinant zoster vaccination, early antiviral therapy, neuropathic pain pharmacotherapy, and, in refractory cases, trigeminal ganglion–targeted interventions. This chapter reviews the anatomical basis, clinical features, diagnosis, and current management of TG-PHN within a neurostomatological framework.

Keywords Trigeminal postherpetic neuralgia, Trigeminal ganglion, Varicella-zoster virus, Neuropathic facial pain

The trigeminal ganglion, also known as the Gasserian or semilunar ganglion, is a key sensory structure unique in its anatomy and function compared to other primary afferent ganglia. It serves as the primary relay for sensory input from nearly all intracranial and extracranial facial structures through its three major divisions: the ophthalmic (V1), maxillary (V2), and mandibular (V3) nerves. Situated in Meckel's cave in the middle cranial fossa, it processes somatosensory modalities such as pain, temperature, and touch [39]. Lesions or dysfunction of the trigeminal ganglion can result from diverse causes, including viral infections (most commonly varicella-zoster virus [VZV]), trauma, intoxication, tumors, or vascular issues. Viral etiologies, particularly VZV reactivation leading to herpes zoster (shingles) and subsequent postherpetic neuralgia (PHN) in the trigeminal distribution—termed trigeminal postherpetic neuralgia (TG-PHN)—are among the most frequent [40].

After primary VZV infection (chickenpox), the virus establishes lifelong latency in sensory ganglia, including the trigeminal ganglion, without being eradicated [41]. Reactivation, often in older adults (>50 years) or immunocompromised individuals, causes herpes zoster, with the ophthalmic division (V1) most commonly affected, manifesting as herpes zoster ophthalmicus with a vesicular rash in the forehead and periorbital region. If pain persists >90–120 days after rash resolution, it becomes PHN, featuring chronic neuropathic pain with burning, allodynia, hyperalgesia, and intermittent lancinating episodes [40,42]. Risk factors include advanced age, severe acute zoster pain, extensive rash, and immunosuppression; TG-PHN incidence following trigeminal zoster ranges from 10% to 20% in various cohorts [42].

Clinically, trigeminal postherpetic neuralgia (TG-PHN) is characterized by persistent, predominantly unilateral facial pain that continues for more than three months following an acute herpes zoster eruption in the trigeminal distribution. The pain is typically described as a constant dull aching or burning sensation, often punctuated by intermittent lancinating or electric shock–like exacerbations. It radiates across the hemiface and most frequently involves the ophthalmic (V1) and maxillary (V2) divisions of the trigeminal nerve, although the mandibular (V3) branch may also be affected. Sensory disturbances are common and include hypoesthesia, hyperesthesia, paresthesia, allodynia, and dysesthesia within the corresponding dermatomes. Dynamic or static mechanical allodynia—pain evoked by light touch, chewing, or even air movement—may significantly impair daily activities such as speaking, shaving, or oral hygiene [43].

Figure 8.2 shows the clinical appearance of acute herpes zoster ophthalmicus affecting the right ophthalmic (V1) branch, characterized by prominent grouped vesicles and erythema on the right forehead extending to the hairline.

Motor branch involvement is distinctly uncommon because herpes zoster predominantly affects sensory ganglia; however, when present, it may result in mild masticatory weakness attributable to involvement of the mandibular division. Electromyographic studies can demonstrate denervation changes in the muscles of mastication in such rare cases. In patients with V1 involvement, ocular complications constitute a critical concern. Keratitis, conjunctivitis, episcleritis, and uveitis may develop, and corneal hypoesthesia increases the risk of neurotrophic keratopathy and secondary infection. Prompt ophthalmologic evaluation is therefore essential to prevent corneal scarring or irreversible vision loss.

The diagnosis of TG-PHN is primarily clinical, based on a documented history of herpes zoster affecting the trigeminal distribution and persistent neuropathic pain in the same territory. Neurological examination typically reveals sensory deficits or hypersensitivity in the affected dermatomes.

Magnetic resonance imaging (MRI) is recommended to exclude secondary causes of facial pain, such as neoplasms, vascular compression, or demyelinating disorders including multiple sclerosis. According to the *International Classification of Headache Disorders, 3rd edition* (ICHD-3), TG-PHN is defined as persistent pain in the trigeminal nerve territory lasting longer than three months after the onset of herpes zoster [44].

Prevention via recombinant zoster vaccine (Shingrix) dramatically reduces herpes zoster incidence (>90%) and PHN risk (66%–89%) in adults ≥50 years [45]. For

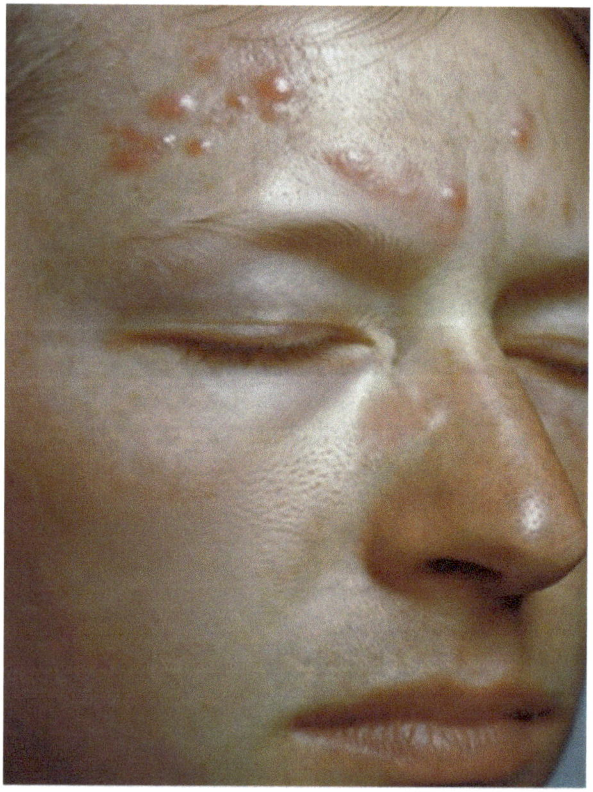

Fig. 8.2 Clinical image of acute herpes zoster ophthalmicus in the right V1 branch. Prominent grouped vesicles and erythema on the right forehead extending to the hairline.

established TG-PHN, first-line pharmacotherapy targets neuropathic mechanisms: gabapentin (900–3600 mg/day) or pregabalin (150–600 mg/day) reduce neuronal hyperexcitability, achieving meaningful relief in 30%–50% of patients, though side effects include dizziness and sedation [46]. Adjunctive options include tricyclic antidepressants (e.g., nortriptyline 25–150 mg/day) or SNRIs (duloxetine 60–120 mg/day) for central pain modulation, plus topical treatments like 5% lidocaine patches or 8% capsaicin cream for localized relief with low systemic risk [47].

Refractory cases benefit from interventional procedures targeting the trigeminal ganglion. CT-guided pulsed radiofrequency (PRF) of the Gasserian ganglion is minimally invasive and effective, with high-voltage, long-duration protocols yielding 70%–80% pain reduction lasting 6–12+ months, often combined with ganglion blocks (local anesthetics + corticosteroids) to enhance outcomes and prevent chronicity in acute/subacute phases [48,49]. A multicenter analysis showed ~67% effectiveness at 3 months post-block with minimal complications (e.g., transient numbness) [49]. Neurolytic approaches, such as Adriamycin (doxorubicin) injection, have demonstrated sustained relief in intractable cases, though evidence remains limited to case reports and requires further validation [50]. Surgical microvascular decompression is less suitable for inflammatory PHN but offers 64–79% long-term relief in select drug-resistant trigeminal neuralgia cases [51].

Multimodal management incorporates physical therapy, cognitive-behavioral therapy, and emerging neuromodulation (e.g., peripheral nerve stimulation). Early antiviral treatment during acute zoster (e.g., valacyclovir) reduces PHN risk by 20%–30%. Complications from interventions are infrequent but include facial hypoesthesia (10%–20%), corneal anesthesia, or infection. Ongoing research explores botulinum toxin, gene-targeted antivirals, and personalized strategies to improve prognosis [40].

In conclusion, TG-PHN is a challenging sequela of trigeminal zoster, best addressed through vaccination, evidence-based pharmacotherapy, and advanced ganglion-targeted interventions in a multidisciplinary framework.

8.3 Burning Mouth Syndrome

Abstract Burning mouth syndrome (BMS) is a chronic idiopathic orofacial pain disorder characterized by persistent intraoral burning or dysesthetic sensations recurring daily for more than 2 h over longer than 3 months, without clinically evident mucosal lesions or other identifiable causes. Current classifications, including ICHD-3 and ICOP, define BMS as a diagnosis of exclusion and distinguish primary BMS from secondary burning symptoms related to systemic or local disorders. The condition occurs predominantly in peri- and postmenopausal women and commonly affects the anterior tongue, lips, palate, and buccal mucosa, often with accompanying xerostomia, dysgeusia, and sensory disturbances. Evidence supports a multifactorial pathogenesis involving peripheral and central neuropathic dysfunction, hormonal influences, psychological factors, and nociplastic mechanisms. Management is primarily symptomatic and multidisciplinary, combining exclusion of secondary causes, patient education, pharmacologic therapy, and supportive non-pharmacological interventions. Despite improved classification and growing pathophysiological insight, BMS remains a diagnostically challenging and therapeutically complex condition with a substantial impact on quality of life.

Keywords Burning mouth syndrome, Chronic orofacial pain, Neuropathic pain, Nociplastic pain, Xerostomia, Dysgeusia, Postmenopausal women, Multidisciplinary management

Definition Burning mouth syndrome (BMS) represents a complex chronic orofacial pain disorder characterized by persistent intraoral burning or dysesthetic sensations in the absence of identifiable local or systemic causes [52]. This condition poses significant diagnostic and therapeutic challenges for clinicians, as affected patients frequently describe unrelenting discomfort that profoundly impairs daily functioning, social interactions, and overall quality of life [53]. Historically, BMS has been referred to by numerous terms, including glossodynia, stomatodynia, glossopyrosis, stomatopyrosis, sore mouth, sore tongue, and oral dysesthesia [54]. These varied and overlapping terminologies have contributed to considerable confusion in clinical practice, research communication, and accurate prevalence estimation, while also complicating etiological investigations and cross-study comparisons [55].

The lack of standardized nomenclature persisted for decades, but contemporary classifications have sought to provide greater clarity and consistency. Current definitions underscore the idiopathic or primary nature of BMS, distinguishing it from secondary forms where an underlying cause can be identified and potentially treated [56]. According to the *International Classification of Headache Disorders, 3rd edition* (ICHD-3), BMS is defined as an intraoral burning or dysaesthetic sensation recurring daily for more than 2 h per day over more than 3 months, without clinically evident causative lesions [57]. This definition emphasizes the burning quality of the pain, its superficial location in the oral mucosa, and the essential requirement for exclusion of visible mucosal abnormalities or other explanatory conditions through thorough clinical examination and appropriate investigations [58].

Similarly, the International Association for the Study of Pain (IASP) describes BMS as a chronic intraoral burning sensation without identifiable local or systemic causes, positioning it within the spectrum of neuropathic or nociplastic pain mechanisms [59]. This alignment highlights the potential involvement of altered pain processing pathways, including peripheral and central sensitization, rather than overt tissue damage [60]. The International Classification of Orofacial Pain (ICOP, 1st edition) further refines this by classifying BMS as an idiopathic orofacial pain, with subtypes distinguished by the presence or absence of somatosensory changes, reinforcing its neuropathic underpinnings [58].

Recent expert consensus has prompted proposals to evolve the terminology further. A 2021 international Delphi study involving orofacial pain specialists recommended renaming the condition from "burning mouth syndrome" to "burning mouth disorder" (BMD) [61]. This shift aims to better reflect its multifaceted pathophysiology, which integrates neuropathic, endocrine, psychosocial, and possibly inflammatory elements, while avoiding the implication of a purely syndromic presentation without a clear disease entity [62]. The proposed reclassification also suggests refining diagnostic criteria in systems like the International Classification of Diseases-11 to exclude emotional factors from core requirements and expand descriptions of affected intraoral sites, thereby improving diagnostic precision and facilitating targeted research [61].

Patients with BMS typically report a burning pain that predominantly affects the anterior two-thirds or tip of the tongue, though it may extend to the lips, hard palate, buccal mucosa, or other oral regions [63]. The sensation is often described as scalding, tingling, or raw, and it tends to be bilateral and symmetric in primary cases [64]. Accompanying complaints frequently include subjective xerostomia (a perceived dry mouth despite normal salivary flow in many instances), dysgeusia (altered or diminished taste perception), and paresthesia or numbness in the affected areas [65].

These sensory disturbances contribute to significant distress, with patients sometimes avoiding eating, speaking, or social activities due to exacerbation by food, temperature, or stress [66]. Symptoms usually persist throughout the day, with variable intensity; in some cases, discomfort may lessen temporarily upon eating or drinking, while in others it worsens progressively [67]. The chronicity and resistance to conventional analgesics further underscore the debilitating impact of BMS on patients' well-being [68].

In summary, BMS is increasingly recognized as a primary neuropathic disorder requiring exclusionary diagnosis and multidisciplinary management. Ongoing refinements in terminology and classification reflect advancing understanding of its complex etiology and promise improved clinical approaches in the future.

Epidemiological data on BMS have improved significantly with the widespread adoption of more stringent diagnostic criteria, such as those from the ICHD-3 and the International Classification of Orofacial Pain (ICOP), which has helped reduce variability in reported prevalence estimates across studies [57, 58]. A landmark 2022 systematic review and meta-analysis, incorporating 18 articles and over 26,000 individuals from diverse geographic regions, estimated the global pooled prevalence of BMS at 1.73% (95% CI: 1.30%–2.20%) in the general population [69]. This figure rises substantially to 7.72% (95% CI: 5.50%–10.10%) among clinical dental patients, reflecting selection bias toward symptomatic individuals who actively seek professional care [69].

The analysis highlighted notable regional differences: in population-based studies, prevalence was higher in Europe (5.58%) and North America (1.00%) compared to Asia (1.05%), whereas clinical prevalence appeared elevated in Asia (8.96%) [69]. These variations may stem from differences in diagnostic awareness, healthcare access, cultural reporting of symptoms, or underlying risk factor distributions across populations.

BMS disproportionately affects women, with a female-to-male ratio typically ranging from 3:1 to 7:1, and it peaks in perimenopausal and postmenopausal individuals, most commonly between 50 and 70 years of age [70]. Gender-specific subgroup analysis from the 2022 meta-analysis confirmed a higher prevalence in females (1.15%) than males (0.38%) in the general population [69]. Age is another key demographic factor; prevalence increases with advancing age, often reaching up to 18%–33% in postmenopausal women according to various reviews and cohort studies [71].

Recent investigations continue to affirm this pattern, with BMS manifesting predominantly around age 65 in some large cohorts and showing strong associations with menopausal hormonal transitions [72]. For instance, a 2024 cross-sectional study in Egypt involving dental clinic attendees reported a 7.9% prevalence, predominantly among females over 50 years old who also had comorbidities such as diabetes mellitus and hypertension [73]. Similarly, a Brazilian retrospective analysis identified a 14.91% prevalence in clinical samples, frequently linked to long-term medication use and depression [74]. These clinical setting estimates align closely with broader meta-analytic findings but underscore the higher burden observed when patients present for care.

Risk factors for BMS encompass demographic, psychosocial, and medical elements. Advancing age and female sex remain the most consistent predictors, with postmenopausal status amplifying vulnerability through estrogen decline and related mucosal or neural changes [65]. Psychosocial stressors play a prominent role; comorbidities including anxiety, depression, and other chronic pain conditions (such as fibromyalgia or irritable bowel syndrome) are highly prevalent among BMS patients and contribute to poorer prognosis and reduced quality of life [75]. Longitudinal observations indicate that spontaneous remission is uncommon, with complete resolution occurring in only 3%–4% of cases over 5–6 years in earlier studies, though more recent cross-sectional surveys suggest partial or complete pain improvement in up to 68%–70% with varied interventions, while persistent symptoms affect a substantial proportion despite treatment [76]. These low spontaneous remission rates emphasize the chronic and burdensome nature of BMS, often leading to prolonged suffering and healthcare utilization.

Diagnostic delays remain a critical issue, averaging 2–3 years from symptom onset to confirmation, largely due to the exclusionary nature of diagnosis and overlap with other orofacial conditions [77]. These epidemiological insights highlight the necessity for heightened awareness and multidisciplinary screening—incorporating dental, neurological, psychological, and endocrinological evaluations—in at-risk populations, particularly postmenopausal women and those with chronic pain or mood disorders. Early identification and targeted management could substantially alleviate the significant personal and societal impact of this underrecognized condition.

Etiopathogenesis of Burning Mouth Syndrome BMS is a chronic orofacial pain disorder whose etiopathogenesis is increasingly understood as multifactorial, involving intricate interactions between neuropathic, hormonal, psychological, nutritional, metabolic, endocrine, and emerging biological factors [52]. Although the precise mechanisms remain partially elusive, contemporary research has shifted away from viewing BMS as purely psychogenic or enigmatic, instead emphasizing its classification as a nociplastic pain condition characterized by altered nociception without evident tissue damage or neural lesions, as per the International Association for the Study of Pain (IASP) criteria [59]. This perspective integrates peripheral and central nervous system dysfunctions with systemic influences, providing a more comprehensive framework for understanding disease onset and progression.

A critical distinction in BMS etiopathogenesis is between primary (idiopathic or essential) and secondary forms [60]. Primary BMS, comprising the majority of cases (50%–70%), lacks identifiable local or systemic causes and is primarily attributed to neuropathic alterations [78]. In contrast, secondary BMS arises from underlying pathological conditions, such as oral infections, nutritional deficiencies, hormonal imbalances, autoimmune disorders, or medication side effects, which can be targeted therapeutically once identified [56]. This classification, endorsed by diagnostic

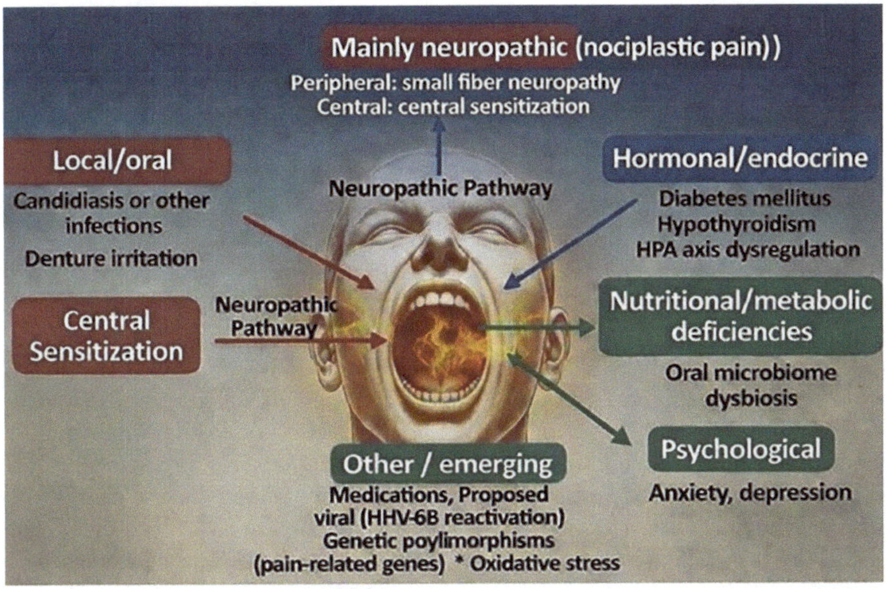

Fig. 8.3 Etiopathogenesis of burning mouth syndrome

frameworks like the ICHD-3 and the International Classification of Orofacial Pain (ICOP), underscores the need for exclusionary diagnostics to differentiate the two, as misclassification can delay appropriate management [57, 58]. Progression in both forms is typically chronic, with symptoms persisting daily for months to years, often exacerbated by comorbidities, and low spontaneous remission rates (3%–4% over 5–6 years) [76]. The multifactorial nature implies that no single factor dominates; rather, a confluence of vulnerabilities—often described as a "perfect storm"—triggers and sustains the condition, particularly in susceptible populations like postmenopausal women [79]. The multifactorial etiopathogenesis of burning mouth syndrome involves complex interactions among peripheral and central neuropathic mechanisms, hormonal imbalances, nutritional deficiencies, psychological factors, and other contributing elements, as illustrated in Fig. 8.3.

Neuropathic Mechanisms

Neuropathic pathways form the cornerstone of primary BMS etiopathogenesis, supported by neurophysiological, histopathological, and imaging evidence [65]. Peripheral neuropathy, especially small fiber neuropathy (SFN), is implicated in 50%–65% of cases, manifesting as axonal degeneration in lingual terminal nerve fibers and sensory abnormalities in heat, cold, taste, and nociceptive stimuli [80].

Quantitative sensory testing (QST) reveals elevated detection thresholds for thermal and mechanical stimuli on the tongue, indicative of hypoesthesia or hyperalgesia, while corneal confocal microscopy demonstrates reduced nerve fiber density and length [81]. Histopathological studies, including tongue biopsies, show decreased intraepidermal nerve fiber density, increased expression of transient receptor potential vanilloid 1 (TRPV1) channels, and purinergic P2X3 receptors, which heighten capsaicin sensitivity and contribute to burning sensations [60]. Electrogustometry further highlights elevated taste thresholds, suggesting involvement of small Aδ afferents in the chorda tympani nerve [65]. Local anesthetics, such as lingual nerve blocks with lidocaine, can temporarily alleviate pain in a subset of patients, confirming peripheral contributions [82].

Central neuropathy complements these peripheral changes, involving altered nociception and pain modulation in the brain [60]. Functional magnetic resonance imaging (fMRI) and positron emission tomography (PET) studies reveal structural deficits, including reduced gray matter volume in the thalamus, hippocampus, and prefrontal cortex, alongside functional hypoactivation in dopaminergic pathways of the basal ganglia, akin to Parkinson's disease [65]. Brainstem reflex recordings, such as the blink reflex, demonstrate abnormal habituation and higher stimulus thresholds for pain-related components, pointing to central sensitization [60]. Heart rate variability (HRV) analyses indicate autonomic instability, with reduced HRV correlating with chronic pain amplification [65]. Trigeminal somatosensory evoked potentials (TSEPs) show earlier spike potentials and lower pain thresholds, further evidencing central dysregulation [60]. These central alterations may explain the diurnal symptom patterns, with pain often worsening throughout the day, and the resistance to conventional analgesics [55].

Hormonal and Endocrine Factors

Hormonal imbalances, particularly those associated with menopause, play a significant role in BMS susceptibility, affecting up to 86% of postmenopausal women [83]. The sharp decline in gonadal steroids during perimenopause alters neuroactive steroid synthesis, leading to neurodegenerative changes in small nerve fibers and brain regions involved in somatic sensation processing [62]. This hormonal "deficit" disrupts salivary composition, mucosal integrity, and neural signaling, exacerbating peripheral neuropathy [60]. Endocrine disorders like hypothyroidism and diabetes mellitus contribute to secondary BMS through microvascular damage and neuropathy; for instance, uncontrolled hyperglycemia impairs nerve conduction and promotes oxidative stress [70]. Recent studies link BMS to neuroendocrine dysfunction, including dysregulation of the hypothalamic-pituitary-adrenal (HPA) axis, which amplifies stress responses and pain perception [78]. Circadian rhythm disturbances, such as altered melatonin secretion, may further influence symptom chronobiology, with pain peaking in the evening due to disrupted sleep-wake cycles [60].

Psychological and Psychogenic Components

Psychological factors are integral to BMS etiopathogenesis, often interplaying with neuropathic mechanisms rather than acting in isolation [84]. Systematic reviews report high prevalence of anxiety (up to 70%) and depression (34%–50%) in BMS patients, assessed via tools like the Hospital Anxiety and Depression Scale (HADS) and Beck Depression Inventory (BDI) [75]. Alexithymia, the inability to identify and express emotions, is significantly elevated (79% vs. 7% in controls), correlating with pain intensity and disability [85]. Personality traits, including high neuroticism, harm avoidance, and perceived stress, exacerbate symptoms, as measured by the Perceived Stress Questionnaire (PSQ) and Temperament and Character Inventory (TCI) [84]. Poor sleep quality, common in 60%–80% of cases, forms a vicious cycle with pain, mediated by central sensitization [86]. The bidirectional relationship—where chronic pain heightens psychological distress, and vice versa—positions BMS as a biopsychosocial disorder, with cytokines like IL-6 linking inflammation to depressive symptoms [75]. Neuroimaging confirms this interplay, showing altered connectivity in emotional processing regions (e.g., amygdala, anterior cingulate cortex) that overlap with neuropathic changes [65].

Nutritional and Metabolic Factors

Nutritional deficiencies contribute to secondary BMS, particularly in cases with identifiable metabolic etiologies [56]. Deficiencies in vitamin B12 (cobalamin), iron, zinc, and B-complex vitamins (e.g., B6, B9) are linked to macrocytic anemia, glossitis, and neuropathic symptoms [56]. Vitamin B12 deficiency, often subclinical, affects nerve myelination and can mimic primary BMS; serum levels detect only 50% of cases, necessitating more sensitive markers like methylmalonic acid (MMA) and homocysteine [56]. Iron and zinc inadequacies impair mucosal repair and taste perception, while metabolic conditions like diabetes induce oxidative stress and microvascular neuropathy [70]. Recent evidence suggests microbiome dysbiosis as an emerging metabolic factor, with altered oral bacterial profiles (e.g., reduced alpha-diversity, overgrowth of Proteobacteria) promoting inflammation and sensory disturbances [60].

Emerging Etiologies

Beyond traditional factors, emerging mechanisms include viral infections, genetic predispositions, and environmental triggers [87]. Human herpesvirus-6B (HHV-6B) has been proposed in BMS pathogenesis, with reactivation potentially causing neuroinflammation, trigeminal neuropathy, and sensory impairments (e.g., dysgeusia, olfactory deficits) [88]. HHV-6B's neurotropism targets cranial ganglia, inducing cytokine storms (TNF, IL-6) that disrupt taste bud renewal and amplify pain [88]. Genetic polymorphisms in pain-related genes (e.g., COMT, SCN9A) may heighten susceptibility, interacting with environmental stressors [81]. Laryngopharyngeal reflux and allergic reactions represent local triggers in secondary cases, while oxidative stress-mediated apoptosis signaling offers a novel hypothesis for cellular damage in the oral mucosa [60]. Bibliometric analyses of recent trends highlight interdisciplinary approaches, integrating neurology, endocrinology, and microbiology for a holistic understanding [79].

The etiopathogenesis of BMS thus reflects a complex interplay of these factors, where peripheral triggers (e.g., nutritional deficiencies) may initiate symptoms, amplified by central sensitization and psychological modulation [60]. Progression often involves a feedback loop: neuropathic pain heightens anxiety, which exacerbates hormonal dysregulation and inflammation [75].

This multifactorial model explains symptom variability and treatment resistance, emphasizing the need for personalized, multidisciplinary interventions [67]. Future research, including biomarkers like neuron-specific enolase (NSE) for neuronal damage and genetic profiling, promises to refine etiological insights and therapeutic targets [52]. In summary, while primary BMS is predominantly neuropathic, secondary forms highlight treatable systemic causes, underscoring the importance of comprehensive diagnostic exclusion to guide management and improve patient outcomes [56].

Clinical Features of Burning Mouth Syndrome BMS is a chronic orofacial pain condition characterized by persistent intraoral burning, dysesthetic, or altered sensations in the absence of detectable mucosal lesions or identifiable systemic disease. The symptoms are highly variable, often fluctuating in intensity and quality, and patients commonly find it difficult to precisely describe their experiences. The cardinal feature is a burning discomfort of the oral mucosa, most frequently described as scalding, tingling, stinging, scorching, prickling, or numb. This sensation typically arises spontaneously without clear triggering events and must be present daily for more than two hours over at least three months to meet current diagnostic thresholds [96].

BMS is classified as a nociplastic pain disorder under the ICOP and ICD-11 frameworks, reflecting altered nociceptive processing without definitive tissue damage or peripheral neuropathy in most cases, although neuropathic mechanisms are increasingly recognized in certain subtypes [96]. The condition has a multifactorial etiology involving central and peripheral sensitization, neuroendocrine changes (especially in peri- and postmenopausal women), psychological factors, and occasionally subtle inflammatory or microbial contributions [102].

The primary symptom—oral burning—is superficial, poorly localized, and most commonly bilateral, though unilateral presentations are documented in a subset of patients and must be differentiated from focal neuropathies or other localized conditions [90]. Pain intensity ranges from mild to severe and characteristically worsens progressively throughout the day, often peaking in the late afternoon or evening and being least bothersome upon awakening. Exacerbating factors include emotional stress, spicy or acidic foods, prolonged speaking, and certain medications, while temporary relief is frequently reported during eating, drinking cool beverages, or oral stimulation (e.g., chewing gum) [91].

Associated symptoms are common and contribute significantly to the clinical picture. Subjective xerostomia (a sensation of dry mouth) is reported by 46%–78.7% of patients, typically without objective reduction in salivary flow rate, and is frequently linked to psychological distress, anxiety, or depression rather than primary salivary gland dysfunction [108].

Dysgeusia or persistent taste disturbance (most often bitter, metallic, or mixed) affects 49%–70% of individuals, with alterations occurring at both detection and recognition thresholds, potentially related to chorda tympani dysfunction or central sensory processing changes [93]. Other sensory complaints include intraoral paresthesia, itching, a coated or filmy sensation on the mucosa, and occasional extraoral dryness (lips, eyes, skin, genitalia) [94].

The anatomical distribution of symptoms in BMS shows a clear predilection for certain intraoral sites. The tongue is the most commonly and severely affected area in the vast majority of cases, with involvement predominantly of the anterior portion (especially the tip), dorsum, and lateral borders/sides. Recent clinical studies across diverse populations consistently identify the tongue as the primary site, with reported involvement ranging from 67.6% to 92.6% of cases; for example, one large cohort of 834 patients found tongue involvement in 92.6%, including the tip in 50.2%, sides in 46.8%, and dorsum in 30.2% [90].

The lips (labial mucosa), anterior hard palate, and buccal mucosa are also frequently involved, though typically less commonly as primary or isolated locations. Involvement of the palate is reported in around 20%–24% in some studies, while buccal mucosa often appears in combined patterns or lower frequencies. Other sites, such as the throat, floor of the mouth, and gingiva, are less commonly primary areas of involvement.

Symptoms in BMS are characteristically diffuse rather than sharply localized, often affecting multiple intraoral areas simultaneously (multi-site distribution) in many patients, and they are typically bilateral [105]. This variable, widespread pattern—most prominently featuring the tongue and buccal mucosa, followed by the palate and lips—underscores the poorly localized and neuropathic-like nature of the persistent burning or dysesthetic sensation in BMS.

Psychological and systemic comorbidities form a core part of the BMS clinical spectrum. Anxiety and depression are present in up to 80% of patients, frequently predating symptom onset and exhibiting bidirectional relationships with pain perception [98]. Sleep disturbances affect approximately 78.8% of individuals, with poor sleep quality strongly correlated with increased symptom severity [99]. Personality traits associated with cluster C disorders (e.g., avoidant, dependent), somatization tendencies, and medically unexplained extraoral symptoms (irritable bowel syndrome, fibromyalgia) are overrepresented [100, 109]. Cognitive complaints, sometimes termed "Burning Fog," involving difficulties with attention, working memory, and executive function, have been linked to chronic pain-induced neuroplastic changes [88]. Stress-related biomarkers, including elevated salivary cortisol and α-amylase, further support a prominent psychoneuroendocrine component [89].

Systemic associations include hypertension, hypercholesterolemia, gastroesophageal reflux disease, thyroid dysfunction, and nutritional deficiencies (vitamin B12, iron, folate). Symptom improvement following treatment of GERD with proton pump inhibitors has been documented in some cohorts [101]. Post-COVID-19 onset or exacerbation has also been reported, potentially mediated by increased psychological stress, social isolation, or direct viral effects on sensory pathways [102].

Diagnosis remains one of exclusion, requiring thorough clinical examination and targeted investigations to rule out local (candidiasis, lichen planus, geographic tongue) and systemic causes (anemia, diabetes, Sjögren syndrome, medication adverse effects). Modern diagnostic approaches incorporate quantitative sensory testing to identify somatosensory subgroups, blood panels (including neuron-specific enolase for neuronal injury markers), salivary flow measurements, and validated psychological screening tools (e.g., Hospital Anxiety and Depression Scale, Pittsburgh Sleep Quality Index) [103]. The ICOP-2020 criteria emphasize the absence of primary mucosal disease and the presence of daily burning for ≥3 months, while advocating for multidisciplinary evaluation [96]. Emerging proposals suggest reclassifying the condition as "Burning Mouth Disorder" to better encompass its heterogeneous sensory, affective, and cognitive features [104].

In conclusion, the clinical presentation of BMS is dominated by chronic intraoral burning pain, sensory disturbances (dysgeusia, xerostomia), and profound psychological and systemic comorbidities that collectively impair quality of life, eating behavior, sleep, and social functioning. Management remains symptomatic and multimodal, with ongoing research into microbiota alterations, circadian rhythm disruptions, central sensitization pathways, and novel pharmacotherapies offering hope for improved outcomes [105].

Management of Burning Mouth Syndrome The management of BMS starts with a comprehensive diagnostic workup to exclude reversible local or systemic causes, including oral infections (e.g., candidiasis), mucosal diseases, nutritional deficiencies (e.g., vitamin B12, iron), allergies, salivary dysfunction, medications (e.g., ACE inhibitors), endocrine disorders (e.g., diabetes, hypothyroidism), and habits like smoking or excessive alcohol use. Where possible, these factors should be eliminated or treated. Patient education and reassurance are fundamental: clinicians should affirm that the symptoms are legitimate, neuropathic in origin, and not indicative of cancer or a life-threatening condition. This validation reduces fear and the secondary anxiety amplification of pain. Given BMS's multifactorial etiology—neuropathic small-fiber changes, central sensitization, psychological factors, and hormonal influences in postmenopausal women—a multidisciplinary team (oral medicine/oral pathology specialists, neurologists, psychologists, and endocrinologists) is essential for coordinated care.

No curative treatment exists, but symptom improvement occurs in 50%–80% of patients with evidence-based, individualized strategies. Management falls into three main categories: topical therapies, systemic medications, and non-pharmacological interventions, often used in combination for superior outcomes [106–108].

Topical therapies target peripheral mechanisms with low systemic risk. Topical clonazepam (0.5–1 mg tablet dissolved in the mouth three times daily) enhances GABAergic inhibition and has shown consistent short-term pain reduction (VAS decrease of 2–4 points) in 60%–80% of patients, with better tolerability than systemic forms and minimal side effects like mild drowsiness [94]. Capsaicin (0.025% cream or rinse) depletes substance P to desensitize nociceptors, providing moderate relief in some cases, though initial burning limits adherence in ~40% of users [89]. Low-level laser therapy (LLLT, also called photobiomodulation; 810–980 nm, 0.5–3 J/cm², sessions two to three times weekly for 4–8 weeks) reduces inflammation, promotes neural repair, and yields significant immediate and sustained pain relief (standardized mean difference ~ −0.87 on pain scales) plus quality-of-life gains, often outperforming or equaling pharmacotherapy with no adverse effects [99].

Emerging compounded topicals (e.g., amitriptyline/gabapentin gels) offer localized analgesia for refractory cases. Topical agents suit mild-to-moderate BMS as first-line or adjunctive options [106].

Systemic medications address central pain processing and comorbidities (e.g., anxiety/depression in ~50% of cases). Oral clonazepam (0.5–2 mg/day) extends topical benefits, with meta-analyses confirming meaningful VAS reductions and symptom control in ~70% over 4–12 weeks [91]. Pregabalin (starting 75 mg twice daily, titrated to 150–300 mg) modulates calcium channels and shows strong efficacy in neuropathic subtypes, often superior to gabapentin with fewer side effects like dizziness [92]. Alpha-lipoic acid (ALA, 600 mg/day) provides antioxidant neuroprotection, with updated reviews indicating positive outcomes in many RCTs (e.g., higher proportion of responders vs. placebo), though pooled pain reduction may be non-significant due to study heterogeneity; it remains a safe adjunct with mild gastrointestinal effects [93]. Antidepressants like vortioxetine (10–20 mg/day) target serotonergic/noradrenergic pathways, reducing pain (~40%) and mood symptoms without a strong anticholinergic burden [94]. Systemic options require slow titration, monitoring, and combination use for optimal results, especially in refractory or comorbid cases [106, 107].

Non-pharmacological approaches emphasize behavioral and physical modalities with excellent safety profiles. Cognitive behavioral therapy (CBT; 8–12 sessions) reframes pain cognition, teaches coping skills, and achieves moderate-to-significant reductions in burning intensity (VAS −1.5 to −2.5), anxiety, and quality-of-life impairment, with sustained benefits and no side effects [95]. LLLT/photobiomodulation, as noted earlier, excels here for rapid, non-invasive analgesia [96,97,110].

Other supportive measures include tongue protectors to reduce mechanical irritation and mindfulness/stress management to mitigate psychogenic contributors. Multimodal regimens (e.g., LLLT + CBT + clonazepam) yield the highest response rates (up to 80%), highlighting personalized, stepped-care plans [106].

Reassurance is crucial throughout, as BMS is chronic but not progressive or malignant. Spontaneous complete remission is rare (~3%–4% within 5 years), though partial improvement occurs in 20%–30% without specific treatment. Long-term follow-up with periodic reassessment allows therapy adjustment. Emerging research explores biomarkers, neuromodulation (e.g., transcranial magnetic stimulation), and novel agents for better targeting.

Overall, a holistic, evidence-based strategy focused on validation, multidisciplinary input, and combined interventions significantly alleviates symptoms and improves well-being in most patients [106].

8.4 Sluder's Syndrome

Abstract Sluder's syndrome is a historically described but diagnostically controversial craniofacial pain syndrome characterized by unilateral facial pain with ipsilateral cranial autonomic symptoms, including nasal congestion, rhinorrhea, lacrimation, and conjunctival injection. Pain usually begins in the nasal root, periorbital, or maxillary region and may radiate to the temple, occiput, teeth, neck, or shoulder. Because of substantial overlap with cluster headache, migraine, trigeminal neuralgia, sinusitis, and rhinogenic contact point headache, it is frequently misdiagnosed, and its status as a distinct clinical entity remains uncertain. Current literature suggests that many cases may represent secondary rhinogenic or contact point–related facial pain rather than a primary neuralgic disorder. Proposed mechanisms include nasal or paranasal inflammation, anatomical variations, irritation of the anterior ethmoidal nerve or sphenopalatine ganglion, and secondary sensitization. Diagnosis is primarily clinical and exclusion-based, supported by otolaryngologic assessment, imaging, and response to sphenopalatine ganglion block in selected cases. Management is individualized and multidisciplinary, incorporating pharmacological therapy, topical nasal treatment, sphenopalatine ganglion block, interventional pain procedures, neuromodulation, and correction of contributory rhinologic abnormalities in carefully selected patients.

Keywords Sluder's syndrome, Sphenopalatine ganglion neuralgia, Cranial autonomic symptoms, SPG block

Definition Sluder's syndrome, also known as sphenopalatine ganglion (SPG) neuralgia or anterior ethmoidal nerve syndrome, is a rare craniofacial pain disorder characterized by severe, unilateral facial pain often accompanied by autonomic symptoms such as nasal congestion, rhinorrhea, lacrimation, conjunctival injection, and occasionally altered gustation. The pain typically originates near the root of the nose, eye, or maxilla and radiates to areas including the orbit, zygoma, temple, occiput, teeth, or even the neck and shoulder. Episodes can be episodic or persistent, lasting from hours to days, and are frequently triggered by environmental factors like barometric pressure changes, cold air exposure, or allergens.

Due to significant overlap with other headache disorders such as cluster headaches, migraines, and trigeminal neuralgia, Sluder's syndrome is often misdiagnosed, leading to debates about its classification as a distinct entity. Recent literature suggests it may be better categorized as a secondary headache under trigeminal autonomic cephalalgias (TACs) or rhinogenic contact point headache (RCPH), emphasizing its nasal and inflammatory origins rather than a simple variant of cluster headache [111]. This reclassification highlights the need for multidisciplinary approaches involving otolaryngologists and neurologists to address its underrecognition and improve patient outcomes.

The prevalence of Sluder's syndrome is poorly defined due to its rarity, diagnostic ambiguity, and frequent misattribution to more common conditions like sinusitis or migraines. It is estimated to account for a small subset of craniofacial pain cases, with only limited case series and reports available in contemporary literature.

A 2022 review indicates that it may represent up to 5%–10% of atypical facial pain referrals in specialized clinics, though global epidemiological data is sparse [112]. Median age of onset is typically in the fourth or fifth decade (around 40–50 years), with a slight male predominance (male-to-female ratio approximately 1.3:1), although both genders are affected. Risk factors include a history of nasal trauma, endoscopic sinus surgery, chronic rhinosinusitis, or anatomical nasal variations, which contribute to underreporting and diagnostic delays. In regions with high allergen exposure or variable climates, incidence may be higher due to environmental triggers exacerbating symptoms.

Recent studies from 2025 underscore the importance of advanced imaging in identifying predisposing anatomical subtypes, potentially aiding in better prevalence estimates through prospective cohorts [113]. Without standardized diagnostic criteria in the *International Classification of Headache Disorders* (ICHD-3), epidemiological tracking remains challenging, necessitating further research to quantify its burden and demographic patterns.

Etiopathogenesis of Sluder's Syndrome The etiopathogenesis of Sluder's syndrome remains complex and multifactorial, involving an interplay of inflammatory, anatomical, neurological, and potentially vascular mechanisms. Although the exact cause is not fully elucidated, contemporary research from the last five years has shifted focus from Sluder's original 1908 description of sphenopalatine ganglion irritation to more precise pathways centered on the anterior ethmoidal nerve and mucosal contact points.

This evolution reflects advances in neuroimaging and endoscopic techniques, which have revealed structural and inflammatory contributors as primary drivers, with secondary neural sensitization amplifying symptoms. At its core, Sluder's syndrome is often linked to inflammatory processes within the paranasal sinuses, particularly the posterior ethmoid and sphenoid regions. Irritation of the pterygopalatine ganglion (also known as the sphenopalatine ganglion) arises secondary to these inflammatory events, leading to unilateral persistent headache that initiates lateral to the nose or near the eye and radiates across the face, maxilla, and beyond [112]. This inflammation may stem from recurrent sinus infections, allergic rhinitis, or non-infectious mucosal edema, which transmits irritants or pressure to the ganglion via contiguous anatomical pathways.

The ganglion, located in the pterygopalatine fossa, serves as a relay station for parasympathetic, sympathetic, and sensory fibers, making it susceptible to such disturbances. When inflamed, it triggers autonomic manifestations like rhinorrhea and lacrimation through parasympathetic overactivation, while sensory fibers convey the neuralgic pain.

Recent case reports from 2025 highlight how chronic inflammation can lead to ganglion demyelination in idiopathic cases, perpetuating a cycle of episodic exacerbations without obvious triggers [114]. This inflammatory hypothesis is supported by clinical observations where resolution of sinus pathology via medical or surgical intervention often alleviates symptoms, suggesting a direct causal link.

Anatomical variations play a pivotal role in the etiopathogenesis, acting as predisposing factors that facilitate mechanical irritation or altered nasal aerodynamics. Rhinogenic contact point headache (RCPH), increasingly viewed as synonymous or overlapping with Sluder's syndrome, arises from mucosal contact points in the nasal and sinus cavities without overt inflammation, polyps, or tumors [113]. Key variations include nasal septal deviation (defined as an angle greater than 5° from the midline), septal spurs (bony or cartilaginous protrusions), and concha bullosa (pneumatization of the middle turbinate creating an air-filled cavity enclosed by bone).

These anomalies create persistent contact between the nasal septum and lateral nasal wall, leading to localized pressure and neurogenic stimulation. For instance, septal spurs can impinge on sensitive mucosal areas, inducing headache through mechanical stress or altered airflow patterns that promote mucosal drying and irritation. Concha bullosa exacerbates this by unevenly distributing contact forces, potentially worsening outcomes post-intervention due to residual asymmetry. A 2025 study utilizing preoperative computed tomography (CT) scans classified these subtypes and found that patients with concha bullosa had higher postoperative pain scores (Visual Analog Scale [VAS] of 4.0) compared to those with isolated septal deviations (VAS 1.8), attributing this to incomplete resolution of contact points [113]. This underscores how anatomical subtypes influence pathogenesis, with CT imaging emerging as a non-invasive tool for predicting symptom severity and therapeutic response.

Neurologically, the condition involves the trigeminal nerve system, particularly the anterior ethmoidal nerve, a branch of the nasociliary nerve within the ophthalmic division (V1). Irritation at this level, often due to anatomical or inflammatory factors, activates nociceptive pathways and produces the characteristic deep, aching pain [114]. Small-diameter afferent fibers transmit signals to the trigeminal nucleus caudalis, where central sensitization may amplify pain perception and contribute to referred symptoms in areas such as the teeth or occiput. In parallel, the sphenopalatine ganglion (SPG) participates in autonomic and sensory integration through its connections with the maxillary nerve (V2), parasympathetic fibers from the greater petrosal nerve, and sympathetic fibers from the superior cervical ganglion. This interaction may trigger reflex lacrimation and nasal hypersecretion, thereby linking trigeminal nociceptive activation with the cranial autonomic features observed in some patients.

Emerging evidence from 2022 reviews suggests a neuroplastic component, where chronic irritation induces cognitive dissonance or heightened central processing, explaining why symptoms persist even after anatomical correction in certain patients

[115]. This neuroplasticity may involve maladaptive changes in pain modulation pathways, similar to those in other chronic headache disorders, and highlights the potential role of psychological factors in symptom perpetuation.

Vascular mechanisms, while less emphasized in recent literature, contribute through vasodilatory responses mediated by peptides within SPG neurons. Vasoactive intestinal polypeptide (VIP), nitric oxide synthase, and other mediators promote mucosal vasodilation and increased blood flow, potentially exacerbating inflammatory changes and pain [112]. Animal models have shown that SPG stimulation leads to nasal mucosa hyperemia, mirroring the vascular component in migraine-like symptoms observed in Sluder's syndrome. However, unlike primary migraines, the vascular changes here are secondary to ganglion irritation rather than central triggers.

Additionally, secondary forms of Sluder's syndrome may arise from structural lesions, such as tumors, cysts, or post-traumatic scarring affecting the SPG or ethmoidal regions. Idiopathic cases, lacking obvious pathology, could involve subtle demyelination or genetic predispositions, though genomic studies are limited due to the condition's rarity. Environmental triggers, including barometric shifts or allergens, may lower the threshold for symptom onset by aggravating mucosal sensitivity.

Overall, the etiopathogenesis of this condition integrates these key elements into a comprehensive unified model: anatomical variations create a predisposition to neurovascular contact points, local inflammation directly irritates sensitive neural structures, and central/peripheral neurological sensitization perpetuates and sustains the chronic pain cycle. Ongoing debates continue regarding its nosological independence from trigeminal autonomic cephalalgias (TACs), with several experts proposing a subdivision into idiopathic forms (primarily driven by demyelination processes) and secondary forms (related to identifiable compressive or lesional pathology) [112]. Further high-quality research, particularly prospective studies employing advanced imaging techniques, remains essential to better elucidate these underlying mechanisms, clarify diagnostic boundaries, and identify reliable biomarkers that could enable the development of more targeted and effective therapeutic strategies.

The principal peripheral and central mechanisms implicated in the etiopathogenesis of Sluder's syndrome are schematically summarized in Fig. 8.4.

Clinical Features of Sluder's Syndrome Sluder's syndrome, also referred to as sphenopalatine ganglion neuralgia or anterior ethmoidal nerve syndrome, presents with a distinctive array of symptoms that primarily revolve around unilateral craniofacial pain accompanied by autonomic manifestations. First described in the early twentieth century, the condition is characterized by pain that originates in the nasal or periorbital region and can significantly impair daily functioning. Unlike more common headache disorders, Sluder's syndrome often stems from irritation of the sphenopalatine ganglion or anterior ethmoidal nerve, leading to a symptom profile that blends neuralgic elements with parasympathetic overactivity [112].

Clinical recognition is crucial, as misdiagnosis is frequent due to overlaps with entities like cluster headache, migraine, and sinusitis. Recent reviews emphasize the

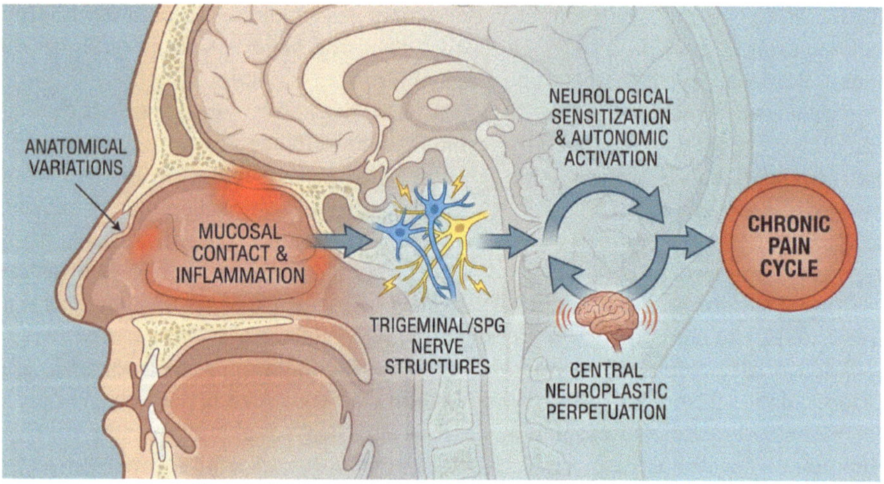

Fig. 8.4 Etiopathogenetic mechanisms of Sluder's syndrome

importance of detailed symptom mapping to differentiate it, highlighting its moderate severity and prolonged duration as key discriminators [112].

Patients typically report onset in mid-adulthood, with a median age of 41.8 years and a slight male predominance (male-to-female ratio of 1.3:1), though it affects both genders [113]. The pain is nearly always unilateral (99% of cases), without side-shifting, and can be either persistent or episodic, underscoring the need for thorough history-taking to capture its variability.

The hallmark symptom of Sluder's syndrome is unilateral facial pain, which patients often describe as burning, dull, aching, or pressure-like. This pain typically initiates at the root of the nose or near the eye, remaining ipsilateral and radiating to adjacent areas such as the maxilla, orbit, upper teeth, temple, ear, mastoid, occiput, neck, or even the shoulder and upper arm in more extensive cases [111]. In severe instances, radiation may extend further to the axilla, forearm, or hand, though this is less common and should prompt evaluation for overlapping neuropathies.

The intensity is generally moderate, rated as severe by approximately 75% of patients and extreme (10/10 on the Visual Analog Scale [VAS]) by 25%, but it lacks the excruciating quality seen in other neuralgias [116]. Episodes can last from hours to days, with 72% of reported cases exceeding 180 min, and attacks may occur daily without the structured clustering or remission periods typical of other disorders.

This prolonged, nagging character distinguishes it from shorter, more intense pains, contributing to chronic disability if untreated. Environmental factors often exacerbate symptoms, including changes in barometric pressure, exposure to cold air, allergens, or nasal irritants, which may lower the threshold for pain initiation by aggravating mucosal sensitivity [117]. Emotional stress has been identified as a trigger in some instances, further complicating the clinical picture and potentially linking to psychosomatic amplification.

Autonomic symptoms are prominent and ipsilateral, adding to the diagnostic specificity of Sluder's syndrome. These include nasal congestion or obstruction, rhinorrhea (often serous), lacrimation, conjunctival injection or redness, and occasionally ptosis or eyelid drooping [118].

Hypersecretion from the nasal mucosa can lead to a sensation of fullness or pressure, mimicking sinus-related issues, while ocular symptoms like tearing and light sensitivity may cause discomfort in bright environments. In rare cases, patients report altered gustation, such as distorted taste perception, or increased salivation on the affected side, reflecting involvement of parasympathetic pathways [119].

These features arise from activation of the sphenopalatine ganglion, which integrates sensory, parasympathetic, and sympathetic inputs, resulting in reflex responses that amplify the pain experience. Unlike purely inflammatory conditions, these autonomic signs occur without overt signs of infection, such as purulent discharge or fever, emphasizing the neurogenic origin. Recent studies note that these symptoms are less pronounced than in related trigeminal autonomic cephalalgias (TACs), providing a subtle but important differentiation point [120].

Associated signs and symptoms extend beyond core pain and autonomic features, occasionally including sensory or motor disturbances. Some individuals experience hyperesthesia or tingling in the affected facial regions, while others report itchiness in the upper extremities or transient weakness in the shoulder and arm that resolves with pain subsidence [121].

These atypical manifestations are infrequent but highlight the potential for referred phenomena due to trigeminal convergence in the brainstem. Dizziness or nausea, though rarely documented, may occur in comorbid scenarios but are not intrinsic to the syndrome; their presence often signals alternative diagnoses like vestibular migraine [122].

The overall symptom complex can lead to significant functional impairment, with patients avoiding triggers such as outdoor activities in variable weather, resulting in social isolation or occupational challenges. Quality-of-life assessments in recent cohorts reveal high VAS scores at presentation (mean 6–9), underscoring the chronic burden and the need for prompt intervention [123].

Variability in presentation is a notable aspect of Sluder's syndrome, influenced by underlying etiological factors such as anatomical nasal variations or inflammatory triggers. For instance, in cases linked to rhinogenic contact point headache (RCPH)—often considered overlapping or synonymous—the pain may correlate with specific structural anomalies like nasal septal deviation, septal spurs, or concha bullosa, leading to more localized nasal pressure and less radiation [124].

Preoperative evaluations in such patients show a mean headache frequency of 9–10 episodes per month, with intensity varying based on the subtype of anatomical variation; concha bullosa cases tend to have higher residual symptoms post-treatment compared to isolated deviations [123]. Episodic forms may mimic cyclical patterns but lack the circadian rhythmicity of other headaches, while persistent variants present as daily discomfort without clear exacerbations. Gender differences are subtle, with females potentially reporting more autonomic ocular symptoms, though data are limited due to the condition's rarity [125]. Pediatric presentations are exceedingly rare, with most cases emerging in adulthood, possibly due to cumulative exposure to nasal irritants or trauma.

To illustrate, consider a case from recent literature: A 45-year-old male presented with persistent right-sided facial pain starting at the nasal root, radiating to the orbit and maxilla, accompanied by ipsilateral rhinorrhea and lacrimation. Symptoms worsened with cold exposure and emotional stress, lasting 4–6 h daily, with a VAS of 8/10. No side-shifting occurred, and autonomic features were mild, distinguishing it from more agitated presentations [124].

Another example involves a 52-year-old female with episodic pain linked to septal deviation, featuring pressure-like discomfort in the periorbital area, nasal congestion, and occasional taste distortion, resolving partially with environmental avoidance [124]. These vignettes highlight the spectrum, from moderate daily pain to intermittent flares, and emphasize the role of triggers in modulating severity.

Differential considerations are integral to understanding clinical features, as Sluder's syndrome often overlaps symptomatically with cluster headache, yet key distinctions exist. In cluster headache, pain is excruciating, orbital/temporal, lasting 15–180 min with 4–12-week clusters followed by remissions, accompanied by marked agitation and restlessness—features absent in Sluder's [125].

Demographics differ too: cluster affects younger males predominantly, while Sluder's has later onset and balanced gender distribution. Migraine presents with throbbing, bilateral pain, photophobia, and nausea, exacerbated by movement, contrasting Sluder's unilateral, non-throbbing nature without vestibular symptoms [126].

Trigeminal neuralgia involves brief, electric-shock pains triggered by touch, lacking autonomic components. Sinusitis includes inflammatory signs like fever or purulent discharge, resolved by antimicrobials, unlike Sluder's neurogenic profile [126]. RCPH, closely related, focuses on mucosal contacts without ganglion emphasis, but shares pain localization and benefits from similar diagnostics like endoscopic confirmation [111]. Accurate delineation relies on symptom duration (prolonged in Sluder's), severity (moderate vs. extreme), and absence of restlessness or periodicity.

In summary, the clinical features of Sluder's syndrome encompass a unique blend of moderate unilateral facial pain, prominent ipsilateral autonomic nasal and ocular symptoms (such as congestion, rhinorrhea, lacrimation, and conjunctival injection), and variable radiating discomfort that may extend to the temple, occiput, neck, or shoulder. This symptom complex is often triggered or exacerbated by environmental factors like barometric pressure changes, cold air, allergens, or stress, reflecting the condition's strong rhinogenic and neurovascular basis. Pain is typically described as aching, burning, or pressure-like rather than excruciating, with episodes lasting hours to days and lacking the short, clustered, restless attacks characteristic of cluster headache. Variability arises from underlying anatomical factors (e.g., septal deviation or concha bullosa) and inflammatory triggers, leading to episodic or persistent presentations. Enhanced awareness through recent studies and improved diagnostic tools, such as nasal endoscopy and targeted sphenopalatine ganglion blocks, facilitates earlier recognition, significantly reducing misdiagnosis with sinusitis, migraine, or

Category		Specific Symptoms	Details
Facial Pain	😟	Severe, unilateral pain	Pain around the eye, nose, cheek, or upper jaw, often described as aching or burning.
Nasal Symptoms	😢	Nasal congestion, runny nose	Stuffy or runny nose on the affected side, mimicking sinus issues.
Eye Symptoms	👁️	Tearing (lacrimation), redness, sensitivity to light	Excessive tearing, red eye (conjunctival injection), and discomfort in bright light.
Headaches	😐	Localized headaches	Pain often concentrated on one side of the head, typically near the eye or temple.
Radiating Pain	❎	Pain spreading to ear, neck, or shoulder	Discomfort may extend beyond the face, affecting nearby areas.
Other Symptoms	⚠️	Occasional dizziness or nausea	Rarely, patients may experience dizziness or nausea due to nerve involvement, though not core features.

Fig. 8.5 Clinical features of Sluder's syndrome

other trigeminal autonomic cephalalgias and thereby improving patient trajectories and quality of life.

The characteristic clinical symptoms of Sluder's syndrome, including unilateral facial pain with associated autonomic features, are summarized in Fig. 8.5.

Future research should prioritize the development of standardized diagnostic criteria, prospective cohort studies to better define prevalence and subtypes, and validation of symptom-based scoring systems to further refine these presentations and guide evidence-based management.

Management of Sluder's Syndrome The treatment of Sluder's syndrome focuses on relieving unilateral facial and headache pain linked to sphenopalatine ganglion (SPG) irritation, often from nasal/sinus issues. Conservative approaches include medications such as anticonvulsants (e.g., carbamazepine or gabapentin) for neuropathic pain, NSAIDs for inflammation, and topical nasal corticosteroids or decongestants to reduce mucosal irritation [112]. Diagnostic and therapeutic SPG blocks using local anesthetics (e.g., transnasal lidocaine application) provide rapid relief and confirm the diagnosis. For refractory cases, interventional options like pulsed or thermal radiofrequency ablation of the SPG offer longer-term pain reduction [117]. Phenolization (88% phenol application to the SPG) has been reported as effective, though temporary, in older and select recent cases [113]. Emerging or adjunctive therapies may include minimally invasive endoscopic nasal surgery to address contact points or structural causes. A multidisciplinary approach (neurology, ENT, pain management) is recommended, with emphasis on excluding mimics like cluster headache or rhinogenic pain before escalating interventions. Evidence remains limited due to the condition's rarity and diagnostic controversy.

Pharmacological Management

Pharmacological approaches primarily target neuropathic pain control, autonomic symptom mitigation, and modulation of trigeminal-parasympathetic reflexes. These strategies are frequently adapted from established trigeminal autonomic cephalalgia (TAC) treatment guidelines, reflecting shared pathophysiological mechanisms between the conditions.

Anticonvulsants form the cornerstone for neuralgic components. Carbamazepine (starting at 200 mg/day, titrated to 800–1200 mg/day) or oxcarbazepine (600–1800 mg/day) stabilize neuronal membranes and suppress ectopic discharges, demonstrating efficacy in TAC-like syndromes and case reports of SN or SPG-related pain [128]. Gabapentin (900–3600 mg/day) and pregabalin (150–600 mg/day) target voltage-gated calcium channels, offering moderate relief in refractory facial neuralgias or persistent pain, though side effects like dizziness, sedation, and weight gain necessitate careful titration and monitoring [129]. In resistant cases, these agents may be combined or switched based on tolerability.

Nonsteroidal anti-inflammatory drugs (NSAIDs), especially indomethacin (75–225 mg/day, often with gastroprotection), serve both diagnostic and therapeutic roles, particularly when paroxysmal hemicrania or other indomethacin-responsive headaches overlap with SN presentations. Adjunctive therapies for autonomic features include topical nasal corticosteroids (e.g., fluticasone) or decongestants to reduce mucosal swelling that may irritate the SPG [116]. Emerging options target neurogenic inflammation via calcitonin gene-related peptide (CGRP) pathways. Monoclonal antibodies like galcanezumab (300 mg monthly subcutaneous for episodic CH; off-label lower doses explored) have shown promise in reducing TAC attack frequency and severity, with preclinical data supporting trigeminal nociceptive inhibition applicable to SPG-related pain [130]. Real-world observational studies in refractory TACs report >50% attack reduction in a substantial proportion of patients, with good tolerability (primarily mild injection-site reactions) [131]. Other CGRP inhibitors (fremanezumab, eptinezumab) exhibit similar preventive benefits in open-label TAC data, though SN-specific evidence remains limited.

Verapamil (240–480 mg/day, with ECG monitoring) provides autonomic stabilization as a first-line preventive in CH but has sparse dedicated data for SN-like presentations. For mild or episodic cases, start with oral agents; escalate to biologics or combinations (e.g., verapamil plus galcanezumab) in refractory scenarios. Efficacy tracking via headache diaries, visual analog scales (VAS), or HIT-6 questionnaires is essential. Evidence levels are moderate (primarily TAC-extrapolated), underscoring the need for dedicated SN research amid diagnostic ambiguity.

Interventional Procedures: SPG Blocks and Radiofrequency Ablation

Interventional techniques targeting the sphenopalatine ganglion (SPG) provide dual benefits: diagnostic confirmation through a positive response to local anesthetic blockade, and therapeutic relief by interrupting excessive parasympathetic outflow while modulating nociceptive signaling pathways. These minimally invasive procedures—such as transnasal blocks, radiofrequency ablation, or neurostimulation—are particularly valuable for achieving rapid acute symptom control in cluster headache or migraine attacks, as well as offering preventive management in medically refractory cases.

Transnasal sphenopalatine ganglion blocks (SPGB) continue to stand out as a highly accessible, minimally invasive, and low-risk interventional option for managing various headache disorders [127]. The procedure typically involves applying a local anesthetic—most commonly 4% lidocaine, though bupivacaine is also used—directly to the mucosal area near the sphenopalatine ganglion. This is achieved using cotton-tipped applicators soaked in the anesthetic or specialized catheters inserted along the middle turbinate to ensure proper contact with the target mucosa (see Fig. 8.6).

Recent systematic reviews and clinical studies have underscored the technique's favorable profile, including a rapid onset of action, frequently within 15–30 min after application, with pain relief often becoming noticeable as early as 15 min in responsive patients. The duration of benefit commonly extends to 24–48 h following a single block, though repeated administrations can yield longer-lasting effects, sometimes spanning several days or contributing to cumulative preventive benefits over time

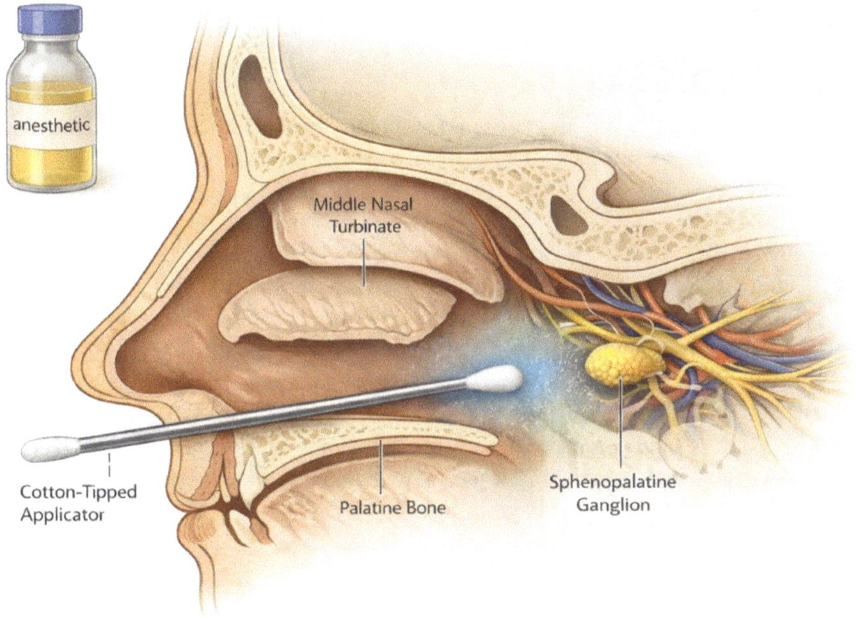

Fig. 8.6 Transnasal SPG block technique: Two cotton-tipped applicators soaked in 4% lidocaine are advanced along the superior border of the middle turbinate in each nostril until contacting the mucosa overlying the SPG. Maintain position for 10–15 min; repeat as clinically indicated

[131]. SPGB demonstrates broad utility across multiple headache and pain conditions, encompassing acute migraine attacks, cluster headache (CH) episodes, chronic or refractory migraine, trigeminal autonomic cephalalgias, and various forms of facial pain such as persistent idiopathic facial pain or trigeminal neuralgia. In primary care and outpatient settings, repeated blocks—whether performed in-clinic or through self-administered protocols using simplified techniques—have been associated with meaningful reductions in attack frequency, severity, and overall headache burden, particularly in cohorts with medically refractory symptoms unresponsive to standard pharmacotherapies. The safety profile remains excellent, with complications occurring infrequently and generally being mild and transient. These include minor epistaxis (nasal bleeding), temporary nasal discomfort or congestion, throat numbness, bitter taste, lightheadedness, or brief episodes of nausea. Serious adverse events are exceedingly rare, further reinforcing SPGB as a safe and practical initial interventional step before considering more invasive alternatives.

In select cases where precise targeting is challenging due to anatomical variations, the addition of image guidance—such as fluoroscopy or ultrasound—can enhance procedural accuracy and optimize outcomes without significantly increasing risk or complexity. Overall, transnasal SPGB offers an effective bridge for acute symptom control while supporting longer-term management strategies in difficult-to-treat headache populations.

Evidence for SPG interventions is strong in CH (Level A/B for blocks/stimulation), moderate for broader facial pain including SN proxies (case series and reviews). Precision via imaging and staged procedures (diagnostic block before ablation) optimizes safety and efficacy.

For longer-lasting relief, radiofrequency ablation (RFA) of the sphenopalatine ganglion (SPG) effectively denervates parasympathetic and nociceptive fibers. Pulsed RFA, a non-thermal technique using 42°C pulses, selectively modulates nerve activity while preserving sensory function, delivering 6–12 months of meaningful pain relief in conditions such as atypical facial pain and trigeminal autonomic cephalalgias (TACs) [132]. In contrast, conventional thermal RFA—typically performed at 80°C via a CT-guided suprazygomatic approach—provides more durable results, with over 50% pain reduction sustained at 12 months in patients with refractory headaches, accompanied by significant improvements in quality-of-life measures [133]. Adverse events, including transient bradycardia due to parasympathetic stimulation and temporary palatal dysesthesia, remain uncommon, occurring in less than 10% of procedures (see Fig. 8.7). In selected complex cases, combined interventions—such as SPG RFA paired with trigeminal branch ablation—may further enhance therapeutic outcomes for challenging craniofacial pain syndromes.

Neurostimulation and Neuromodulation

Neuromodulation provides non-pharmacological alternatives for chronic or refractory SN-like pain by modulating SPG or related pathways.

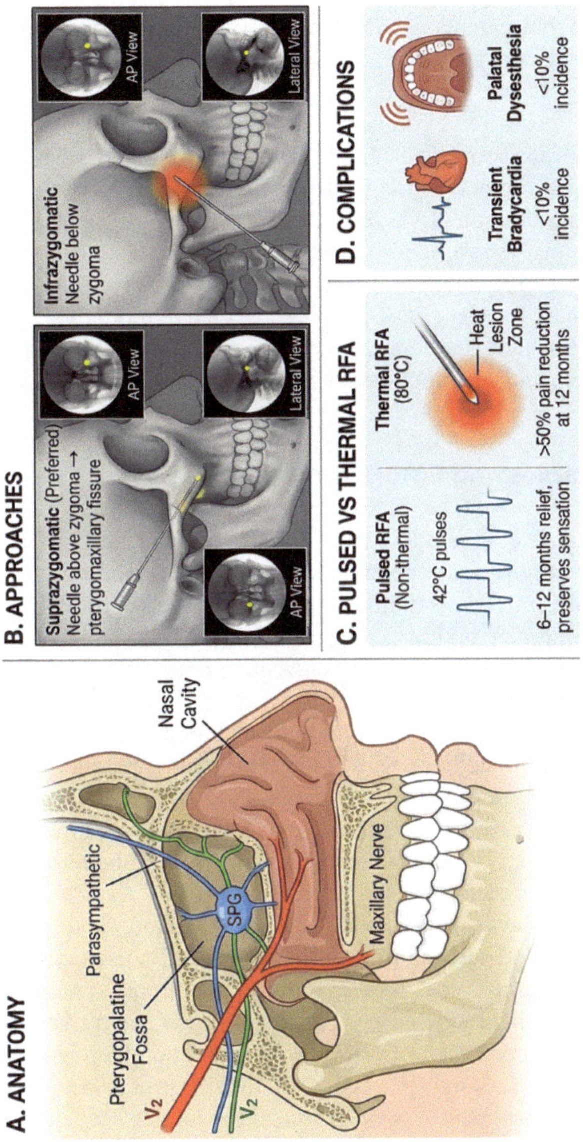

Fig. 8.7 Sphenopalatine ganglion radiofrequency ablation (RFA)

Implantable SPG stimulation (SPGS) uses microstimulators for on-demand pulses via suprazygomatic placement. Retrospective analyses in refractory craniofacial pain (including TAC overlaps) demonstrate acute attack abortion and preventive frequency reduction, with trial phases predicting responders [134]. Combined SPG-occipital nerve stimulation yields synergistic long-term benefits in chronic TAC syndromes [135]. Noninvasive transcutaneous SPGS (external devices) offers home-based safety and efficacy in facial pain case series, expanding accessibility [136].

Noninvasive vagus nerve stimulation (nVNS), such as with the gammaCore device, activates central antinociceptive pathways and has demonstrated preventive efficacy in reducing trigeminal autonomic cephalalgia (TAC) attacks, as supported by real-world clinical evidence [137]. Emerging adjunctive noninvasive modalities, including transcranial magnetic stimulation (TMS) and transcutaneous electrical nerve stimulation (TENS), show promising results for managing refractory headache disorders when standard treatments prove insufficient

Evidence is robust for CH (FDA-approved devices), moderate for SN proxies. Advantages include pharmacotherapy sparing, though implantation risks (infection <5%), costs, and patient selection remain considerations.

Emerging Therapies and Future Directions

Innovative approaches continue to evolve. Cryoneuromodulation (transnasal cryotherapy) induces temporary fiber denervation with prolonged relief in facial pain. Botulinum toxin SPG injections inhibit acetylcholine release, providing 3–6 months of benefit in TAC/migraine reviews [138]. Preclinical explorations of gene therapies targeting CGRP receptors suggest future preventive potential.

Multidisciplinary management of trigeminal autonomic cephalalgias (TACs) and related headache syndromes effectively combines pharmacological treatments and various interventional procedures with essential non-pharmacological approaches. These include carefully implemented lifestyle modifications—such as consistent trigger avoidance and optimized sleep hygiene—alongside comprehensive psychological support, particularly cognitive-behavioral therapy tailored to enhance coping mechanisms for chronic pain.

Patient education plays a central role, empowering individuals with knowledge about their condition and self-management strategies. Evidence demonstrates that well-coordinated integrated care models significantly improve clinical outcomes, quality of life, and treatment adherence in patients with TACs and associated disorders.

Looking ahead, key research priorities encompass conducting dedicated randomized controlled trials focused specifically on Sluder's syndrome to better delineate diagnostic boundaries and reduce overlap with other entities. Concurrently, ongoing investigations are exploring the efficacy of low-frequency sphenopalatine ganglion stimulation (SPGS), next-generation advanced biologics targeting novel pathways, and minimally invasive rhinologic procedures—such as endoscopic decompression or relief of mucosal contact points—in carefully selected cases where a rhinogenic contribution is suspected.

Although SN remains diagnostically ambiguous and understudied, evidence-based, stepwise management—starting with conservative pharmacological options, progressing to interventional and neuromodulatory therapies for refractory presentations—significantly enhances patient function and well-being. Clinicians must emphasize excluding established diagnoses (e.g., CH, secondary rhinogenic pain) while applying SPG-targeted strategies judiciously.

8.5 Charlin's Syndrome

Abstract Charlin's syndrome, also known as nasociliary neuralgia or anterior ethmoidal nerve syndrome, is a rare neuropathic facial pain disorder caused by irritation or dysfunction of the nasociliary nerve (ophthalmic division of the trigeminal nerve). First described by Carlos Charlin in 1931, it features severe, unilateral paroxysmal pain in the ocular, retro-orbital, nasal bridge, and facial areas, often with autonomic signs including ipsilateral lacrimation, rhinorrhea, and conjunctival injection. Pain episodes, triggered by local tactile stimulation (e.g., nasal contact points), last minutes to over an hour and differ from trigeminal neuralgia (briefer attacks) and cluster headache (circadian patterns). Etiopathogenesis is poorly defined due to limited evidence (mostly case reports and small series), primarily involving peripheral nerve irritation from intranasal anatomical variations (e.g., deviated septum with spurs, concha bullosa), post-traumatic damage (e.g., post-rhinoplasty neuroma or fibrosis), inflammatory causes (e.g., herpes zoster), or idiopathic factors. Historical overlap exists with entities like Sluder's neuralgia or contact-point headaches. Diagnosis depends on clinical features (sharp unilateral pain from inner canthus/nasal root, variable autonomic signs, hyperesthesia), nasal endoscopy for contact points, imaging to exclude central lesions, and temporary relief from local anesthetic blocks or application. Management, extrapolated from trigeminal neuralgia due to scarce specific data, includes anticonvulsants (e.g., carbamazepine, oxcarbazepine), gabapentinoids, or anti-inflammatories; interventional options like nasociliary nerve blocks or pulsed radiofrequency; and, rarely, endoscopic surgery for structural anomalies. Multidisciplinary care is advised.

Keywords Charlin's syndrome, Nasociliary neuralgia, Anterior ethmoidal nerve syndrome, Neuropathic facial pain, Trigeminal nerve, Autonomic cephalalgia differential, Paroxysmal ocular pain

Definition Charlin's syndrome, also known as nasociliary neuralgia or anterior ethmoidal nerve syndrome (AENS), is a rare neuropathic facial pain disorder characterized by severe, paroxysmal pain primarily localized in the ocular, retro-orbital, nasal, and facial regions. First described by Chilean ophthalmologist Carlos Charlin in 1931, the condition arises from irritation or dysfunction of the nasociliary nerve, a key sensory branch of the ophthalmic division (V1) of the trigeminal nerve. This syndrome is marked by its distinctive pain distribution and associated autonomic symptoms, which often overlap with other craniofacial pain disorders such as cluster headaches, migraines, or sinogenic headaches, leading to frequent misdiagnosis and delayed treatment. The nasociliary nerve supplies sensory innervation to critical structures including the cornea, iris, ciliary body, nasal mucosa, and the skin of the external nose, making its involvement central to the syndrome's clinical manifestations [139].

Due to its rarity, comprehensive epidemiological data on Charlin's syndrome remain limited, primarily derived from isolated case reports and small-scale observational studies rather than large population-based surveys. Recent studies indicate that it predominantly affects middle-aged adults, with a mean age of onset around 36 years, ranging from 24 to 53 years in reported cohorts. A slight female predominance has been observed, with approximately 60% of cases involving women, though this may reflect selection or reporting biases rather than true population differences [139].

In clinical practice, Charlin's syndrome accounts for a minor proportion of headache consultations, often underdiagnosed and mistaken for more prevalent conditions like trigeminal neuralgia or chronic rhinosinusitis. Increased awareness, facilitated by advancements in diagnostic tools such as high-resolution nasal endoscopy and computed tomography (CT) imaging, has led to better recognition of underlying nasal anatomical abnormalities contributing to the syndrome [140].

For instance, in a prospective study of 31 patients with persistent craniofacial pain, anterior ethmoidal nerve irritation was confirmed in cases associated with nasal septal spurs, highlighting the syndrome's link to structural nasal variants [140]. Recent literature emphasizes the importance of distinguishing Charlin's from similar entities to improve patient outcomes, noting that misdiagnosis can lead to unnecessary treatments and prolonged suffering [141].

Historically, Charlin's syndrome has been occasionally grouped under broader categories of craniofacial neuralgias, but it is distinct from trigeminal autonomic cephalalgias (TACs) such as cluster headache, as it lacks the characteristic rhythmic, seasonal, or nocturnal patterns typical of TACs. Instead, it is classified as a neuropathic pain syndrome within orofacial pain disorders, emphasizing its peripheral nerve origin.

Differentiation from related but separate entities is crucial: for example, Sluder's neuralgia (sphenopalatine ganglion neuralgia) involves pain irradiating to the cheek and palate, whereas Charlin's focuses on nasociliary structures with prominent ocular and nasal symptoms [141]. This distinction avoids conflation, as seen in some older literature, and underscores the need for precise anatomical correlation in diagnosis. Recent reviews have revisited these syndromes to enhance diagnostic accuracy, pointing out that Charlin's often presents with unilateral symptoms tied to nasociliary irritation [141].

Anatomically, the nasociliary nerve originates from the trigeminal ganglion as part of the ophthalmic division (V1) of the trigeminal nerve (cranial nerve V). It enters the orbit through the superior orbital fissure, courses along the medial orbital wall, and gives off several branches. Key among these is the anterior ethmoidal nerve, which passes through the anterior ethmoidal foramen into the cranial cavity, ascends over the cribriform plate, and then descends into the nasal cavity via a slit lateral to the crista galli. In the nasal cavity, it divides into internal (medial and lateral) nasal branches and the external nasal branch. The internal nasal branches innervate the anterior nasal septum and lateral nasal wall, while the external nasal branch emerges between the nasal bone and upper lateral cartilage to supply the skin of the nasal tip and ala.

The nasociliary nerve also provides a sensory root to the ciliary ganglion but does not originate from it; the ciliary ganglion primarily handles parasympathetic functions for pupillary constriction and accommodation [139]. This intricate pathway renders the nerve vulnerable to compression or irritation at multiple sites, from the orbit to the nasal cavity, particularly at points of anatomical narrowing or contact. Case reports illustrate how infections like herpes zoster can affect this nerve, leading to complications such as keratitis or iridocyclitis, further complicating the clinical picture [142].

The nasociliary nerve's proximity to the ethmoid sinuses and its role in sensory innervation explain the syndrome's overlap with sinonasal pathologies. Irritation at the anterior ethmoidal nerve can trigger referred pain to orbital and facial areas, often mimicking inflammatory or vascular headaches. In differential classification, Charlin's syndrome is recognized as a peripheral neuropathic disorder requiring multidisciplinary evaluation by otolaryngologists, neurologists, and ophthalmologists. Recent insights emphasize its underdiagnosis in primary care settings, where it may be overlooked in favor of more common headache types.

For example, in reviews of facial neuralgias, nasociliary neuralgia is highlighted as a rare but treatable cause of unilateral ocular pain, distinct from glossopharyngeal or occipital neuralgias [143]. Accurate identification relies on correlating clinical symptoms with anatomical findings, avoiding misattribution to central mechanisms. Studies on related neuropathies, such as trigeminal neuropathy, provide context for understanding sensory dysfunction in Charlin's, though direct overlaps are limited [144].

In summary, Charlin's syndrome represents a niche within neuropathic facial pains, driven by nasociliary nerve dysfunction. Its rarity, combined with symptomatic overlap, necessitates heightened clinical suspicion and targeted diagnostics to prevent prolonged morbidity. Advances in the last five years, including prospective studies linking it to specific nasal anomalies, have refined our understanding, positioning it as an under-recognized yet manageable entity in craniofacial pain medicine [140].

Etiopathogenesis of Charlin's Syndrome The etiopathogenesis of Charlin's syndrome is multifaceted but centers on irritation, compression, or inflammation of the nasociliary nerve or its branches, resulting in aberrant neuropathic pain signaling. A primary mechanism involves mechanical irritation from intranasal contact points, where anatomical variations cause chronic friction between the nasal septum (medial wall) and turbinates (lateral wall).

Common structural culprits include deviated nasal septum (DNS) with or without septal spurs, concha bullosa (pneumatized middle turbinate), paradoxically curved middle turbinate, enlarged ethmoidal bullae, or superior turbinate impinging on the septum.

These anomalies compress the terminal branches of the anterior ethmoidal nerve, which innervate the middle turbinate and septum, leading to neuralgic pain radiating to orbital and facial regions [139].

Trauma significantly contributes, especially iatrogenic injury from nasal procedures such as septorhinoplasty or turbinectomy. During such surgeries, shearing forces between the mobile upper lateral nasal cartilage and fixed nasal bone can damage the external nasal branch, potentially forming neuromas or entrapping the nerve in scar tissue. Post-traumatic cases often exhibit delayed onset, as fibrosis and neuroma development progress over months.

Inflammatory causes encompass infections like herpes zoster involving the nasociliary nerve or chronic sinusitis inducing mucosal edema and secondary compression. In rare instances, ciliary ganglionitis may occur, precipitated by infections or vascular insufficiency, manifesting the syndrome's autonomic features [142]. Idiopathic variants persist, possibly linked to central sensitization, though metabolic factors such as diabetes or substance use (e.g., alcohol, tobacco) may heighten nerve susceptibility without direct causation [141].

Pathophysiologically, the syndrome implicates the intricate neural network around the ciliary ganglion, where sensory, sympathetic, and parasympathetic fibers interconnect, enabling cross-irritation. Unlike Sluder's neuralgia, which centers on the pterygopalatine (sphenopalatine) ganglion with cheek irradiation, Charlin's emphasizes nasociliary and ciliary involvement, producing localized ocular redness, nasal hypersecretion, and pain at the inner canthus. Evidence for vascular compression, akin to trigeminal neuralgia's arterial impingement, is scant in Charlin's, with intranasal mechanical factors predominating. Recent studies underscore the role of nasal septal spurs in etiology: in a cohort of 31 patients, 93% had DNS with spurs pressing on the lateral nasal wall, confirming contact-point irritation as a key trigger [140]. This mechanical hypothesis is supported by diagnostic responses to local anesthetics, which temporarily abolish pain by blocking irritated nerve fibers [141].

Functional impairments, such as altered sinus ciliary motility, can exacerbate mechanical compression by inducing pressure differentials and edema. Observational data from recent years identify varied etiologies: DNS with septal spur (most common), concha bullosa, post-traumatic fibrosis, nerve sheath tumors, or oversized ethmoidal bullae. For instance, in a 2025 review of craniofacial pains, nasociliary involvement was differentiated from other neuralgias, emphasizing peripheral compression over central demyelination [141]. Similarly, a 2023 StatPearls update on nummular headache lists nasociliary neuralgia as a differential, highlighting its neuropathic basis without migrainous auras [143]. Case reports of herpes zoster with nasociliary involvement illustrate inflammatory pathways, where viral reactivation leads to nerve damage and secondary bacterial superinfections, adding layers to the etiopathogenesis [142].

The syndrome's multifactorial nature highlights the interplay of anatomical, traumatic, and inflammatory elements. In prospective analyses, females show higher susceptibility, possibly due to hormonal influences on mucosal edema, though robust evidence is lacking. Systemic diseases like multiple sclerosis or tumors must be excluded as secondary causes. Literature on trigeminal neuropathy offers similar and analogous insights, indicating that sensory loss or even pain resulting from nerve damage may mimic Charlin's syndrome specifically in its distribution patterns [144].

Overall, due to the rarity of Charlin's syndrome, it demands a thorough and meticulous evaluation to properly differentiate primary mechanical irritation from other mimicking conditions, with recent literature advocating early endoscopic assessment to identify treatable structural anomalies [140]. This methodical approach not only elucidates the underlying pathogenesis but also guides the application of targeted interventions, ultimately reducing diagnostic delays in this often under-recognized disorder [141].

A schematic overview of the multifactorial etiopathogenesis of Charlin's syndrome is presented in Fig. 8.8.

Clinical Features of Charlin's Syndrome The clinical presentation of Charlin's syndrome is typified by acute, unilateral paroxysms of sharp, stabbing, or burning pain in the ocular or retro-orbital region, often radiating ipsilaterally to the forehead, nasal bridge, maxilla, and upper jaw. Pain episodes feature rapid onset and peak intensity, averaging 45–60 min but varying from minutes to hours, evolving from throbbing to intense stabbing.

A hallmark trigger is tactile stimulation, such as palpation of the lateral nostril, nasal bridge, or intranasal contact points, which reliably reproduces symptoms. Distinct from migraines, there is generally no associated nausea, vomiting, or primary photophobia, though secondary photophobia may arise from ocular irritation [139].

Autonomic manifestations are prominent, including ipsilateral rhinorrhea, nasal congestion, lacrimation, sneezing, conjunctival hyperemia, and scleral injection. Ocular signs like redness at the inner canthus and eye irritation are pathognomonic, reflecting ciliary ganglion involvement.

Additional features encompass cutaneous hyperesthesia, forehead erythema, mild upper eyelid edema, and tenderness at nasal contact points. Pain localizes over the nasal root or bridge, with no reported bilateral cases in recent series. In chronic presentations, pain may become persistent, with exacerbations triggered by environmental changes, stress, or infections [141]. The pain is often described as deep, dull, or pressure-like, worsening with barometric shifts or cold exposure, distinguishing it from other neuralgias [141].

The principal clinical characteristics of Charlin's syndrome are summarized in Table 8.2.

Diagnostic evaluation relies on clinical criteria: unilateral pain, contact-point tenderness, absence of migrainous auras, and symptomatic relief with topical decongestants via mucosal shrinkage. Nasal endoscopy identifies contact points (e.g., turbinate-septum impingement), while CT imaging delineates anatomical variants,

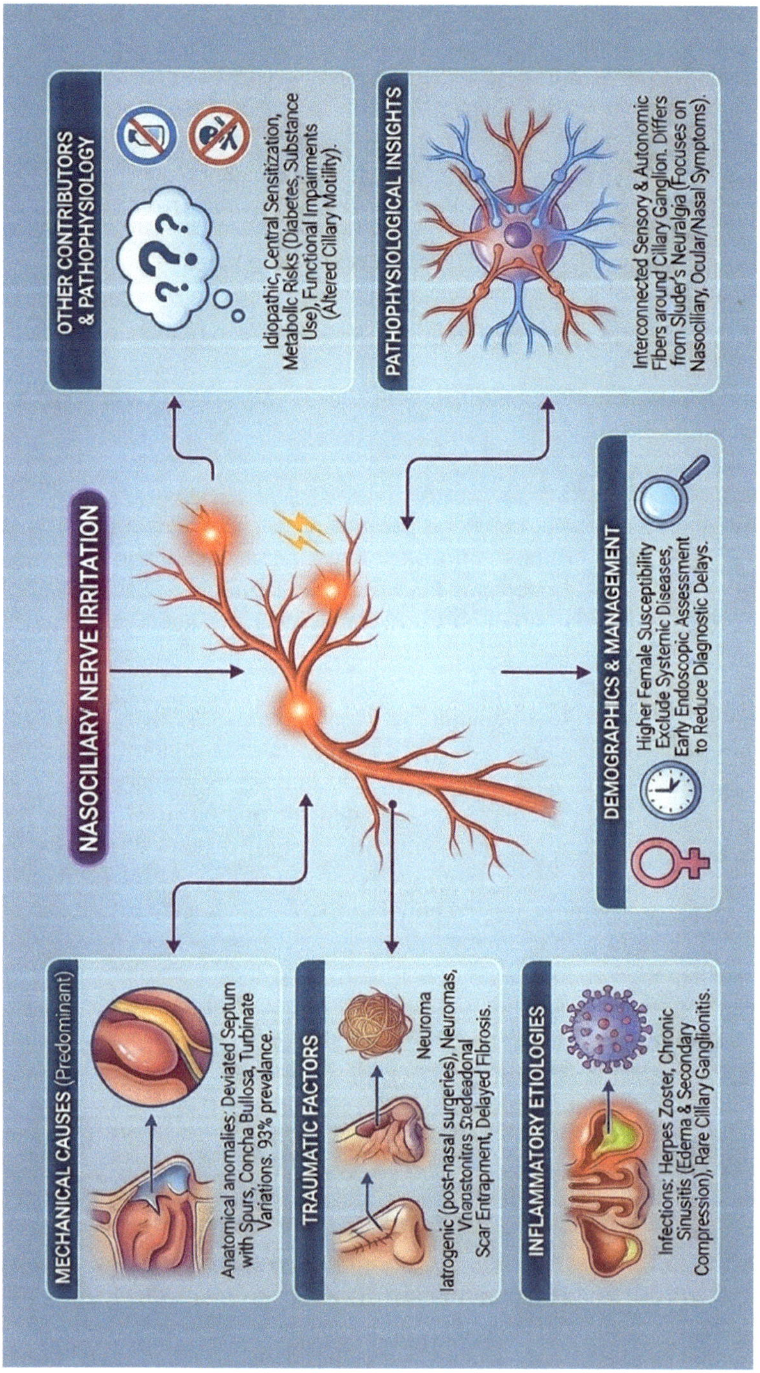

Fig. 8.8 Etiopathogenesis of Charlin's syndrome

Table 8.2 Clinical features of Charlin's syndrome

Feature	Description
Pain location	Ocular/retro-orbital, radiating to forehead, nose, maxilla
Pain quality	Sharp, stabbing, burning; paroxysmal or continuous
Duration	45–60 min typical; seconds to hours
Triggers	Palpation of lateral nostril, nasal bridge touch
Autonomic symptoms	Ipsilateral rhinorrhea, lacrimation, nasal congestion, photophobia
Ocular signs	Scleral injection, iridocyclitis, keratitis, corneal ulcer
Other	Sneezing, forehead erythema, eyelid swelling, hyperesthesia
Diagnostic test	Pain relief with local anesthetic on nasal mucosa
Course	Paroxysms with long remissions (months–years)

albeit with potential false positives from edema. A confirmatory test involves applying 4% lidocaine to the anterior nasal mucosa or contact point, yielding temporary pain abolition. Long remissions (months to years) are common, but relapses follow trauma or infection [140].

Differential diagnosis encompasses cluster headache (lacks tactile triggers), trigeminal neuralgia (briefer attacks), persistent idiopathic facial pain, chronic rhinosinusitis, dental pathologies, or atypical migraines [145]. Recent case series reveal variability: post-rhinoplasty persistence for years or trauma-onset in middle-aged females. Underdiagnosis arises from symptomatic mimicry, necessitating high suspicion in refractory headaches. In a 2023 review, nasociliary neuralgia was differentiated from nummular headache by its autonomic features and nasal triggers [143]. Inflammatory cases, such as those with herpes zoster, add ocular complications like keratitis, emphasizing the need for prompt recognition [142]. Related neuropathies, like trigeminal, share pain patterns but differ in etiology and response [144]. In conclusion, Charlin's syndrome poses diagnostic challenges due to its rarity and mimicry, requiring integrated approaches [141].

Management of Charlin's Syndrome Management strategies for Charlin's syndrome remain challenging due to its rarity and limited specific evidence. A stepwise, multidisciplinary approach is advised, involving neurologists, pain specialists, otolaryngologists, and psychologists, with emphasis on early intervention to mitigate central sensitization risks. Patient education on trigger avoidance (e.g., nasal contact, cold air) and psychosocial support are crucial, as chronic pain impacts quality of life [146]. However, outcomes vary, with case reports suggesting potential self-resolution in post-traumatic cases, underscoring the need for long-term follow-up [147].

Diagnostic evaluation is essential for tailored therapy. Confirmation often relies on positive response to nasociliary nerve block, which provides both diagnostic clarity and temporary relief [148]. Etiological factors, such as sinusitis or herpes zoster, warrant targeted interventions like antibiotics (e.g., amoxicillin-clavulanate 875 mg twice daily for 7–10 days) or antivirals (e.g., acyclovir 800 mg five times daily for 7 days)

[149]. Imaging, particularly MRI, helps exclude central causes like multiple sclerosis (MS), which rarely complicates Charlin's but increases recurrence if present [150].

Differential diagnoses include anterior ethmoidal neuralgia (AEN), which overlaps with Charlin's in symptoms like nasal bridge pain but may involve middle turbinate-septum contact points; distinguishing requires endoscopic evaluation [151]. Unlike TN, Charlin's lacks chronobiological patterns or alcohol triggers, and sphenopalatine blocks are ineffective, highlighting its peripheral nature [152]. Functional assessments, such as rhinomanometry for nasal obstruction, aid in identifying structural anomalies [153].

Pharmacological management draws from TN evidence but must account for Charlin's inflammatory component. Sodium channel blockers are prioritized to reduce ectopic firing. Carbamazepine, starting at 200 mg and titrating to 1200 mg daily, is first-line, with meta-analyses in TN showing 75%–91% pain reduction, though evidence in Charlin's is anecdotal from case series [154]. Monitoring for hyponatremia, dizziness, and hematologic issues via regular blood tests is vital [155]. Oxcarbazepine (600–1800 mg daily) offers similar efficacy with better tolerability and fewer drug interactions, making it preferable for older patients [156].

For partial response or intolerance, gabapentinoids like gabapentin (900–3600 mg daily) or pregabalin modulate calcium channels, suppressing neurotransmitter release; case reports in Charlin's indicate 200–600 mg gabapentin effective [157]. Combination therapy, such as carbamazepine with gabapentin, may enhance control but risks polypharmacy [158]. Short-term corticosteroids (e.g., prednisone 40–60 mg tapered over 10–14 days) address inflammation, though limited data suggest caution for hyperglycemia [159]. Antidepressants like nortriptyline (25–75 mg nightly) provide dual analgesic and mood benefits via serotonin-norepinephrine reuptake inhibition, useful for associated anxiety or depression [160]. Opioids are avoided due to poor neuropathic efficacy and addiction potential, with tramadol reserved for acute flares [161]. Emerging agents like vixotrigine show TN promise but lack Charlin's data [162]. Pharmacogenomics could personalize therapy, but research gaps remain [163].

If symptoms persist after 4–8 weeks of optimized pharmacotherapy, interventional procedures target the peripheral nerve. Nasociliary blocks with 1–2 mL 2% lidocaine plus 40 mg triamcinolone via medial orbital approach offer weeks-to-months relief; repeat blocks maintain control, with anatomical precision minimizing orbital hematoma [164]. The administration of a nasociliary nerve block via the medial orbital approach is illustrated in Fig. 8.9.

Pulsed radiofrequency ablation (PRFA) non-destructively modulates signaling, with TN trials showing sustained relief applicable to Charlin's [165]. Chemical neurolysis using 6%–10% phenol or alcohol prolongs effects but risks sensory loss or neuroma [166]. Cryoneuroablation, cooling the nerve controllably, is minimally invasive with low recurrence in facial neuralgias [167]. These procedures are supported by TN data, but Charlin's-specific efficacy relies on small series, with variable outcomes [168]. Cost-effectiveness and procedural risks, like infection, should be weighed, especially in elderly patients [169].

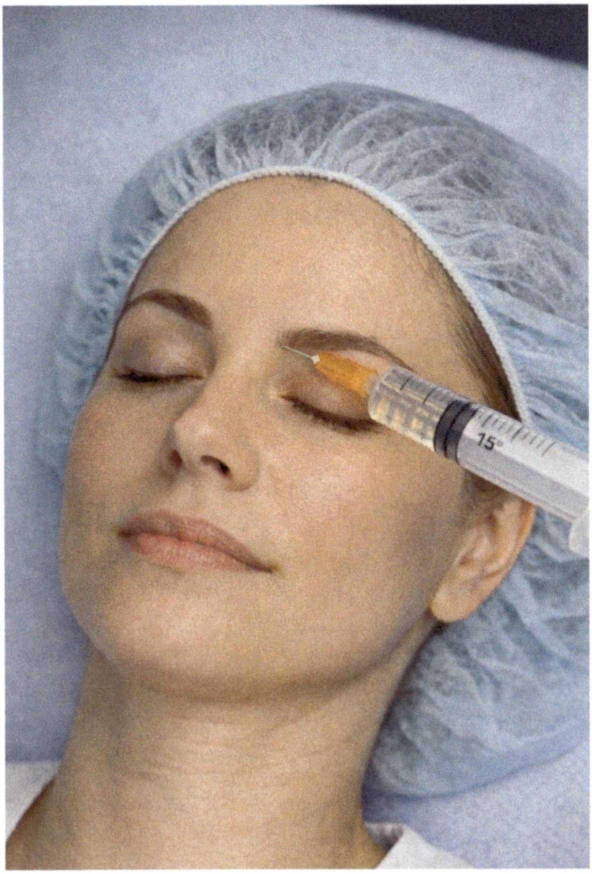

Fig. 8.9 Nasociliary nerve block injection procedure

For refractory cases, surgical options emphasize peripheral intervention, as central TN procedures like microvascular decompression are unsuitable. Transection or electrocoagulation of the nasociliary or anterior ethmoidal nerve provides permanent relief, particularly post-trauma, though dysesthesia or trophic changes may occur [170]. Endoscopic approaches reduce scarring in rhinoplasty-related cases [171]. In MS-linked Charlin's, combining disease-modifying therapies with neuralgia interventions improves outcomes, per TN reviews [172]. Long-term monitoring for recurrence or complications, using tools like SF-36 for quality-of-life assessment, is recommended [173].

Innovative therapies adapt TN advancements. Botulinum toxin A (50–100 units) at trigger sites inhibits nociception; refractory TN studies support off-label use in Charlin's, with 55%–85% relief lasting months [170]. Peripheral nerve stimulation implants modulate pain, reducing intensity in chronic facial syndromes [166].

Cannabinoids and ketamine are investigational, showing adjunctive neuropathic potential but no Charlin's-specific trials [165]. Gene therapies targeting ion channels represent future directions [174].

Outcomes are limited by rarity; case series suggest 60%–93% control with pharmacotherapy and blocks, but 20%–40% require escalation, with higher variability than TN [159]. Central sensitization risks necessitate early cognitive-behavioral therapy [160]. Holistic care addresses ocular complications and distress. In summary, Charlin's management adapts TN strategies with peripheral focus, but evidence gaps demand cautious extrapolation and further research for personalization.

8.6 Ramsay Hunt Syndrome

Abstract Ramsay Hunt syndrome (RHS), or herpes zoster oticus, is caused by reactivation of varicella-zoster virus in the geniculate ganglion of the facial nerve. It classically presents with the triad of ipsilateral peripheral facial paralysis, otalgia, and vesicular eruption of the auricle or external auditory canal, although atypical forms may occur. Adjacent cranial nerve involvement, particularly of the vestibulocochlear nerve, may produce hearing loss, tinnitus, vertigo, and taste disturbance. RHS is an uncommon but important cause of peripheral facial palsy and generally carries a poorer prognosis than Bell's palsy if treatment is delayed. Early diagnosis and prompt treatment with antiviral agents and corticosteroids are essential to improve recovery and reduce complications, including postherpetic neuralgia, synkinesia, and persistent hearing impairment.

Keywords Ramsay Hunt syndrome, Herpes zoster oticus, Varicella-zoster virus, Geniculate ganglion, Facial nerve palsy, Peripheral facial paralysis, Postherpetic neuralgia

Definition Ramsay Hunt syndrome (RHS), also known as herpes zoster oticus, is a neurological disorder resulting from the reactivation of the varicella-zoster virus (VZV) in the geniculate ganglion, the sensory ganglion of the facial nerve (cranial nerve VII). This reactivation leads to inflammation and subsequent clinical manifestations, primarily involving ipsilateral facial paralysis, ear pain, and a vesicular rash in the auricle or external auditory canal. It is crucial to recognize that the syndrome is distinct from other forms of herpes zoster (shingles) due to its specific involvement of the geniculate ganglion, which can also affect adjacent cranial nerves, such as the vestibulocochlear nerve (cranial nerve VIII), leading to auditory or vestibular symptoms. Unlike earlier misconceptions that attributed similar symptoms to inflammation of the trigeminal ganglion (Gasserian ganglion) or its auriculotemporal branch, RHS is now understood to originate from VZV latency in the geniculate ganglion [175].

Furthermore, the geniculate ganglion does not "contain" multiple cranial nerves but is anatomically proximate to them, allowing viral spread in severe cases. This polyneuropathy aspect was historically described as a form of viral cranial polyneuritis, but contemporary definitions emphasize the triad of symptoms: facial palsy, otalgia, and herpetic vesicles in a dermatomal distribution. It is important to note that RHS represents a rare but serious complication of VZV reactivation, accounting for approximately 12% of all peripheral facial nerve palsies, with a poorer prognosis compared to Bell's palsy if untreated. Early recognition is critical, as delays can lead to permanent sequelae, highlighting the need for prompt medical intervention to mitigate long-term effects [176].

Etiopathogenesis of Ramsay Hunt Syndrome The etiopathogenesis of RHS begins with primary VZV infection, commonly manifesting as chickenpox (varicella) in childhood. During this acute phase, the virus spreads via respiratory droplets, causing a widespread vesicular rash and systemic symptoms. Following resolution, VZV establishes latency in sensory ganglia, including the dorsal root ganglia, cranial nerve ganglia, and specifically the geniculate ganglion for RHS.

Reactivation occurs under conditions of immunosuppression, physiological stress, or age-related immunosenescence, allowing the virus to replicate and travel along sensory nerve fibers to the corresponding dermatome. In RHS, reactivation primarily affects the facial nerve, but proximity to other nerves can lead to involvement of cranial nerves V (trigeminal), VIII (vestibulocochlear), IX (glossopharyngeal), or X (vagus) in rare instances [177].

Immunocompromise plays a pivotal role; factors such as advanced age, HIV, malignancy, or iatrogenic immunosuppression (e.g., post-organ transplant or corticosteroid use) increase risk. A recent case highlighted RHS triggered by local intra-articular steroid injections for knee osteoarthritis, suggesting even localized immunosuppression may precipitate reactivation in susceptible individuals. This underscores the importance of monitoring patients undergoing such treatments for potential viral reactivation [178]. The virus's neurotropism causes ganglionitis and neuritis, resulting in pain, vesicular eruptions due to keratinocyte infection, and nerve dysfunction. Molecularly, VZV reactivation involves downregulation of cell-mediated immunity, particularly T-cell responses, which wane with age.

This explains the higher incidence in older adults, where latent VZV in ganglia evades immune surveillance until a triggering event, emphasizing the role of immune health in preventing such occurrences [179]. RHS is uncommon, comprising less than 1% of all herpes zoster cases, but its incidence rises with age due to declining immunity.

The etiopathogenesis of Ramsay Hunt syndrome, from primary varicella-zoster virus infection to reactivation and nerve involvement, is visually summarized in Fig. 8.10.

The lifetime risk of herpes zoster is approximately one-third, with RHS specifically affecting about 5 per 100,000 individuals annually in the general population. Rates are higher in immunocompromised groups, reaching up to 50 per 100,000 in those

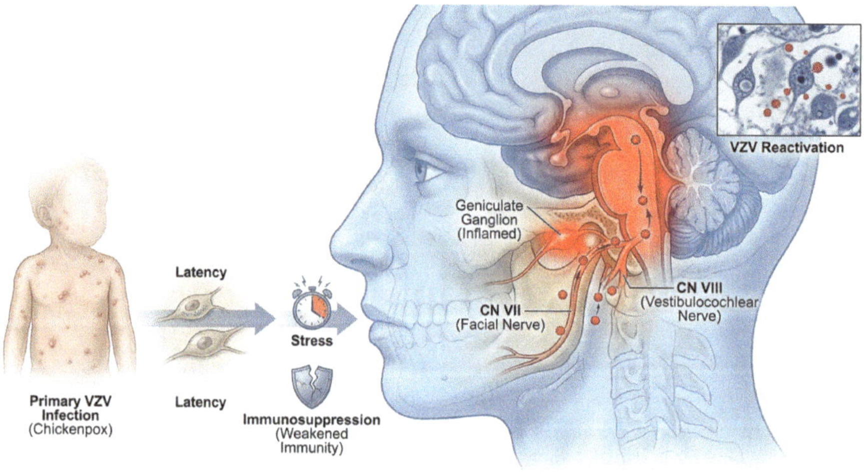

Fig. 8.10 Etiopathogenesis of Ramsay Hunt syndrome

over 60 years. Pediatric cases are rare, representing only 2%–3% of RHS diagnoses, often linked to prior chickenpox rather than vaccination, with a median onset age of 12 years and a 7-year median latency from primary infection. Males predominate slightly (60%), and unilateral presentation is standard, though bilateral cases have been reported in severely immunocompromised patients [180].

The COVID-19 pandemic influenced RHS epidemiology indirectly; studies noted a potential increase in facial palsies, including RHS, post-vaccination or infection, possibly due to immune modulation. However, large-scale data suggest no definitive causal link, emphasizing the need for vigilance in post-pandemic surveillance to better understand these patterns. Global variations exist, with higher rates in regions with lower varicella vaccination coverage, underscoring prevention's role and the benefits of widespread immunization programs [181].

Clinical Features of Ramsay Hunt Syndrome Clinical presentation of RHS typically evolves over days, beginning with prodromal symptoms such as flu-like malaise, headache, or fever. Pain (otalgia) is a hallmark, often severe and localized to the ear, progressing from a dull ache to dysesthetic neuritic discomfort due to ganglionitis. The characteristic vesicular rash appears as red, fluid-filled blisters on the pinna, external auditory canal, or occasionally the soft palate and tongue, reflecting the facial nerve's sensory distribution.

Facial weakness or paralysis follows, usually unilateral, manifesting as inability to close the eye (lagophthalmos), drooling, or asymmetry in smiling. Progression to peak severity occurs within 72 h, contrasting with slower evolution in some differentials, which aids in differential diagnosis [182].

Involvement of cranial nerve VIII can cause auditory symptoms like tinnitus (48%), hyperacusis, or sensorineural hearing loss (24%), which may persist permanently in 5% of cases. Vestibular dysfunction leads to vertigo, nystagmus, or nausea. Rare symptoms include dysgeusia (altered taste on the anterior two-thirds of the tongue), dry eyes (reduced lacrimation), nasal congestion, or dysarthria from oropharyngeal involvement. In children, presentations mirror adults but with higher rates of prior chickenpox history (44%) and complications like meningitis (12%).

Atypical forms lack vesicles or involve disseminated zoster, complicating diagnosis and requiring heightened clinical suspicion [183]. Severity is graded using the House-Brackmann scale, with grades IV-VI indicating poor prognosis if untreated. Complications extend beyond acute symptoms; postherpetic neuralgia (PHN) affects 10%–20%, causing chronic pain. Ocular issues like keratitis arise from incomplete eye closure, while rare extensions include encephalitis or cerebrovascular events. In immunocompromised patients, dissemination increases mortality risk, making comprehensive evaluation essential [184].

Diagnosis is primarily clinical, based on the triad of ipsilateral facial palsy, otalgia, and auricular vesicles. Differential diagnoses include Bell's palsy (idiopathic facial palsy), acoustic neuroma, Lyme disease, or otitis media. Laboratory confirmation via polymerase chain reaction (PCR) on vesicular fluid or saliva detects VZV DNA with high sensitivity (95%). Serology (IgM/IgG) supports reactivation but is less specific. Imaging, such as gadolinium-enhanced MRI, reveals geniculate ganglion enhancement in 70-80% of cases, aiding in atypical presentations where symptoms may not fully align with the classic triad [185].

Electroneurography (ENoG) and electromyography (EMG) assess nerve degeneration; >90% amplitude reduction predicts poor recovery. Audiometry evaluates hearing loss, while vestibular testing (e.g., videonystagmography) quantifies balance issues. In delayed cases, healed rashes pose challenges, requiring high suspicion, especially in elderly or post-steroid patients. Early EMG may show hyperacute lesions with reduced motor recruitment, guiding prognosis and helping clinicians tailor follow-up care [186].

For a comprehensive overview of the primary manifestations and potential sequelae associated with Ramsay Hunt syndrome, refer to Table 8.3, which summarizes key symptoms and complications.

Management of Ramsay Hunt Syndrome Management of RHS addresses viral replication, inflammation, pain, and complications. Early initiation within 72 h of rash onset optimizes outcomes, with combination therapy of antivirals and corticosteroids as first-line. Antivirals like acyclovir (800 mg five times daily for 7–10 days), valacyclovir (1 g three times daily for 7 days), or famciclovir (500 mg three times daily for 7 days) inhibit VZV replication.

A meta-analysis showed a 30% relative reduction in non-recovery with antivirals, though statistical significance was marginal (OR 0.68, 95% CI 0.37–1.27). Combination with systemic corticosteroids (e.g., prednisone 1 mg/kg/day tapered over 10 days) enhances recovery rates to 70%–90%, superior to monotherapy, and is recommended to improve overall patient outcomes [187].

Table 8.3 Key Symptoms and Complications of Ramsay Hunt Syndrome.

Category	Details 🩺
🩻 **Facial Nerve Palsy**	✅ Unilateral facial weakness/paralysis
	✅ Inability to close the eye (lagophthalmos)
	✅ Drooping of the mouth, leading to drooling
	✅ Loss of facial expressions
🔥 **Vesicular Rash**	✅ Painful blisters in the ear, external auditory canal, or tympanic membrane
	✅ May also appear on the tongue or hard palate
	✅ Rash often follows severe pain
🎧 **Otologic Symptoms**	✅ Severe ear pain (otalgia)
	✅ Hearing loss (sensorineural)
	✅ Tinnitus (ringing in the ears)
	✅ Vertigo & dizziness
👅 **Other Symptoms**	✅ Altered taste sensation (chorda tympani involvement)
	✅ Dry mouth & dry eyes (due to salivary & lacrimal gland dysfunction)
	✅ Hyperacusis (increased sound sensitivity)
⚠️ **Complications**	❌ Permanent facial weakness (if untreated)
	❌ Postherpetic neuralgia (chronic nerve pain)
	❌ Corneal damage (exposure keratitis from incomplete eyelid closure)

For severe cases or hospitalization, intravenous acyclovir (10 mg/kg every 8 h for 7–10 days) is preferred, especially in immunocompromised patients. Pain management follows a stepwise approach: non-opioids (acetaminophen, NSAIDs) for mild pain; gabapentin (starting 300 mg, titrated to 3600 mg/day) or pregabalin for neuropathic components; opioids (e.g., morphine) for breakthrough throbbing pain. Topical capsaicin is FDA-approved for PHN. Sympathetic nerve blocks (stellate ganglion) may alleviate acute pain and prevent ischemia-induced nerve damage, though evidence is limited, warranting further research [188].

Supportive care includes eye protection (artificial tears, ointments, patches) to prevent corneal abrasions. Physical therapy targets facial muscle rehabilitation, reducing synkinesia risk (up to 45%). In children, acyclovir (intravenous or oral) plus prednisolone yields high recovery (78%), with early therapy crucial. Botulinum toxin injections manage synkinesia or hyperlacrimation. Surgical decompression is rarely indicated, reserved for refractory cases. Monitoring for complications like PHN (treated with anticonvulsants) or hearing loss is essential, ensuring holistic patient care [189].

Prognosis varies; with early treatment, full recovery occurs in 70% within 6–12 months, but untreated cases drop to 50%. Factors worsening outcomes include age >50, House-Brackmann grade >III, and delayed therapy (>72 h). PHN persists in

10-15%, causing debilitating pain. Permanent hearing loss affects 5%–10%, while facial synkinesia or contractures occur in 20%–30%. In pediatrics, partial recovery is seen in 22%, with persistent paralysis in 44% of those. Rare sequelae include stroke or myocardial infarction from vasculopathy. Long-term follow-up (3–6 months) assesses recovery via EMG and clinical grading, allowing for timely interventions if needed [190].

Prevention centers on vaccination against VZV. The recombinant zoster vaccine (RZV, Shingrix) is recommended for adults ≥50 years, demonstrating >90% efficacy against HZ and complications like RHS, sustained >70% at 10 years. It reduces HZ burden and interference with daily activities by over 90%. In immunocompromised individuals, RZV is safe and effective. Childhood varicella vaccination lowers primary infection rates, indirectly reducing RHS risk. High-risk groups (e.g., post-steroid therapy) warrant preemptive vaccination. Public health strategies emphasizing vaccination could halve HZ incidence, promoting healthy aging and reducing healthcare burdens [191].

RHS is a multifaceted medical syndrome that carries significant morbidity if it remains unrecognized. Prompt diagnosis through the classic clinical triad, along with early administration of combined antiviral and corticosteroid therapy, effectively mitigates associated risks. This management strategy underscores the value of interdisciplinary care, which involves collaboration among neurologists, otolaryngologists, and ophthalmologists. Vaccination serves as a robust preventive tool, particularly effective in aging populations. Future research should prioritize optimizing therapeutic approaches for atypical or pediatric cases, with the goal of further improving outcomes and ensuring a better quality of life for affected individuals.

8.7 Tolosa–Hunt Syndrome

Abstract Tolosa–Hunt syndrome (THS) is a rare idiopathic granulomatous inflammatory disorder characterized by unilateral periorbital or retro-orbital pain associated with ipsilateral ophthalmoplegia due to involvement of the cavernous sinus, superior orbital fissure, or orbital apex. Classified in the *International Classification of Headache Disorders, 3rd edition* (ICHD-3) among painful cranial neuropathies and other facial pains, THS remains a diagnosis of exclusion because its clinical and radiological presentation may mimic neoplastic, vascular, infectious, and systemic inflammatory disorders. The syndrome typically presents with severe unilateral orbital pain followed by paresis of cranial nerves III, IV, and/or VI, sometimes accompanied by sensory disturbance in the ophthalmic division of the trigeminal nerve. High-resolution contrast-enhanced magnetic resonance imaging is central to diagnosis, although careful laboratory evaluation, exclusion of secondary causes, and longitudinal follow-up are essential to confirm the diagnosis. Current evidence supports an immune-mediated granulomatous inflammatory process, while recent reports suggest possible associations with autoimmune conditions and post-infectious immune dysregulation in selected cases. Corticosteroids remain the first-line treatment and usually produce rapid pain relief with gradual recovery of ocular motor deficits. Prognosis is generally favorable, although relapses may occur and warrant prolonged monitoring or steroid-sparing immunotherapy in recurrent cases. This chapter reviews the contemporary understanding of THS, including its definition, epidemiology, etiopathogenesis, clinical features, diagnostic approach, differential diagnosis, treatment, and prognosis.

Keywords Tolosa–Hunt syndrome, Painful ophthalmoplegia, Cavernous sinus, Superior orbital fissure, MRI, Corticosteroids

Definition Tolosa-Hunt syndrome (THS) is a rare idiopathic granulomatous inflammatory disorder characterized by unilateral painful ophthalmoplegia due to involvement of the cavernous sinus, superior orbital fissure, or orbital apex [192]. First described by Eduardo Tolosa in 1954 on the basis of surgical observations and further delineated by William Hunt and colleagues in 1961 through detailed clinical and angiographic correlations, the syndrome has evolved from a purely clinical observation into a precisely defined entity that now mandates neuroimaging confirmation for definitive diagnosis [193]. It remains a diagnosis of exclusion, frequently mimicking a wide array of more sinister conditions such as tumors, aneurysms, infections, or other inflammatory processes [198]. This diagnostic challenge underscores the critical importance of applying stringent diagnostic criteria together with close multidisciplinary evaluation involving neurologists, ophthalmologists, and neuroradiologists to avoid potentially harmful interventions. Patients typically experience severe, boring periorbital or retro-orbital pain that is often described as relentless and debilitating, followed by paresis of the ocular motor cranial nerves, resulting in diplopia, ptosis, or restricted gaze that significantly impairs daily function and quality of life [194]. Prompt recognition is critical because THS responds dramatically to corticosteroids, often with pain relief occurring within hours to days and ophthalmoplegia resolving over subsequent weeks, thereby clearly distinguishing it from neoplastic or infectious mimics that may actually worsen with steroid administration [194].

Recent literature from 2020 to 2025 has substantially refined our understanding of THS through high-quality systematic reviews, well-documented case series, and insightful reports of atypical presentations, including post-viral triggers and previously under-recognized overlaps with autoimmune disorders [195]. A 2024 systematic review of clinical profiles synthesized data from multiple international cohorts, providing robust pooled estimates, while dedicated pediatric-specific analyses have illuminated important differences in disease expression among younger patients [195]. Advances in high-resolution magnetic resonance imaging (MRI) have markedly improved detection of even subtle granulomatous changes that were previously overlooked, and emerging associations with conditions such as neuromyelitis optica spectrum disorder (NMOSD) and post-COVID-19 inflammation have meaningfully broadened the etiological discussion without altering the core idiopathic nature of classic THS [196].

According to the *International Classification of Headache Disorders, 3rd edition* (ICHD-3), Tolosa-Hunt syndrome (THS) is formally classified under the section on painful cranial neuropathies and other facial pains. The ICHD-3 diagnostic criteria for Tolosa-Hunt syndrome are summarized in Table 8.4. This classification highlights the syndrome's hallmark feature of painful involvement of the cranial nerves. The condition is strictly idiopathic by definition. Consequently, any cases associated with identifiable underlying causes—including infections, systemic granulomatous diseases (e.g., sarcoidosis), malignancies, or vascular lesions—are deliberately excluded from the THS diagnosis and are instead reclassified as secondary painful ophthalmoplegia [197].

Historically, the syndrome has been referred to by several alternative terms such as "painful ophthalmoplegia syndrome" or "superior orbital fissure syndrome." However, in modern practice, the designation "Tolosa-Hunt syndrome" is reserved exclusively for the idiopathic form that fulfills the strict modern diagnostic criteria. Recent case series have provided valuable validation of these criteria while carefully noting their limitations in specificity. Specifically, these investigations have shown that up to 20–30% of patients initially suspected of having THS are ultimately found to have alternative diagnoses upon extended clinical and radiological follow-up [199].

THS is exceedingly rare, with an estimated annual incidence of approximately 1 case per million population based on population registries and systematic reviews [192]. No consistent geographic, ethnic, or racial predilections have been identified, and the condition affects males and females equally. The mean age of onset is 41–50 years in adults, although pediatric cases are increasingly recognized, comprising up to 10%–15% of reported series. A 2025 systematic review identified 61 unique pediatric patients, with a mean age of onset around 10–12 years and a slight male predominance in children [200].

Bilateral involvement occurs in 5%–9% of cases, typically asynchronous. Recurrence rates have been reported as high as 40%–50% in older literature, but

Table 8.4 Diagnostic Criteria for Tolosa-Hunt Syndrome (ICHD-3)

Criterion	Description
🎯 **A. Headache**	✅ Unilateral orbital or periorbital headache
🧠 **B. Inflammation & Nerve Involvement**	✅ **Both must be present:** 1️⃣ Granulomatous inflammation in the cavernous sinus, superior orbital fissure, or orbit (confirmed by MRI or biopsy). 2️⃣ Ipsilateral paresis of cranial nerves: • III (oculomotor) – Controls eye movement & eyelid • IV (trochlear) – Control superioroblique muscle • VI (abducens) – Controls lateral eye movement
🔗 **C. Evidence of Causation**	✅ **Both must be present:** 1️⃣ Headache is ipsilateral to the inflammation. 2️⃣ Headache precedes or coincides with nerve paresis (≤2 weeks).
❌ **D. Exclusion of Other Causes**	✅ Not explained by another ICHD-3 diagnosis (e.g., tumor, infection, vascular disease).

contemporary data from a large 2024 cohort indicate a lower rate of 27.5%, with recurrences more common in younger patients and during steroid tapering [194]. Improved MRI availability has increased recognition, particularly in resource-limited settings where underdiagnosis previously prevailed. Emerging associations include autoimmune conditions such as systemic lupus erythematosus (SLE) and NMOSD, as well as post-infectious triggers following COVID-19, potentially expanding the epidemiological footprint [196, 201].

Etiopathogenesis of THS The precise etiology of THS remains idiopathic, characterized by nonspecific granulomatous inflammation featuring lymphocytic and plasma cell infiltrates, epithelioid cells, occasional multinucleated giant cells, and fibroblast proliferation within the cavernous sinus or adjacent structures [192]. Histopathology shows dural thickening and compression of traversing cranial nerves without significant necrosis or vasculitis. No specific infectious agent or autoantibody has been consistently identified, although immune-mediated mechanisms are strongly suspected.

Recent reports have documented overlaps with autoimmune disorders, including aquaporin-4 antibody-positive NMOSD and SLE, suggesting THS may occasionally represent an initial manifestation of broader systemic inflammation [196, 202]. Post-viral immune dysregulation has gained attention, with documented cases following COVID-19 infection attributed to molecular mimicry or cytokine storm effects [201]. Infectious mimics, such as bacterial sphenoid sinusitis causing recurrent alternating THS, challenge the strict "idiopathic" classification and have prompted proposals for an "inflammatory THS" subcategory [203]. Traumatic, aneurysmal, or neoplastic processes are important mimics rather than causes and must be rigorously excluded [193]. Pathogenesis involves mass effect from granulomatous tissue on cranial nerves III, IV, VI, and the ophthalmic division of V, with occasional involvement of sympathetic fibers leading to Horner syndrome. Intracranial extension is uncommon but documented in recent imaging series [198]. Molecular and immunological studies remain limited, supporting a steroid-responsive, non-infectious inflammatory process without a single definitive trigger.

Clinical Features of THS THS typically presents acutely or subacutely with severe, unilateral periorbital or retro-orbital pain described as boring, stabbing, or throbbing, often radiating to the frontal or temporal regions. Pain precedes ophthalmoplegia by up to 2 weeks in most cases, though simultaneous onset occurs [204]. Ophthalmoplegia arises from variable involvement of cranial nerves: oculomotor (III) in ~70%–80%, abducens (VI) in ~60%–70%, trochlear (IV) in ~25%–30%, and sensory disturbances in the ophthalmic division of trigeminal (V1) in ~30%. Symptoms include diplopia, ptosis, restricted eye movements, and forehead hypoesthesia.

Less common manifestations include optic neuropathy with visual loss (potentially irreversible if untreated), Horner syndrome (~15%–20%), mild proptosis, or extension to V2/V3 divisions [198]. Pediatric cases may feature more frequent bilateral or recurrent episodes, systemic symptoms such as fatigue or fever, and slightly different nerve involvement patterns [200]. Elderly presentations can be subtler, with prominent ptosis or headache predominance [205]. The natural history is relapsing-remitting; untreated attacks last weeks to months, with intervals between recurrences ranging from months to years. Associated features may include nausea or fatigue, but focal neurological deficits beyond cranial nerves are absent. Recent case reports emphasize phenotypic variability across age groups and the value of serial neuroimaging [206].

Diagnosis integrates clinical presentation, neuroimaging, and therapeutic response per ICHD-3 criteria (Table 8.4) [197]. High-resolution contrast-enhanced MRI is the cornerstone, typically revealing enlargement and avid enhancement of the cavernous sinus, superior orbital fissure, or orbital apex, with possible dural tail or extension along the tentorium. Thin-slice (≤1 mm) fat-suppressed sequences improve sensitivity. Follow-up imaging demonstrates resolution with treatment, aiding confirmation. Biopsy is reserved for atypical or steroid-refractory cases to exclude malignancy or infection. Limitations of ICHD-3 include potential over-reliance on imaging and reduced specificity in early disease; reappraisal studies recommend prolonged clinical and radiological surveillance [199, 207].

THS must be differentiated from a broad spectrum of painful ophthalmoplegia causes. Neoplastic mimics include meningioma, lymphoma, metastasis, or pituitary apoplexy. Vascular etiologies encompass intracavernous carotid aneurysm, cavernous sinus thrombosis, or carotid-cavernous fistula. Infectious causes include bacterial/fungal sinusitis with extension, tuberculosis, Lyme disease, or herpes zoster. Inflammatory mimics comprise sarcoidosis, IgG4-related disease, granulomatosis with polyangiitis, and orbital pseudotumor. In children, rhabdomyosarcoma or Langerhans cell histiocytosis must be considered. Comprehensive workup includes MRI/MRA, CT angiography, cerebrospinal fluid analysis, serologies (ANA, ANCA, aquaporin-4, ACE), and chest imaging as indicated. Steroid response alone is insufficient for diagnosis [208].

Management of THS High-dose glucocorticoids remain first-line therapy, typically prednisone 1 mg/kg/day (or intravenous methylprednisolone pulses in severe cases) with a slow taper over 3–6 months to minimize relapse. Pain relief occurs within 24–72 h, and ophthalmoplegia resolves over 1–4 weeks in most patients [208]. For recurrent or steroid-dependent disease, steroid-sparing agents such as methotrexate, azathioprine, mycophenolate mofetil, or infliximab are employed, with growing evidence for adalimumab in pediatric refractory cases [216]. Radiotherapy is rarely used for truly refractory disease. Supportive care includes analgesics, eye patching for diplopia, and regular ophthalmologic monitoring. All patients require close follow-up with serial MRI to confirm resolution and exclude evolving alternative diagnoses.

With appropriate treatment, prognosis is generally favorable, with complete or near-complete recovery in >80% of cases. Residual mild cranial nerve deficits occur in <20%. Recurrence affects approximately 27.5% of patients, more frequently in younger individuals, yet long-term functional outcomes remain excellent, with all patients achieving remission in recent cohorts [210]. Pediatric cases may have higher relapse rates but similar ultimate recovery [216]. Long-term monitoring (at least 2 years) is recommended to detect late recurrences or unmasked underlying conditions. Mortality is negligible, and quality of life returns to baseline in the majority.

Tolosa-Hunt syndrome exemplifies a treatable cause of painful ophthalmoplegia when diagnostic rigor is applied. Updated criteria, advanced imaging, and recognition of autoimmune/post-infectious overlaps have enhanced diagnostic accuracy and management. Continued research into biomarkers and targeted immunotherapies holds promise for reducing recurrences. Clinicians should maintain a high index of suspicion in appropriate presentations while systematically excluding mimics to optimize patient outcomes.

8.8 Frey Syndrome

Abstract Frey syndrome, also known as auriculotemporal syndrome, is a gustatory autonomic disorder characterized by unilateral sweating, flushing, warmth, and occasionally burning or pruritus in the preauricular, temporal, and adjacent cheek regions during eating or in response to food-related stimuli. It most commonly develops after injury to the auriculotemporal nerve, particularly following parotid surgery, although trauma, infection, and other procedures in the parotid region may also precipitate the condition. The accepted pathophysiological mechanism is aberrant regeneration of postganglionic parasympathetic secretomotor fibers, which become misdirected to cutaneous sweat glands and blood vessels, thereby producing sweating and vasodilation during gustatory stimulation. Clinically, Frey syndrome usually appears months after surgery or trauma and ranges from mild subjective discomfort to socially distressing symptoms that impair quality of life. Diagnosis is based primarily on a characteristic history and clinical presentation and may be confirmed by Minor's starch-iodine test, which remains the standard method for objectively demonstrating and mapping gustatory sweating. Management is best approached in a stepwise manner. Preventive intraoperative barrier techniques during parotidectomy can significantly reduce postoperative incidence. In established cases, topical antiperspirants and anticholinergic agents may provide limited benefit, whereas botulinum toxin type A is considered the most effective first-line symptomatic treatment because it offers high rates of improvement with minimal morbidity. Auriculotemporal nerve block may be useful in selected patients, particularly when pain is prominent, while surgical intervention is reserved for severe refractory cases. Overall, Frey syndrome is a benign but often distressing complication of parotid-region nerve injury whose burden can be substantially reduced through accurate diagnosis, preventive surgical strategies, and effective minimally invasive treatment.

Keywords Frey syndrome, Auriculotemporal syndrome, Gustatory hyperhidrosis, Parotidectomy, Auriculotemporal nerve, Minor's starch-iodine test

Definition Frey syndrome (FS), also known as auriculotemporal syndrome, Baillarger's syndrome, Dupuy's syndrome, or gustatory hyperhidrosis, is a neurological disorder characterized by unilateral facial sweating, flushing, warmth, and occasionally burning or itching in the preauricular, temporal, and cheek regions triggered by gustatory stimuli or even the thought of food [208]. It arises from aberrant regeneration of parasympathetic nerve fibers following injury to the auriculotemporal nerve, typically near the parotid gland. Although uncommon in the general population, FS is one of the most frequent complications of parotid surgery [208, 209].

The condition was first noted by Baillarger in 1853 as facial sweating after parotid abscess drainage. In 1923, Polish neurologist Lucja Frey provided the classic description in a patient after parotid trauma, coining the term "auriculotemporal syndrome" and linking it to nerve injury [210]. Subsequent reports confirmed its occurrence after parotidectomy, trauma, infection, and other insults. The eponym "Frey syndrome" became standard, while synonyms reflect historical contributors [210].

FS predominantly follows parotid-region surgery or trauma, with no clear age or sex predilection [208]. The incidence after parotidectomy varies widely (4%–96%) due to differences in diagnostic criteria, follow-up duration, and surgical technique. Subjective (symptomatic) rates range from 10%–50%, while objective testing (Minor's starch-iodine test) reveals up to 80% positivity [208, 211]. Recent series report clinically bothersome FS requiring treatment in only 2%–3% of cases in some regions, reflecting improved prevention [212].

Salivary gland neoplasms account for 3%–6% of all head and neck tumors, with 80–85% occurring in the parotid gland; of these, 75%–80% are benign (primarily pleomorphic adenoma) [213, 214]. Parotidectomy for tumor resection remains the leading iatrogenic cause in adults. Other etiologies include submandibular gland surgery, radical neck dissection, temporomandibular joint surgery, facelift procedures, direct trauma, infection, burns, and forceps delivery in infants [208, 210].

Etiopathogenesis of Frey Syndrome FS results from injury to the auriculotemporal nerve or its parotid branches during surgery or trauma. The auriculotemporal nerve, a branch of the mandibular division of the trigeminal nerve (V3), carries sensory fibers to the skin over the temple and ear, parasympathetic secretomotor fibers to the parotid gland (via the otic ganglion from the glossopharyngeal nerve), and sympathetic fibers (postganglionic from the superior cervical ganglion) to local sweat glands and vessels [215].

The predominant theory is aberrant regeneration: severed parasympathetic axons regrow aberrantly along sympathetic pathways that normally innervate cutaneous sweat glands and blood vessels. Because both use acetylcholine at muscarinic receptors, gustatory stimuli (which normally activate parotid secretion) now trigger sweating and vasodilation in the skin [208]. Symptoms typically appear 6–18 months post-injury as regeneration occurs. Alternative hypotheses include parasympathetic hypersensitivity or ephaptic transmission, but aberrant reinnervation best explains the classic presentation [208, 210].

Clinical features of Frey Syndrome The clinical manifestations of Frey syndrome are highly characteristic. Patients typically report abnormal gustatory sweating, flushing, and burning or tingling sensations in the affected area while eating or even thinking about food [208, 209]. These symptoms usually develop weeks to months after parotid surgery, trauma, or infection, although onset may occasionally be delayed for years [208]. The affected skin is most commonly located in the preauricular, temporal, or mandibular region. Severity ranges from mild (noticed only by the patient) to socially debilitating, causing embarrassment during meals. Rarely, patients experience gustatory neuralgia with more pronounced burning pain. Numbness or paresthesia in the distribution of the greater auricular nerve may coexist as a result of surgical trauma [208, 209]. The affected area is usually confined to the skin overlying the resected parotid bed but can vary [210]. The main clinical features of Frey syndrome are summarized in Table 8.5.

Diagnosis is primarily clinical, based on history of parotid-region insult and characteristic gustatory symptoms [208]. The gold-standard confirmatory test is Minor's starch-iodine test (described by Victor Minor in 1928): the affected skin is painted with iodine solution, allowed to dry, and dusted with starch powder. A gustatory stimulus (lemon juice, sour candy) is given. Sweat turns the starch-iodine mixture dark blue or black within minutes, mapping the hyperhidrotic area [208, 216]. Infrared thermography or quantitative sudomotor axon reflex testing can provide additional objective quantification but is less commonly used [217]. No routine imaging or laboratory tests are required.

FS must be distinguished from other causes of facial sweating: diabetic gustatory sweating (bilateral, often with autonomic neuropathy), idiopathic hyperhidrosis, auriculotemporal neuralgia (pain without sweating), or flushing from carcinoid syndrome or menopause [208]. History of parotid surgery strongly favors FS. Bilateral presentation is rare and suggests systemic causes.

Table 8.5 Clinical Features of Frey Syndrome

🔷 Feature	🔥 Description
🥵 **Gustatory Sweating**	Abnormal sweating in the **preauricular, temporal, or mandibular region** while eating or even thinking about food.
🌡️ **Flushing (Redness)**	Skin turns **red** in the same areas where sweating occurs, especially during meals.
⚡ **Burning/Tingling**	**Warmth, tingling, or discomfort** in the affected region during eating.
⏳ **Onset**	Symptoms appear **weeks to months** after **parotid surgery, trauma, or infection.**
🍴 **Triggers**	**Chewing, eating, smelling, or thinking about food** can induce symptoms.

Management of Frey Syndrome Management is stepwise, starting with prevention (ideally intraoperative), followed by conservative and medical/interventional measures; surgery is reserved for refractory cases [208].

Intraoperative placement of a barrier between the parotid bed and overlying skin flaps is the most effective strategy. Options include superficial musculoaponeurotic system (SMAS) flap, sternocleidomastoid (SCM) muscle flap, temporoparietal fascia (TPF) flap, acellular dermal matrix (ADM/AlloDerm), autologous fat grafting, and fascia lata. A 2021 network meta-analysis of 31 studies found all barrier techniques superior to no barrier, with SCM and SMAS flaps ranking highest for reducing both subjective and objective FS (odds ratios 0.15–0.30) [218]. ADM reduces objective incidence by ~85% and subjective by ~68% [208]. Preservation of the parotidomasseteric fascia also significantly lowers risk [213]. These techniques add minimal operative time and donor-site morbidity while improving contour and preventing FS.

Medical and Interventional Treatment

Topical therapies include aluminum chloride antiperspirants or glycopyrrolate wipes/creams (0.5%–2%), which inhibit eccrine sweating but require daily application and may cause skin irritation [219]. Systemic anticholinergics (e.g., oxybutynin) carry side effects (dry mouth, blurred vision) and are rarely used long-term.

Botulinum toxin type A (BTA) injection is the first-line medical therapy with the strongest evidence [208, 220]. Doses of 1.9–2.5 U/cm² (total 20–100 U) are injected intradermally into the affected area mapped by Minor's test. Onset occurs within 4–7 days, with 80%–98% complete or near-complete resolution of sweating and flushing. Effects last 6–24 months (average 12–15 months); repeat injections maintain efficacy without tachyphylaxis. Side effects (temporary facial weakness, dry mouth) are mild and transient [221]. Quality-of-life scores improve markedly [208].

Auriculotemporal nerve block is a valuable interventional option, particularly for patients with prominent pain (gustatory neuralgia), focal symptoms, or situations in which intradermal injections are technically difficult or poorly tolerated. Local anesthetic, such as 3–5 mL of 0.5% bupivacaine, is injected at the point where the auriculotemporal nerve crosses the zygomatic arch or just anterior to the tragus (see Fig. 8.11). Analgesic and symptom relief typically begins within minutes and may persist for several hours to several days, making the procedure useful for both diagnostic confirmation and short-term therapeutic management. Ultrasound guidance can further improve anatomical localization, procedural accuracy, and safety by helping avoid adjacent vascular or soft-tissue structures [222]. In selected cases, botulinum toxin A may also be administered proximally through this nerve-block approach to achieve a longer-lasting effect, although diffuse intradermal injection remains the preferred technique for patients with widespread gustatory sweating or hyperhidrosis [222, 225]. When performed correctly, the technique is generally safe and may help identify suitable candidates for permanent surgical intervention.

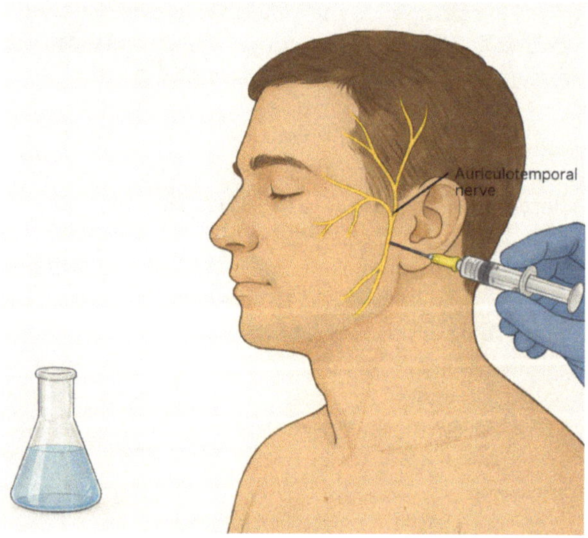

Fig. 8.11 Auriculotemporal nerve block for Frey syndrome

Surgical Treatment

Surgical intervention is considered only after failure of conservative measures, botulinum toxin type A (BTA) injections, and auriculotemporal nerve-block therapies in severely symptomatic patients [223]. Several surgical options exist for these refractory cases. These include tympanic (Jacobson's) neurectomy performed via a transcanal approach, which achieves a success rate of 50%–82% although recurrence remains possible; auriculotemporal nerve section, which is technically difficult and carries a risk of sensory loss; re-exploration of the surgical site with placement of thicker barriers such as temporoparietal fascia (TPF) flap, fascia lata, or acellular dermal matrix (ADM); and autologous fat grafting, which is minimally invasive and simultaneously improves facial contour while providing an effective barrier. A 2025 scoping review of 53 patients reported that surgical interventions achieved symptom reduction or resolution in 94% of cases, with 51% of patients attaining complete relief [224]. Potential complications include hematoma, infection, or temporary facial weakness. No single technique is universally superior; the choice depends on individual patient anatomy and history of prior surgery [224].

FS is chronic but non-progressive. Many patients adapt or remain mildly symptomatic; however, severe cases impair social functioning and dietary habits [208]. With modern prevention, BTA, and targeted nerve blocks, most achieve satisfactory control. Long-term follow-up shows sustained benefit from barriers and repeated interventions [208, 218]. Patient education and multidisciplinary care (otolaryngology, neurology, dermatology, pain management) optimize outcomes.

Frey syndrome is a well-recognized yet preventable iatrogenic condition that arises from aberrant parasympathetic reinnervation following injury to the auriculotemporal nerve. Accurate diagnosis is primarily based on a detailed clinical history and confirmation with Minor's starch-iodine test. Intraoperative barrier techniques have been shown to dramatically reduce its incidence, while botulinum toxin injections and auriculotemporal nerve blocks offer safe and highly effective options for symptom control. For refractory cases, various surgical interventions remain available. Increased awareness among head and neck surgeons, combined with the routine implementation of standardized prevention protocols, will play a crucial role in further minimizing this bothersome sequela.

8.9 Glossopharyngeal Neuralgia

Abstract Glossopharyngeal neuralgia (GPN) is a rare cranial neuralgia characterized by recurrent, unilateral, brief paroxysms of severe stabbing, shooting, or electric shock-like pain in the sensory distribution of the glossopharyngeal nerve, typically involving the posterior tongue, tonsillar fossa, pharynx, deep ear, and region beneath the angle of the mandible. Attacks are abrupt and commonly triggered by swallowing, coughing, speaking, yawning, or chewing. Contemporary classifications distinguish classical, secondary, and idiopathic forms, with classical GPN most often related to neurovascular compression at the root entry zone. Secondary forms should be considered in atypical or progressive presentations. In neurostomatologic practice, GPN is especially important because it may mimic dental, temporomandibular, otologic, pharyngeal, or trigeminal pain disorders, leading to diagnostic delay and unnecessary procedures. Diagnosis is primarily clinical, supported by careful pain characterization and magnetic resonance imaging with vascular assessment to identify neurovascular conflict and exclude secondary causes. Rare vagal manifestations, including bradycardia, hypotension, and syncope, may occur. Management is stepwise and etiology-based, with carbamazepine or oxcarbazepine as first-line therapy, minimally invasive nerve block in selected refractory cases, and microvascular decompression as the preferred definitive treatment for medically refractory classical GPN.

Keywords Glossopharyngeal neuralgia, Cranial neuralgia, Orofacial pain, Neurostomatology, Differential diagnosis, Microvascular decompression, Glossopharyngeal nerve block

Definition Glossopharyngeal neuralgia (GPN) is a rare cranial neuralgia characterized by recurrent unilateral brief paroxysms of severe pain in the sensory distribution of the glossopharyngeal nerve, classically involving the posterior tongue, tonsillar fossa, pharynx, deep ear, and/or the region beneath the angle of the mandible. The pain is abrupt in onset and termination, usually electric shock-like, stabbing, shooting, or sharp, and is commonly precipitated by swallowing, coughing, talking, or yawning. In contemporary classification systems, GPN is subdivided into **classical, secondary**, and **idiopathic** forms. Classical GPN is diagnosed when neurovascular compression of the glossopharyngeal nerve is demonstrated, secondary GPN when an underlying lesion or disease explains the syndrome, and idiopathic GPN when neither neurovascular compression nor another causative disorder is identified. Importantly, ICHD-3 also notes overlap with the auricular and pharyngeal branches of the vagus nerve, which explains the older term *vagoglossopharyngeal neuralgia* [226, 227].

From a neurostomatologic perspective, GPN is clinically important because patients may present initially with throat, tongue, deep ear, mandibular angle, or even oral pain, and may therefore be misdirected into dental, temporomandibular, otolaryngologic, or nonspecific facial pain pathways. The International Classification of Orofacial Pain (ICOP) places cranial neuralgias within the broader orofacial pain taxonomy, reinforcing that dentists, oral medicine specialists, and neurostomatologists must distinguish brief neuropathic paroxysms from pain caused by local odontogenic or mucosal disease [227, 228].

GPN is uncommon. Contemporary reviews describe it as accounting for roughly 0.2%–1.3% of craniofacial or cranial neuralgia syndromes, with an annual incidence generally reported in the range of 0.2–0.8 per 100,000 persons, and some reviews citing about 1% of the frequency of trigeminal neuralgia. Onset is most often in later adulthood. Because the disorder is rare, because its pain may radiate, and because many patients first present outside neurology or neurosurgery clinics, diagnostic delay remains a recurring problem. In neurostomatology, the practical consequence is important: patients may undergo repeated dental consultations, noncontributory imaging, and sometimes unnecessary irreversible procedures before the neuralgic phenotype is recognized [229–231].

Etiopathogenesis of Glossopharyngeal Neuralgia The best-supported mechanism in classical GPN is neurovascular compression at the glossopharyngeal nerve root entry zone, the vulnerable transitional segment where central and peripheral myelin meet. Chronic pulsatile vascular contact at this site is thought to induce focal demyelination, increase membrane instability, and facilitate ectopic impulse generation together with ephaptic transmission between adjacent nerve fibers, thereby producing the characteristic brief, severe, paroxysmal pain phenotype. Among the vascular offenders identified in surgical and imaging series, the posterior inferior cerebellar artery (PICA) is most frequently implicated, although compression by the vertebral artery and more complex or combined vascular conflict patterns are also recognized. Because the underlying mechanism in classical GPN is closely related to this neurovascular relationship, high-resolution posterior fossa MRI with dedicated vascular assessment has become central to contemporary etiologic evaluation and preoperative planning [230, 233].

Secondary GPN should be suspected when the clinical presentation is atypical, progressive, bilateral, associated with major neurologic deficits, or accompanied by persistent nonparoxysmal pain between attacks. ICHD-3 specifically notes reported secondary causes such as multiple sclerosis, tonsillar or regional tumors, cerebellopontine angle tumors, neck trauma, and Arnold–Chiari malformation. This distinction is clinically important because continuous or near-continuous pain is not typical of classical GPN. Instead, such a pattern should prompt consideration of painful glossopharyngeal neuropathy or another secondary disorder rather than being interpreted as a minor variant of classical neuralgia. In practical terms, the etiopathogenetic classification of GPN is therefore not merely descriptive; it directly informs diagnostic reasoning, imaging strategy, differential diagnosis, and subsequent treatment selection [226, 230].

Figure 8.12 illustrates the principal etiopathogenetic mechanisms of glossopharyngeal neuralgia, emphasizing neurovascular compression at the root entry zone in classical cases and the major secondary causes that should be considered in atypical presentations.

Clinical features of Glossopharyngeal Neuralgia The hallmark of GPN is brief, recurrent, unilateral, severe paroxysmal pain localized to the deep throat, tonsillar region, posterior tongue, pharynx, deep ear, or submandibular angle. The pain is typically abrupt in onset and termination, and patients often describe it as sharp, stabbing, shooting, or electric shock-like. Individual attacks usually last from a few seconds up to 2 minutes and may recur many times during the same day, although symptom-free intervals of variable duration may also occur between clusters of attacks. This episodic pattern is one of the most important clinical clues because it helps distinguish GPN from other chronic orofacial pain disorders that are more persistent or diffuse in character. The most characteristic triggers are swallowing, coughing, speaking, yawning, chewing, and contact with the oropharyngeal mucosa. In some patients, even minimal stimulation of the throat region may provoke an attack. In contrast to persistent idiopathic facial pain or painful glossopharyngeal neuropathy, the classic pain of GPN is paroxysmal rather than continuous [226, 227].

Some patients also develop vagal manifestations, including bradycardia, hypotension, syncope, or even convulsive episodes secondary to cardioinhibitory reflexes. Although such autonomic and cardiovascular features are uncommon, they are clinically important because they may render the condition potentially life-threatening. For this reason, the history should not focus solely on pain quality and location but should also actively explore syncopal episodes, presyncope, dizziness, palpitations, or collapse temporally related to painful attacks. Recognition of these associated manifestations is particularly important because they may indicate a more hazardous clinical presentation requiring urgent attention [239].

A particularly important diagnostic pitfall in neurostomatology is pain radiation into trigeminally innervated territory. Patients may report pain extending to the auricle, temple, cheek, mandibular gingiva, or anterior tongue, which can create confusion with

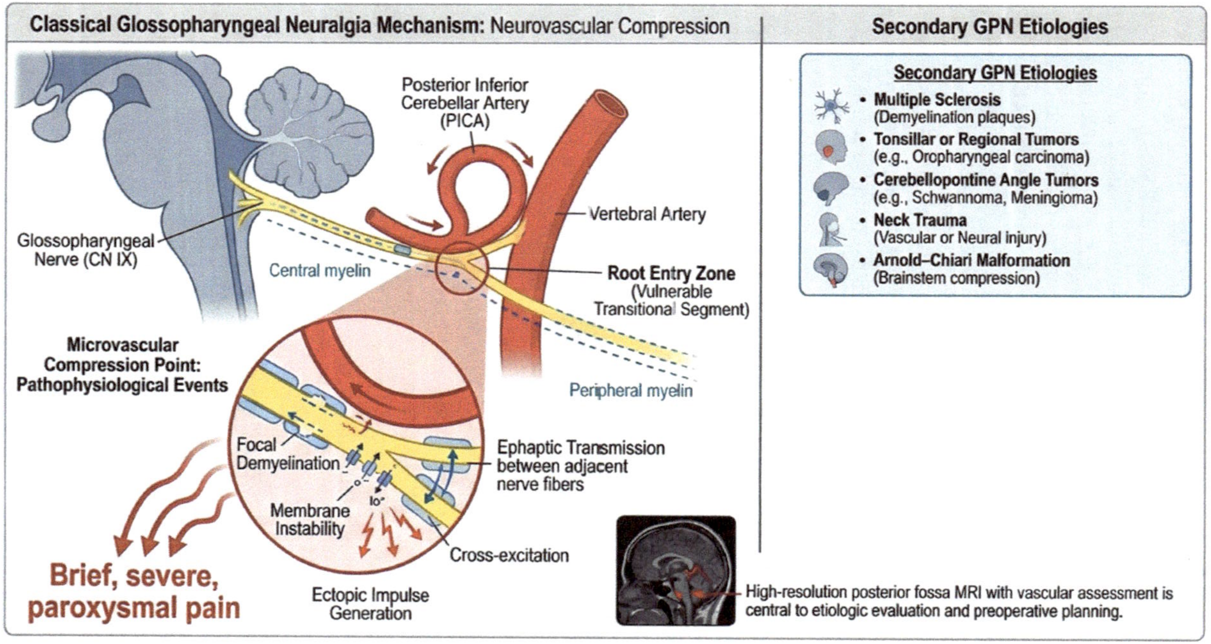

Fig. 8.12 Etiopathogenesis of glossopharyngeal neuralgia

trigeminal neuralgia, dental pain, temporomandibular disorders, or other regional pain syndromes. In one surgical series, swallowing remained the most common trigger, but pain frequently extended into regions ordinarily assigned to the trigeminal system. This overlap may contribute to delayed recognition and inappropriate referral if the characteristic pharyngeal pattern is not carefully elicited during history-taking [232].

Diagnosis is primarily clinical, but structured evaluation is essential. History should establish the exact pain topography, attack duration, trigger profile, laterality, radiation, frequency, and whether pain is entirely paroxysmal or partly continuous. Because many patients first present to dental or oral medicine services, examination should include oral, dental, mucosal, periodontal, temporomandibular, cranial nerve, and otolaryngologic screening. The purpose is not to "find" classical GPN on examination, since physical findings may be minimal, but to exclude common mimics and avoid attributing neuralgic pain to incidental dental findings [227, 228].

MRI of the posterior fossa with vascular assessment is the imaging study of choice. It may demonstrate neurovascular compression in classical GPN and can also help exclude tumors, demyelinating disease, and other structural pathology. When the clinical picture includes syncope, marked bradyarrhythmia, or other cardiorespiratory symptoms, further cardiovascular assessment is warranted. Persistent interictal pain, sensory loss, or gag reflex abnormalities should prompt stronger investigation for neuropathy or secondary causes rather than simple acceptance of "atypical GPN" [226, 230].

The clinical topography and characteristic triggers of glossopharyngeal neuralgia are illustrated in Fig. 8.13, highlighting the unilateral paroxysmal pain distribution, common radiation into adjacent trigeminal territories, and associated vagal/autonomic manifestations.

Table 8.6 summarizes the practical diagnostic criteria for glossopharyngeal neuralgia, adapted from ICHD-3 and aligned with ICOP for use in contemporary neurostomatologic practice [226, 227].

The most important differential diagnosis is trigeminal neuralgia (TN). Misdiagnosis can occur when pain radiates to the mandible, auricle, cheek, or tongue, or when clinicians focus too heavily on a facial pain location while underrecognizing pharyngeal triggers. In the 2024 case series by Hurth and colleagues, mandibular radiation strongly correlated with initial misdiagnosis as TN. The reverse pitfall also exists: a 2023 case report showed that vascular contact of the glossopharyngeal nerve on MRI may mislead clinicians if the actual clinical phenotype is TN rather than GPN [132, 231]. Other relevant differentials in neurostomatology include painful glossopharyngeal neuropathy, superior laryngeal neuralgia, temporomandibular disorders, myofascial orofacial pain, otologic disease, tonsillopharyngeal inflammation, skull base tumors, and other causes of referred throat or ear pain. The practical message is that diagnosis must remain anchored to the pain phenotype and trigger pattern, not to imaging alone and not to the site where the patient first seeks care [226, 228, 232].

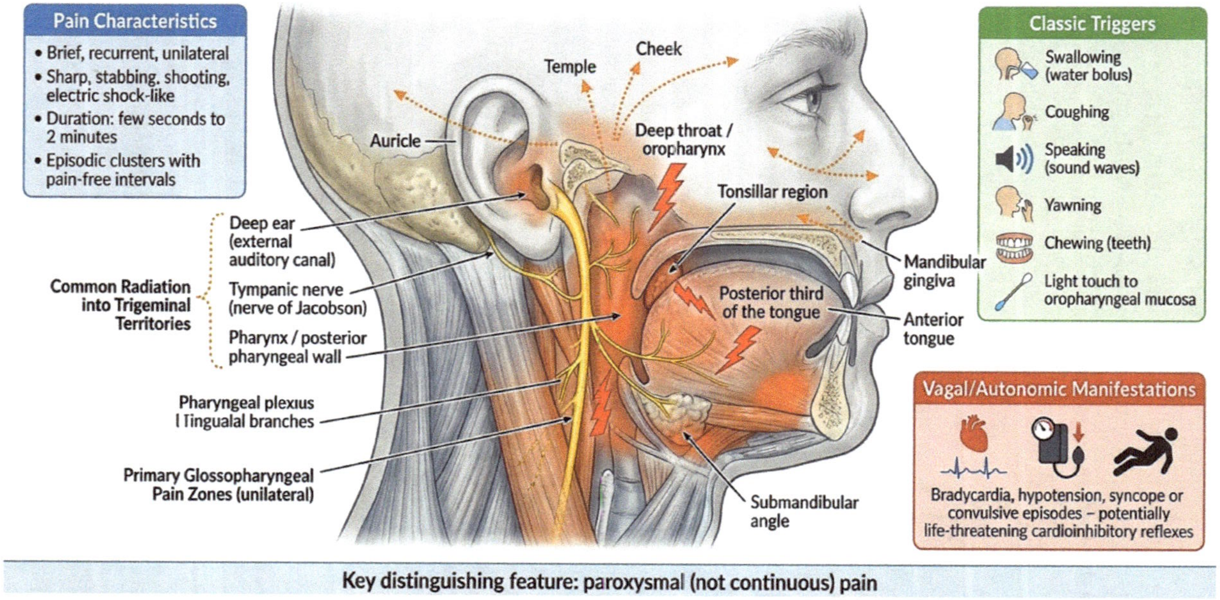

Fig. 8.13 Clinical distribution, trigger profile, and associated vagal manifestations of glossopharyngeal neuralgia

Table 8.6 Practical diagnostic criteria for glossopharyngeal neuralgia

Criterion	Description
A	Recurrent unilateral paroxysms of pain in the distribution of the glossopharyngeal nerve
B1	Attack duration from a few seconds to 2 min
B2	Severe intensity
B3	Electric shock-like, shooting, stabbing, or sharp
B4	Commonly precipitated by swallowing, coughing, talking, or yawning
C	Not better explained by another diagnosis, especially painful glossopharyngeal neuropathy, trigeminal neuralgia, tumor, demyelinating disease, or other secondary causes

Adapted from ICHD-3 and aligned with ICOP

Management of Glossopharyngeal Neuralgia The intensity of the symptoms, the underlying cause, and the patient's reaction to conservative treatment all influence how glossopharyngeal neuralgia is managed. Pharmacologic therapy is the first stage in modern treatment, which advances to interventional or surgical treatments in situations that are resistant or physically defined.

Pharmacologic Treatment

Initial treatment is usually conservative. As in other cranial neuralgias, carbamazepine remains the conventional first-line agent, and oxcarbazepine is a commonly used alternative. Gabapentin and pregabalin may be useful when first-line drugs are ineffective or poorly tolerated. Pharmacotherapy can substantially reduce attack burden, but it does not correct underlying neurovascular conflict in classical disease. Medically refractory disease, medication intolerance, or disabling vagal symptoms should prompt specialist referral rather than indefinite dose escalation [229, 230, 233].

Glossopharyngeal Nerve Block

Glossopharyngeal nerve block deserves more emphasis than it received in the earlier version because it is increasingly relevant in modern interventional management.

Ultrasound-guided glossopharyngeal nerve block is a minimally invasive option that may be considered when medication is insufficient, poorly tolerated, or undesirable, and it may be especially attractive before intracranial surgery or destructive procedures in selected patients. The key anatomical relationships and procedural steps of ultrasound-guided glossopharyngeal nerve block are illustrated in Fig. 8.14, including probe placement, ultrasound identification of the styloid process and adjacent vascular structures, and the intended needle path to the injection site. In the 2024 retrospective study of 43 patients, 81.4% had pain improvement at discharge, 30.2% achieved

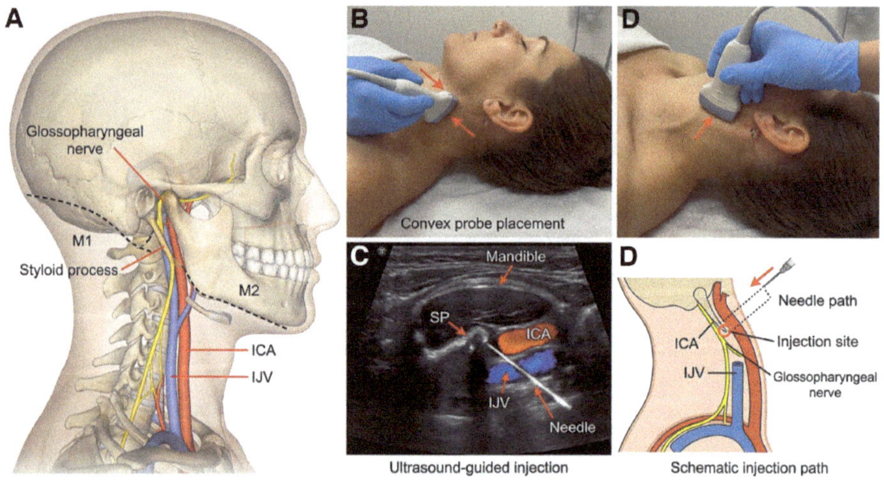

Fig. 8.14 Ultrasound-guided glossopharyngeal nerve block: anatomical landmarks, probe positioning, sonographic view, and schematic needle trajectory

recovery, and recurrence-free survival remained 77.04% at 12 months, 70.31% at 36 months, and 54.66% at 120 months. Among patients with relapse, repeat block improved pain in some cases, and no severe adverse reactions were documented [236].

For neurostomatologic practice, the practical value of this technique is twofold. First, it expands the treatment ladder between oral medication and intracranial surgery. Second, ultrasound-guided approaches allow direct visualization of important submandibular vascular structures during the procedure, which is clinically useful in a region where blind techniques can be anatomically unforgiving. While current evidence is retrospective rather than randomized, glossopharyngeal nerve block is now a reasonable option in carefully selected refractory cases, particularly when a reversible, repeatable, low-morbidity intervention is preferred [236].

Microvascular Decompression and Other Surgical Approaches

For medically refractory classical GPN with neurovascular compression, microvascular decompression (MVD) is the preferred definitive treatment because it addresses the presumed underlying mechanism rather than merely interrupting pain transmission. A 2025 systematic review and meta-analysis including 1009 patients reported pooled rates of 94.2% initial complete pain relief and 92.1% long-term complete relief, with 2.3% recurrence and 1.7% permanent complications, supporting MVD as the gold-standard treatment for classical disease [235].

Other recent evidence supports careful patient selection. In a 2022 retrospective analysis of 97 patients, venous compression and a lower degree of neurovascular conflict were independently associated with shorter pain-free survival after MVD.

Contemporary surgical review data also indicate that MVD alone is associated with fewer adverse effects than combined MVD plus rhizotomy, while rhizotomy carries greater risk of dysphagia and sensory sequelae. Accordingly, rhizotomy is best reserved for selected cases in which decompression is not feasible or the vascular conflict is surgically unfavorable [233, 234].

Radiofrequency Procedures

Radiofrequency-based procedures provide additional options for patients who are poor candidates for craniotomy or who prefer less invasive intervention. In a 2020 multicenter retrospective study, CT-guided pulsed radiofrequency (PRF) achieved an initial effective rate of 93.3%, with durable satisfactory pain-relief survival in many patients and no serious morbidity or mortality reported [237].

More recently, a 2024 study of ultrasound-guided high-voltage, long-duration PRF reported pain reduction, improved quality of life, and minimal complications, supporting its role as a promising neuromodulatory option for selected primary GPN cases [238]. When more destructive approaches are considered, percutaneous radiofrequency thermocoagulation (PRT) remains relevant. In a 2023 comparative study, both PRT and MVD significantly reduced pain and improved sleep, but MVD showed a superior medium-term complete curative effect, whereas PRT was less resource-intensive and may be more attractive in patients unsuited to craniotomy or general anesthesia [239].

Radiosurgery

Stereotactic radiosurgery (SRS) is a reasonable nonopen option for elderly, medically fragile, or surgery-averse patients, and for some recurrent cases after other interventions. The 2024 systematic review by Siempis and colleagues identified 16 studies with 97 patients and found favorable short- and long-term pain outcomes with generally low complication rates. Reported maximal doses ranged from 70 to 88.7 Gy, and the glossopharyngeal meatus was the most common target [240].

The 2025 systematic review and meta-analysis refined these estimates: among 75 patients, pooled short-term complete and acceptable pain-relief rates were 44% and 86%, while long-term complete and acceptable pain-relief rates were 46% and 79%. The pooled adverse radiation effect rate was 1%, but recurrence remained substantial at 39%. These data support SRS as a useful alternative when open surgery is undesirable, while also making clear that it is generally less durable than MVD for classical compressive disease [241]. Prognosis depends on etiology, delay to diagnosis, and treatment selection. Classical GPN with clearly identified neurovascular compression often has an excellent outcome after successful decompression, whereas secondary GPN requires treatment of the underlying lesion. Patients with venous compression, weaker neurovascular conflict, or prior surgery may have less durable surgical results [234, 235].

Follow-up in neurostomatology should document attack frequency, severity, triggers, medication burden, swallowing function, nutritional status, voice change, syncopal or presyncopal symptoms, and any recurrence pattern. It is also useful to monitor whether the patient continues to seek repeated dental opinions or requests irreversible oral procedures despite the absence of local pathology, since this may signal persisting diagnostic uncertainty rather than true dental disease [231, 235].

Glossopharyngeal neuralgia is a rare but highly characteristic cranial neuralgia that should be recognized promptly in neurostomatologic practice because it can mimic dental, otologic, pharyngeal, and trigeminal pain disorders. Accurate diagnosis depends on recognizing the paroxysmal phenotype, the trigger pattern, and the anatomical distribution, while also distinguishing classical GPN from painful glossopharyngeal neuropathy and systematically excluding secondary causes. Contemporary management is stepwise: pharmacotherapy remains first-line, glossopharyngeal nerve block has become an important minimally invasive option, and microvascular decompression remains the most effective definitive treatment for medically refractory classical disease. Radiofrequency procedures and stereotactic radiosurgery provide additional options in selected patients who are poor candidates for open surgery or prefer less invasive care [226, 230, 235, 236, 241].

8.10 Bruxism

Abstract Bruxism is a heterogeneous masticatory muscle behavior that occurs during sleep or wakefulness and should not automatically be regarded as a disease entity. Contemporary consensus distinguishes sleep bruxism, characterized by rhythmic and/or non-rhythmic jaw-muscle activity during sleep, from awake bruxism, which is defined predominantly by repetitive or sustained tooth contact and/or mandibular bracing or thrusting during wakefulness. In neurostomatology, bruxism is clinically important not as an isolated label, but because of its potential association with orofacial pain, jaw-muscle overuse, temporomandibular disorders, tooth and restorative damage, sleep-related comorbidity, and psychosocial stress. Current evidence supports a multifactorial etiopathogenesis involving central neural regulation, sleep-arousal physiology, autonomic activation, behavioral conditioning, and psychosocial as well as biologic-environmental influences, rather than older occlusion-centered concepts. Because no single sign is pathognomonic, assessment requires a multidimensional approach integrating history, circadian phenotype, associated symptoms, clinical consequences, differential diagnosis, and, when indicated, instrumental methods. Tooth wear, muscle tenderness, or restorative failure alone should not be interpreted as proof of active bruxism. Management should be individualized and directed toward clinically relevant consequences rather than complete elimination of jaw-muscle activity. Contemporary care focuses on patient education, behavioral strategies, protection of teeth and restorations, reduction of pain and overload, and identification of relevant sleep or psychosocial contributors. A modern neurostomatologic approach therefore frames bruxism as a behavior with variable clinical significance that requires precise terminology, cautious interpretation, and consequence-based management.

Keywords Bruxism, Sleep bruxism, Awake bruxism, Neurostomatology, Orofacial pain, Tooth wear, Behavioral management

Definition Bruxism is currently defined as a masticatory muscle activity that may occur during sleep or wakefulness and should not automatically be regarded as a disease entity. Contemporary consensus distinguishes two circadian manifestations. Sleep bruxism (SB) occurs during sleep and is characterized by rhythmic (phasic) and/or non-rhythmic (tonic) jaw-muscle activity. Awake bruxism (AB) occurs during wakefulness and is characterized predominantly by repetitive or sustained tooth contact and/or by bracing or thrusting of the mandible. An important conceptual advance is that bruxism is best understood as a behavior with variable clinical significance. Depending on its intensity, frequency, duration, associated symptoms, comorbidities, and consequences, it may be harmless, may act as a risk factor for oral or musculoskeletal complications, or may coexist with other conditions without being the primary cause of symptoms. The earlier qualifier "in otherwise healthy individuals" has been removed from the definitions, acknowledging that bruxism may occur in persons with coexisting oral, medical, neurologic, psychiatric, or sleep-related disorders [242, 243].

This conceptual framework is especially important in neurostomatology. Bruxism should not be reduced to "tooth grinding" alone, because its clinical relevance lies in its possible association with orofacial pain, jaw-muscle overuse, temporomandibular disorders, restorative failure, sleep-related comorbidity, psychosocial stress, and functional impairment. Accordingly, contemporary evaluation is not centered on a binary yes/no diagnosis. Instead, it requires a multidimensional interpretation that considers behavior, evidence level, associated symptoms, clinical consequences, and the broader neurostomatologic context [243, 257].

Bruxism is common, but its prevalence varies substantially according to phenotype definition, age group, population sampled, and method of assessment. A recent systematic review and meta-analysis estimated a global prevalence of approximately 22.2% for bruxism broadly defined, with pooled estimates near 21% for sleep bruxism and 23% for awake bruxism. However, these figures should be interpreted with caution because self-report, clinical examination, ecological momentary assessment, portable electromyography, and polysomnography do not identify the same construct with the same specificity [245]. A 2025 meta-analysis focused on awake bruxism further showed that self-reported "possible" AB yields higher prevalence estimates than clinically based "probable" AB and that prevalence may be substantially higher in selected subpopulations, particularly in individuals with temporomandibular disorders or systemic conditions [246]. The main differences between sleep bruxism and awake bruxism are outlined in Table 8.7.

The most important epidemiologic point is not a single prevalence value, but the recognition that bruxism is common, heterogeneous, and strongly influenced by the method of assessment. Reported prevalence differs between pediatric and adult populations, varies across studies using different thresholds for case definition, and is often inflated when based solely on retrospective self-report.

Thus, prevalence data should never be presented without clarifying whether the estimate refers to SB or AB, which assessment method was used, and whether the studied population represents the general community or a selected clinical sample [245, 246].

Etiopathogenesis of Bruxism The etiopathogenesis of bruxism is multifactorial and not adequately explained by older occlusion-centered models. Contemporary evidence supports an interaction among central neural regulation, sleep-arousal physiology, autonomic activation, psychosocial influences, behavioral conditioning, and comorbid sleep or pain disorders. Current neurobiological work implicates distributed cortical and subcortical motor networks, brainstem pathways, and neurotransmitter systems including dopamine, serotonin, glutamate, and GABA in the modulation of bruxism-related motor activity. These findings support the view that bruxism is not primarily a peripheral dental phenomenon but rather a centrally modulated motor behavior that may be influenced by multiple biologic and environmental inputs [247].

In sleep bruxism, particular attention has been directed toward arousal-related activation during sleep. SB appears to overlap with transient changes in autonomic and motor activity and may coexist with other disturbances of sleep regulation. A systematic review found that SB may be associated with sleep-related disorders, including sleep-disordered breathing, but the available evidence does not justify simple linear

Table 8.7 Comparison of sleep bruxism and awake bruxism

Factor	Sleep bruxism	Awake bruxism
Occurrence	While asleep, mostly during sleep arousal	While awake
Time–intensity	Pain is worst on waking, improves during the day	Pain worsens throughout the day
Noises	Commonly associated	Rarely associated
Activity	Clenching and grinding	Mostly clenching, sometimes grinding
Stress relationship	Unclear, little evidence	Stronger evidence, but not conclusive
Gender distribution	Equal between genders	More common in females
Heritability	Some evidence	Unclear

Adapted from Verhoeff et al. [252], Zieliński et al. [255], Stanisic et al. [256], Uchima Koecklin et al. [257], Polmann et al. [258], and Colonna et al. [263], Adapted from Verhoeff et al. [252], Zieliński et al. [255], Stanisic et al. [256], Uchima Koecklin et al. [257], Polmann et al. [258], and Colonna et al. [263]

cause-and-effect assumptions. Instead, the relationship is best interpreted as one of possible comorbidity and partially shared mechanisms [249]. Recent clinical guidance has likewise emphasized that sleep-related bruxism should be assessed within a broader sleep-medicine framework, especially when symptoms suggest obstructive sleep apnea, insomnia, frequent arousals, or fragmented sleep [257].

Psychosocial factors are also relevant. A systematic review and meta-analysis found a positive association between sleep bruxism and self-reported stress symptoms, although heterogeneity across studies and variable definitions limit the certainty of causal interpretation [248]. Awake bruxism appears particularly sensitive to behavioral and psychosocial modulation, including concentration-related tooth contact, stress-linked jaw bracing, and habitual motor patterns reinforced during daily activities [253]. Altogether, current evidence supports describing bruxism as a biopsychophysiological behavior with multiple contributing pathways rather than as a single disorder with one universal cause [247–249]. The multifactorial etiopathogenesis of bruxism is summarized in Fig. 8.15, which highlights the interaction between central neural regulation, sleep-arousal physiology, autonomic activation, psychosocial stress, behavioral conditioning, comorbid sleep disorders, and biologic-environmental influences.

Accurate assessment requires a clear distinction between findings suggestive of bruxism and direct evidence of current bruxism activity. The Standardized Tool for the Assessment of Bruxism (STAB) was developed to support this multidimensional approach by evaluating bruxism status, possible etiologic factors, comorbid conditions, and clinical consequences in a structured manner [243]. This is a major improvement

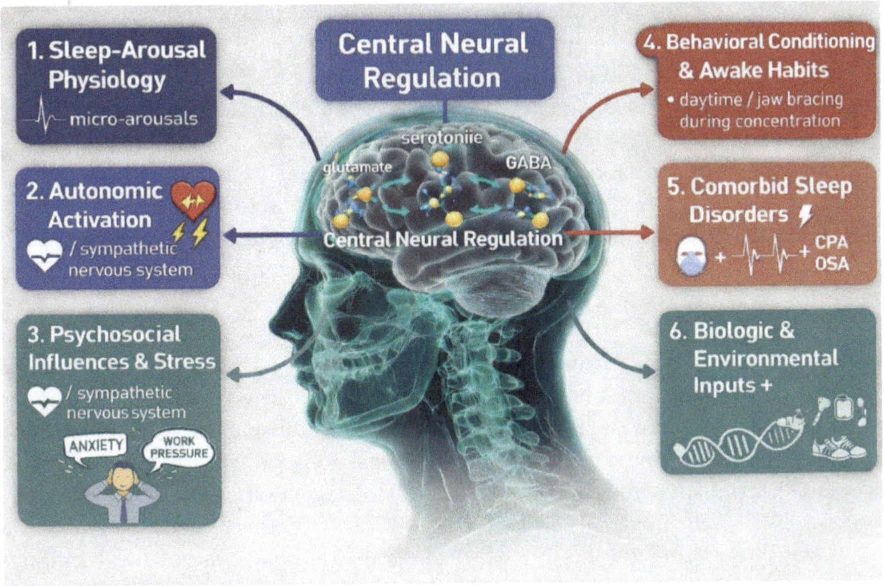

Fig. 8.15 Multifactorial etiopathogenesis of Bruxism

over older approaches that relied excessively on isolated signs such as tooth wear, muscle tenderness, or patient belief.

For sleep bruxism, polysomnography remains the reference standard when high diagnostic specificity is required, while portable electromyographic devices may be useful in selected settings and for supportive assessment in carefully chosen clinical or research contexts. Nevertheless, recent systematic review evidence indicates considerable methodological heterogeneity, variable diagnostic thresholds, inconsistent device performance, and low-certainty evidence for some portable devices. Consequently, instrumental findings should be interpreted cautiously and not detached from the broader clinical context, patient history, symptoms, and examination findings [244]. In parallel, recent sleep-medicine guidance has stressed that dental and sleep classifications do not completely overlap and that pain, tooth wear, or restorative damage should not be mistaken for direct proof of active SB, since these findings may reflect past exposure, coexisting conditions, or multifactorial processes rather than ongoing sleep-related jaw-muscle activity [257].

For awake bruxism, assessment depends more heavily on history, structured self-report, and real-time behavioral monitoring. Because grinding is relatively uncommon during wakefulness, AB more often presents as tooth contact, clenching, or mandibular bracing. Ecological momentary assessment (EMA), especially smartphone-based monitoring, has emerged as a particularly useful method for capturing AB behaviors in real-world conditions while reducing recall bias. A recent scoping review of smartphone-based studies reported frequencies of awake bruxism behaviors that varied by method and study population, highlighting both the promise of EMA and the need for careful interpretation [253]. In addition, a 2024 clinical study found that AB behaviors were more frequent in patients with masticatory muscle myalgia than in healthy controls and that self-reported clenching was especially relevant [252].

Figure 8.15 illustrates the multifactorial etiopathogenesis of bruxism, emphasizing the central role of neural regulation and its interaction with sleep-arousal physiology, autonomic activation, psychosocial stress, behavioral conditioning, comorbid sleep disorders, and biologic as well as environmental influences.

In practice, bruxism assessment should proceed in stages: a careful history, clarification of the sleep or wake phenotype, evaluation of symptoms and consequences, clinical examination for overload-related findings, screening for psychosocial and sleep-related contributors, and selective use of instrumental methods when greater diagnostic certainty is needed [243, 244, 252, 257].

Clinical Features of Bruxism Bruxism has no single pathognomonic presentation. Some individuals exhibit jaw-muscle activity without clinically relevant consequences, whereas others develop substantial oral, muscular, temporomandibular, or restorative complications. Reported features may include morning jaw-muscle fatigue, masticatory stiffness, temporal pain, masseter hypertrophy, tooth fracture, cracked restorations, dentin hypersensitivity, and soft-tissue impressions on the tongue or cheeks. However, none of these findings alone proves active bruxism [243, 257]. One of the most important corrections to older bruxism teaching concerns tooth wear. A 2024 scoping review found that the association between tooth wear and bruxism is often weak or inconsistent, especially when confounding factors are considered. Dental erosion, diet, aging, parafunctions unrelated to bruxism, and other mechanical or chemical influences may all contribute to wear. Therefore, tooth wear alone should not be used as a surrogate marker of current SB or AB activity [251].

The relationship with temporomandibular disorders is more consistent, but it still requires careful interpretation. A systematic review and meta-analysis found a positive association between bruxism and TMD, suggesting that bruxism may increase the likelihood of temporomandibular pain or dysfunction in susceptible individuals [250]. Even so, association does not imply that bruxism is the sole cause of TMD, and many patients with TMD do not have clinically meaningful bruxism. In neurostomatology, this point is essential because facial pain, temple pain, ear symptoms, and jaw dysfunction are often prematurely attributed to "bruxism" on the basis of nonspecific findings. A consequence-based interpretation is more appropriate: the clinician should determine whether the observed jaw-muscle activity is plausibly related to the patient's current symptoms or tissue overload rather than assuming that any sign automatically indicates an active pathologic process [250, 251]. The principal clinical features, associated findings, and possible consequences of bruxism are summarized in Table 8.8.

Table 8.8 Clinical features, associated findings, and possible consequences of bruxism

Domain	Feature	Clinical relevance/interpretation
Patient-reported symptoms	Jaw-muscle fatigue or stiffness	Common, especially on waking or after daytime clenching
	Masticatory muscle pain or tenderness	May reflect muscle overload or coexisting myalgia
	Temporal or facial pain	May occur, but is nonspecific
	Tooth sensitivity	May result from wear, microfracture, or overload
Clinical findings	Tooth wear	*May be present, but does not confirm active bruxism*
	Tooth fracture or cracked restorations	*Possible consequence of repetitive loading*
	Masseter hypertrophy	*May suggest sustained muscle overactivity*
	Cheek or tongue indentations	*Associated finding, but nonspecific*
	Restricted jaw comfort on waking	*May suggest nocturnal muscle activity*
Associated conditions/possible consequences	Temporomandibular symptoms or disorders	May coexist and aggravate symptoms
	Restorative or prosthodontic complications	May increase risk of mechanical failure or overload
	Sleep disturbance or nonrestorative sleep	May reflect associated sleep comorbidity
	Referred otalgia or craniofacial discomfort	May occur with masticatory or temporomandibular dysfunction

A modern neurostomatologic approach requires explicit differential diagnosis. Bruxism-related complaints may overlap with masticatory myalgia, myofascial temporomandibular pain, arthralgia, primary headache disorders presenting as facial or temporal pain, sleep-disordered breathing with morning symptoms, and odontogenic or periodontal pathology. In some patients, the main complaint is not grinding itself, but pain, stiffness, headache, or restorative breakdown, and these consequences may have multifactorial origins. For that reason, bruxism should not be diagnosed merely because the patient has tooth wear, morning discomfort, or a history of clenching [243, 257].

Differential diagnosis is particularly important when pain predominates. Morning facial pain may reflect muscle overuse, but it may also arise from TMD, primary headache disorders, poor sleep quality, or coexisting psychosocial distress. Likewise, a patient with extensive wear may have had previous bruxism but little current activity. Conversely, a patient with active AB may show minimal dental damage but substantial jaw-muscle symptoms. These scenarios illustrate why the neurostomatologic evaluation of bruxism must integrate pain medicine, behavioral assessment, and sleep medicine rather than rely on isolated dental signs [243, 250, 257].

Bruxism has major implications for restorative and prosthodontic care because repetitive loading may compromise teeth, restorations, prostheses, and implants. Cracked teeth, ceramic chipping, restorative debonding, and recurrent mechanical failure may be more likely when clinically meaningful bruxism coexists with high occlusal load or parafunctional habits. This does not mean that restorative care is contraindicated, but it does require risk communication, careful material selection, maintenance planning, and follow-up.

The strongest recent evidence concerns implant therapy. A 2024 systematic review and meta-analysis found that implants placed in probable bruxers have a significantly higher risk of failure than implants placed in non-bruxers [256]. This finding supports incorporating bruxism into prosthodontic risk assessment, especially in extensive rehabilitations or implant-supported restorations. Importantly, the clinical implication is not to deny treatment automatically but to plan restorations more cautiously, monitor them more closely, and address potentially modifiable overload-related factors [256].

Management of Bruxism Management should be individualized, conservative, and directed toward clinically relevant consequences rather than toward the unrealistic goal of eliminating all bruxism activity. At present, no intervention has been shown to suppress sleep bruxism reliably across all patient groups. Therefore, treatment goals should include protection of teeth and restorations, reduction of pain or muscle overload, improvement of jaw function, identification of comorbid sleep or psychosocial factors, and patient education [254, 257].

Occlusal splints remain among the most commonly used interventions. Current systematic review evidence shows substantial heterogeneity across splint designs, study protocols, and outcome measures. Their most consistent role is mechanical protection and, in selected patients, symptom control rather than definitive eradication of bruxism behavior [254]. Irreversible occlusal adjustments should not be promoted as routine primary therapy for bruxism itself, because contemporary evidence does not support a simplistic occlusal-causation model and because irreversible interventions may overtreat patients whose main problem is behavioral, muscular, or sleep-related [243, 254].

For awake bruxism, behavioral management is central. Education, reassurance, self-awareness training, habit-reversal strategies, stress-informed counseling, and smartphone-based monitoring may reduce nonfunctional tooth contact and mandibular bracing during the day. EMA-based approaches may also help patients recognize high-risk contexts such as concentration, emotional stress, or prolonged screen use [252, 253].

For sleep bruxism, additional evaluation is warranted when history suggests coexisting sleep-disordered breathing, insomnia, medication-related exacerbation, frequent arousals, or disproportionate morning symptoms. In such cases, referral or co-management within sleep medicine may be appropriate [249, 257]. Botulinum toxin type A may offer short-term benefit in selected patients with refractory jaw-muscle pain or prominent temporomandibular symptoms associated with sleep bruxism, but current evidence remains preliminary, protocols are heterogeneous, and potential adverse effects include injection-site discomfort, transient masticatory weakness, and cosmetic changes [255]. Pharmacologic treatment should therefore be reserved for carefully selected situations and not presented as standard routine therapy [255].

Bruxism should be approached as a heterogeneous masticatory muscle behavior with two circadian manifestations rather than as a uniform standalone disease. Its clinical importance lies in its possible relationship to pain, functional disturbance, restorative overload, sleep-related comorbidity, and psychosocial context. A modern neurostomatologic approach therefore requires precise terminology, multidimensional assessment, careful differential diagnosis, caution against overdiagnosis from nonspecific signs, and management directed toward meaningful clinical consequences rather than toward complete behavioral eradication [242, 243, 257].

8.11 Painful Post-Traumatic Trigeminal Neuropathy

Abstract Painful post-traumatic trigeminal neuropathy (PTTN), also termed post-traumatic trigeminal neuropathic pain, is a chronic neuropathic orofacial pain disorder caused by injury to the trigeminal nerve or its peripheral branches. It presents with persistent or recurrent facial and/or oral pain in a neuroanatomically plausible trigeminal distribution together with clinical evidence of somatosensory dysfunction. Contemporary diagnostic frameworks, including the *International Classification of Headache Disorders, 3rd edition* and the *International Classification of Orofacial Pain, 1st edition*, emphasize a documented traumatic event, temporal relation to pain onset, and objective confirmation of trigeminal nerve dysfunction. In dental practice, many cases are iatrogenic, particularly after third-molar surgery, implant placement, endodontic procedures, or local anesthetic trauma. Clinically, patients commonly report continuous burning, aching, or dull pain, often with allodynia, hyperalgesia, paresthesia, or numbness, in the absence of ongoing odontogenic pathology. Diagnosis requires careful history taking, neurosensory assessment, and exclusion of other causes of persistent facial or dentoalveolar pain. Management is multidisciplinary and includes early recognition, avoidance of unnecessary invasive procedures, neuropathic pain treatment, and selected surgical intervention in reparable lesions.

Keywords Painful post-traumatic trigeminal neuropathy, Post-traumatic trigeminal neuropathic pain, Trigeminal nerve injury, Neuropathic orofacial pain, Diagnosis, Management

Definition Painful post-traumatic trigeminal neuropathy (PTTN), also termed post-traumatic trigeminal neuropathic pain (PTNP), is a chronic neuropathic orofacial pain disorder caused by injury to the trigeminal nerve or one of its peripheral branches. It presents as unilateral or bilateral facial and/or oral pain distributed within the territory of one or both trigeminal nerves and is accompanied by clinical signs of trigeminal nerve dysfunction [258, 259]. The condition is especially important in dental practice because many cases are iatrogenic, developing after routine dental or oral surgical procedures that unintentionally injure the dental plexuses or related trigeminal branches, thereby producing persistent sensory disturbance and neuropathic pain.

The traumatic event may be mechanical, chemical, thermal, or radiation-related. The parallel criteria in the International Classification of Orofacial Pain, 1st edition (ICOP) require pain persisting or recurring for >3 months, with history and diagnostic confirmation of peripheral trigeminal nerve injury [259].

The odontogenic (dentoalveolar) subtype involves persistent pain after trauma to the superior or inferior dental nerve plexuses and is now classified under the broader PTTN framework. Older terms such as "odontogenic dental plexalgia" appear only in limited historical literature and are no longer recommended for international use.

PTTN is uncommon but carries high clinical and psychosocial burden, especially in dental practice. Reported prevalence after dental procedures ranges from 1.55% to 13%, depending on procedure type, follow-up duration, and diagnostic criteria [260]. Females predominate (60%–82% of cases in most series), with peak incidence in the 40–60-year age group [260]. The mandibular division, particularly the inferior alveolar and lingual nerves, is affected most often because of its vulnerability during third-molar removal, implant placement, and endodontic treatment [260].

Table 8.9 summarizes the diagnostic criteria for 13.1.2.3 Painful post-traumatic trigeminal neuropathies according to the *International Classification of Headache Disorders, 3rd edition* (ICHD-3).

Etiopathogenesis of PTTN In the odontogenic form, injury is predominantly iatrogenic and includes overextension of root-canal materials into the mandibular canal, traumatic extractions (especially multiple or complex third molars), implant

Table 8.9 ICHD-3 diagnostic criteria for painful post-traumatic trigeminal neuropathy

Criterion	Description
A.	Facial and/or oral pain in the distribution(s) of one or both trigeminal nerve(s) and fulfilling criterion C
B.	History of an identifiable traumatic event to the trigeminal nerve(s), with clinically evident positive (hyperalgesia, allodynia) and/or negative (hypaesthesia, hypalgesia) signs of trigeminal nerve dysfunction
C.	Evidence of causation demonstrated by both of the following:
	1. Pain is localized to the distribution(s) of the trigeminal nerve(s) affected by the traumatic event
	2. Pain has developed **<6 months** after the traumatic event
D.	Not better accounted for by another ICHD-3 diagnosis.

compression of the inferior alveolar nerve, local-anesthetic needle trauma, and suboptimal restorative or surgical procedures [260, 261]. These events cause mechanical compression, chemical irritation, thermal damage, or ischemia. Peripheral mechanisms begin immediately after axon injury. Inflammatory mediators (bradykinin, prostaglandins, cytokines IL-1β and TNF-α) lower nociceptor thresholds and promote ectopic firing [261]. Sodium-channel isoforms Nav1.3 and Nav1.8 are upregulated, while potassium channels (Kv7.2, Kv4.1) are downregulated, generating spontaneous action potentials. Satellite glial cells in the trigeminal ganglion become activated, releasing ATP and additional cytokines that amplify neuronal excitability and induce phenotypic switching of large-diameter Aβ fibers to a nociceptive phenotype [261].

Centrally, within the spinal trigeminal nucleus, particularly the subnucleus caudalis, activation of microglia and astrocytes plays a central role in driving central sensitization. Activated microglia release brain-derived neurotrophic factor (BDNF), which downregulates the potassium-chloride cotransporter KCC2 on second-order neurons. This disrupts chloride homeostasis and leads to neuronal disinhibition. At the same time, astrocytes alter the glutamate-glutamine shuttle, resulting in enhanced excitatory synaptic transmission. These neuroglial changes lead to expansion of receptive fields, loss of inhibitory control, and wind-up phenomena, which ultimately produce spontaneous pain, hyperalgesia, and allodynia [261]. Genetic variants affecting ion channels, inflammatory pathways, and descending modulation, together with psychosocial stressors, strongly influence the transition from acute to chronic pain [261].

Clinical Features of PTTN Patients typically report persistent, continuous or near-continuous pain that is strictly confined to the teeth, gingiva, alveolar ridge, or edentulous areas supplied by the injured dental plexus. The quality of pain is most commonly described as burning (approximately 53% of cases), although aching, gnawing, or dull sensations are also frequently reported. These ongoing symptoms are often interrupted by superimposed paroxysmal flares lasting from several minutes up to several hours [260]. Radiation of the pain toward the jaw, ear, or temple area is a common complaint. A range of factors tends to aggravate the pain intensity, including cold stimuli, light touch, oral functions such as chewing or speaking, emotional stress, and fluctuations in weather conditions. Some patients describe partial relief from competing sensory input, for instance while chewing coarse or strongly flavored foods, although current evidence for this phenomenon remains largely anecdotal [260]. Routine clinical dental examination combined with appropriate imaging studies consistently fails to reveal any active pathology. Furthermore, additional restorative, endodontic, or surgical dental procedures do not alleviate the pain and commonly lead to exacerbation of symptoms.

Quantitative sensory testing (QST) reliably detects well-localized somatosensory abnormalities, manifesting as either positive signs (hyperalgesia and allodynia) or negative signs (hypaesthesia or hypalgesia). By comparison, conventional bedside sensory examination frequently shows only subtle deficits [260]. Associated symptoms often include paresthesia, numbness, allodynia, an unpleasant foreign-body sensation, or occasional unexplained swelling in the affected region.

The condition exerts a substantial negative effect on quality of life, commonly resulting in sleep disturbance, increased anxiety and depression, and moderate-to-severe disability in up to 13% of patients [260].

The typical clinical features of painful post-traumatic trigeminal neuropathy in its odontogenic (dentoalveolar) presentation are summarized in Table 8.10.

Diagnosis rests on a documented traumatic event with temporal linkage (<6 months onset), neuroanatomically congruent pain distribution, and objective signs of trigeminal dysfunction confirmed by clinical examination or QST [258, 259]. Cone-beam computed tomography (CBCT) is mandatory to exclude compressive lesions, overfills, or implant malposition. Neurophysiological tests (blink reflex, quantitative sensory testing) and, when indicated, MRI provide supportive evidence.

Key differentials include persistent idiopathic dentoalveolar pain disorder (no clear trauma), classical or idiopathic trigeminal neuralgia (paroxysmal, seconds-long shocks without sensory loss), post-herpetic trigeminal neuropathy, sinus pain, myofascial pain, and residual odontogenic infection. Multidisciplinary evaluation involving orofacial pain specialists prevents misdiagnosis and unnecessary invasive procedures [258, 260].

Table 8.10 Clinical features of painful post-traumatic trigeminal neuropathy

Category	Features
Pain quality	• Burning, dull, aching, gnawing; continuous or near-continuous
Duration	• Persistent (>3 months); paroxysms lasting minutes to hours
Location	• Unilateral, dental plexus distribution (teeth, gingiva, post-extraction sites)
Aggravating factors	• Cold, touch, function, stress, weather changes
Associated findings	• Clear trauma history; no visible pathology; no response to dental treatment; QST abnormalities

Management of PTTN Management is multidisciplinary, conservative, and focused on neuropathic pain modulation. Further invasive dental procedures in the painful region are strictly contraindicated because they may exacerbate central sensitization [260].

Acute Phase (First 7–10 Days)

In the acute phase (first 7–10 days), any ongoing compressive or irritative factor such as overextended material or a malpositioned implant should first be removed. A short course of systemic corticosteroids is recommended to limit inflammation. Typical regimens include prednisone 40–60 mg daily or dexamethasone 12–16 mg daily, tapered over 7–10 days [260].

Chronic Phase Pharmacotherapy

First-line agents include gabapentinoids such as gabapentin or pregabalin titrated to patient tolerance. Tricyclic antidepressants such as amitriptyline 10–75 mg at night are also recommended. Serotonin-norepinephrine reuptake inhibitors such as duloxetine serve as useful alternatives or adjuncts [260]. Topical therapy with lidocaine gel or low-concentration capsaicin ($\leq$0.1%) delivered via custom intraoral trays provides localized relief with minimal systemic adverse effects [260].

Non-Pharmacological Approaches

Cognitive-behavioral therapy, patient education on pain neurobiology, and mindfulness-based stress reduction address the psychosocial dimension. Transcutaneous electrical nerve stimulation (TENS) and low-level laser therapy may be considered as adjuncts, although evidence specific to PTTN remains limited [260].

Surgical Options

Microsurgical repair (neurolysis, neuroma resection, neurorrhaphy, or allograft) is reserved for cases with clear reparable lesions when performed early (within 90–200 days post-injury) [260, 262]. Outcomes are best when compression or partial transection is confirmed by imaging and neurophysiology. Microvascular decompression has no role in this peripheral neuropathic entity.

Prevention relies on preoperative CBCT planning, meticulous surgical technique, informed consent regarding nerve-injury risk, and immediate referral to orofacial pain specialists at the first sign of persistent symptoms. These steps are essential to reduce incidence and severity [260].

PTTN often follows a chronic course. Symptom persistence rates range from 50% to 70% across heterogeneous studies, rising to 86.2% at 2 years in patients with painful PTTN [260, 263]. Most improvement occurs within the first 20 weeks; little further recovery is seen after 60 weeks [263]. Negative prognostic factors include older age, female sex, painful (versus purely sensory) presentation, maxillary or implant-related injury, and thermal or mechanical hyperesthesia [263]. Lingual-nerve injuries and third-molar-related cases have relatively better long-term outcomes [263]. Early intervention (within weeks) limits central sensitization and improves quality of life [260]. Most patients achieve partial relief with multimodal management, although complete resolution remains uncommon [263].

Emerging research is actively investigating several promising therapeutic approaches such as glial-cell modulators, selective ion-channel blockers, and minimally invasive peripheral nerve stimulation [261]. Heightened awareness among general dentists, the adoption of standardized diagnostic protocols, and prompt referral to specialists will be critical in lowering the iatrogenic burden of PTTN.

8.12 Persistent Idiopathic Dentoalveolar Pain

Abstract Persistent idiopathic dentoalveolar pain (PIDAP), previously termed atypical odontalgia, phantom tooth pain, or persistent dentoalveolar pain disorder, is a chronic orofacial pain condition defined by the International Classification of Orofacial Pain (ICOP) as persistent unilateral intraoral dentoalveolar pain recurring daily for more than 2 h per day for longer than 3 months, in the absence of a preceding causative event or any local pathology sufficient to explain the complaint. As a diagnosis of exclusion, PIDAP is clinically important because it is frequently underrecognized and may lead to repeated ineffective or unnecessary dental procedures. It predominantly affects middle-aged women and is increasingly understood within a nociplastic pain framework involving altered central pain processing and trigeminal sensitization. Clinically, patients typically report deep, dull, pressing, pulling, burning, or throbbing pain localized to a tooth or alveolar ridge, often without corresponding clinical or radiographic abnormalities. Mechanical allodynia of the gingival sulcus is a characteristic finding in many patients. Diagnosis requires exclusion of odontogenic disease, painful post-traumatic trigeminal neuropathy, persistent idiopathic facial pain, trigeminal neuralgia, and referred musculoskeletal pain. Management is multidisciplinary and individualized, combining patient education, avoidance of further invasive treatment, pharmacologic neuromodulation, and supportive behavioral strategies. Early recognition is essential to reduce iatrogenic harm and improve long-term outcomes.

Keywords Persistent idiopathic dentoalveolar pain, PIDAP, Atypical odontalgia, Chronic orofacial pain, Nociplastic pain, ICOP

Definition Persistent idiopathic dentoalveolar pain (PIDAP), previously known as atypical odontalgia, phantom tooth pain, or persistent dentoalveolar pain disorder (PDAP), is a chronic orofacial pain disorder characterized by persistent unilateral intraoral pain localized to the dentoalveolar region. According to the International Classification of Orofacial Pain (ICOP), it is defined as persistent unilateral intraoral dentoalveolar pain, rarely occurring in multiple sites, with variable features but recurring daily for more than 2 h per day for more than 3 months, in the absence of any preceding causative event [264]. This condition is strictly a diagnosis of exclusion, meaning that all potential local causes must be thoroughly ruled out before assigning this label.

The full ICOP diagnostic criteria for PIDAP are presented in Table 8.11.

Furthermore, subtypes have been recognized based on the presence or absence of somatosensory abnormalities, which can be detected through quantitative sensory testing (QST). This subclassification helps in better understanding individual patient presentations and may guide more personalized approaches [265].

PIDAP is an uncommon condition, with prevalence estimated at approximately 2% among patients attending specialized orofacial pain clinics and potentially higher rates, ranging between 2.5% and 5%, observed in endodontic referral populations [266]. It predominantly affects middle-aged women, showing a clear female-to-male ratio of approximately 4:1, with a mean age of onset around 49 years [265, 266]. Due to its relatively low prevalence and atypical presentation, the condition is frequently underrecognized by general dental practitioners. This often leads to prolonged patient suffering and multiple cycles of ineffective dental interventions that fail to address the underlying pain mechanism [267].

Etiopathogenesis of PIDAP PIDAP is widely regarded as a nociplastic pain condition, which involves altered nociception in the absence of clear evidence of actual or threatened tissue damage or a distinct somatosensory lesion [265]. In this context, central nervous system mechanisms play the predominant role. These include processes such as central sensitization, dysfunction of descending inhibitory

Table 8.11 ICOP diagnostic criteria for persistent idiopathic dentoalveolar pain

Criterion	Description
A	Persistent unilateral intraoral dentoalveolar pain, rarely affecting multiple sites. The pain may vary in character but recurs daily for **>2 h per day** and persists for **>3 months**
B	No clinical or radiological evidence of dental, periodontal, or other orofacial pathology that could explain the pain
C	The condition is not better accounted for by another diagnosis within the ICOP classification

Adapted from ICOP, 1st edition, and supplemented by recent review literature

pain pathways, and alterations in cortical neuroplasticity specifically within the trigeminal system [266].

Although many patients report a remote history of dental procedures, such as root canal therapy (observed in approximately 69% of cases), no identifiable causative event is present according to the strict diagnostic definition. This key feature clearly distinguishes PIDAP from post-traumatic trigeminal neuropathic pain (PTNP) [265]. Frequent comorbidities, such as fibromyalgia, thyroid disorders, other chronic widespread pain syndromes, anxiety, and depression, are commonly observed. These associations strongly point toward a broader dysregulation of central pain processing mechanisms [266]. Preliminary research has identified elevated serum 3-hydroxybutyrate levels as a potential supportive biomarker possibly linked to altered energy metabolism in chronic pain states; however, additional confirmatory studies are required before widespread clinical validation can be recommended [268]. Genetic predispositions combined with hormonal influences are thought to contribute to the observed female predominance in the condition.

Clinical Features of PIDAP Patients with PIDAP typically describe their pain as a deep, dull, pressing, pulling or dragging, burning, or throbbing sensation. The discomfort is usually localized to a specific tooth or the alveolar ridge and is frequently associated with teeth that have previously undergone root canal treatment. The average pain intensity remains in the moderate range (NRS scores approximately 3–5), although it can fluctuate. A characteristic diurnal pattern is often noted, with pain intensity usually lower upon waking in the morning and progressively increasing as the day advances. Nocturnal pain that awakens the patient from sleep is relatively uncommon, occurring in only about 15% of cases [265].

One of the most reliable clinical hallmark findings is the presence of mechanical allodynia in the gingival sulcus, which can be easily detected through gentle probing with a periodontal probe. This sensitivity is reported in over 90% of patients [265, 267]. When compared with patients experiencing inflammatory dental pain, individuals with PIDAP demonstrate distinctly different quantitative sensory testing profiles. These often include higher mechanical pain thresholds within certain subgroups [267]. Importantly, the pain is not typically provoked or intensified by standard odontogenic tests such as percussion or thermal stimulation when no underlying pathology is present. The sensation may radiate mildly to the jaw or face; however, it does not adhere to strict neural or dermatomal distributions. Beyond the sensory experience, the condition exerts significant negative impacts on patients' quality of life, including difficulties with chewing, speaking, concentration, and overall mood. These effects frequently extend to secondary disturbances in sleep quality and psychological well-being [267].

Reaching an accurate diagnosis of PIDAP requires a systematic and comprehensive approach to exclude other potential sources of pain. This includes ruling out odontogenic causes through thorough vitality testing, conventional radiographs, and cone-beam computed tomography (CBCT) when indicated. It is equally important to differentiate PIDAP from post-traumatic trigeminal neuropathic pain, which is characterized by a clear history of injury or surgery with a close temporal relationship to pain onset. Other conditions to consider include persistent idiopathic facial pain (which presents with extraoral location), trigeminal neuralgia, and myofascial pain disorders.

Key differentiating features of PIDAP include the consistent lack of radiographic or clinical abnormalities, the characteristic pattern of gingival sulcus allodynia, and the specific pain chronology that follows the ICOP criteria [264, 266]. Because of the complexity involved in accurate differentiation, multidisciplinary evaluation involving experienced orofacial pain specialists is often necessary to ensure diagnostic precision and avoid inappropriate treatment pathways [271].

Management of PIDAP The management of PIDAP remains clinically challenging due to its chronic and often refractory nature. The most effective strategy involves a multidisciplinary and highly individualized approach that focuses on central pain modulation, careful management of associated comorbidities, and the deliberate avoidance of additional invasive dental procedures that could potentially exacerbate the condition [266, 269].

Pharmacologic interventions commonly include tricyclic antidepressants such as amitriptyline and serotonin-norepinephrine reuptake inhibitors such as duloxetine, which serve as foundational agents for addressing the neuropathic and nociplastic components. Gabapentinoids may be added for further neuromodulation when needed. It is important to note that the current evidence base for these medications in PIDAP is largely extrapolated from established neuropathic pain treatment guidelines, given the relative rarity of the condition [269].

Adjunctive therapies have also shown value. Topical agents including capsaicin and local anesthetics, along with intraoral injections of onabotulinumtoxinA, have demonstrated positive outcomes in observational studies and case series [270]. Non-pharmacological interventions constitute essential components of care and typically encompass cognitive behavioral therapy, acceptance and commitment therapy, various relaxation techniques, acupuncture, and transcutaneous electrical nerve stimulation (TENS). For patients with refractory symptoms, emerging neuromodulation techniques such as repetitive transcranial magnetic stimulation (rTMS) offer additional therapeutic promise [269]. Throughout the treatment process, strong emphasis must be placed on patient education and validation that their pain is genuine and has a neurobiological basis. Such validation plays a critical role in building trust and enhancing long-term treatment adherence.

Although PIDAP is generally a persistent condition, a considerable proportion of patients can achieve substantial symptom relief and meaningful improvements in daily function when managed with a well-coordinated multimodal therapeutic approach. Early and accurate recognition of the disorder, followed by prompt referral to appropriate specialists, plays a vital role in minimizing iatrogenic harm and optimizing long-term patient outcomes [266]. Regular long-term follow-up remains essential to monitor progress and adjust management strategies as needed.

Looking ahead, several important research priorities have been identified. These include the need for large-scale prospective epidemiological studies, further validation of potential biomarkers such as 3-hydroxybutyrate, the conduct of high-quality randomized controlled trials, and more in-depth exploration of the underlying neuroinflammatory and metabolic mechanisms. Such advancements hold significant potential for the development of more targeted and effective therapies in the future [263, 269].

8.13 Persistent Idiopathic Facial Pain

Abstract Persistent idiopathic facial pain (PIFP) is a chronic idiopathic orofacial pain disorder characterized by persistent facial and/or oral pain recurring daily for more than 2 h per day over more than 3 months, in the absence of a dental, neurologic, or other local lesion sufficient to explain the complaint. Replacing the older term *atypical facial pain*, PIFP is classified in the International Classification of Orofacial Pain among idiopathic orofacial pain disorders and is best understood within a contemporary biopsychosocial and chronic primary pain framework. Clinically, it usually presents as a poorly localized, deep, dull, aching, pressure-like, or burning pain, often unilateral at onset but capable of spreading over time. The pain does not follow the distribution of a peripheral nerve and lacks the brief, stereotyped paroxysms and typical triggerability of trigeminal neuralgia. Its etiopathogenesis is multifactorial and remains incompletely understood, with evidence suggesting interactions among peripheral triggers, altered nociceptive processing, central pain modulation, and psychosocial contributors. Diagnosis depends on recognition of the characteristic clinical pattern together with careful exclusion of competing odontogenic, neuropathic, musculoskeletal, sinus-related, and headache disorders. Management should be conservative, multidisciplinary, and explicitly non-destructive, emphasizing clear patient education, avoidance of unnecessary irreversible dental procedures, rational pharmacotherapy, and attention to sleep, mood, coping, and functional outcomes. In neurostomatology, early recognition of PIFP is essential to reduce diagnostic delay, prevent iatrogenic interventions, and support effective long-term care.

Keywords Persistent idiopathic facial pain, PIFP, Idiopathic orofacial pain, Chronic primary pain, Differential diagnosis, Multidisciplinary management, Neurostomatology

Description Persistent idiopathic facial pain (PIFP) is a chronic orofacial pain disorder defined by persistent facial and/or oral pain recurring daily for more than 2 h per day for more than 3 months, in the absence of a dental, neurologic, or other local lesion that adequately explains the complaint. In contemporary terminology, PIFP replaces the older and less precise label *atypical facial pain*.

Within the International Classification of Orofacial Pain (ICOP), it is classified among idiopathic orofacial pain disorders; at the same time, its clinical behavior is consistent with a modern biopsychosocial and chronic primary pain framework, in which persistent pain and disability cannot be understood solely through structural findings. For neurostomatology, PIFP remains a particularly important diagnosis because patients commonly undergo prolonged diagnostic journeys, repeated imaging, and unnecessary dental procedures before the correct diagnosis is reached [272, 273].

PIFP usually presents as a persistent, poorly localized, deep pain in the face and/or oral region. Patients commonly describe the pain as dull, aching, nagging, heavy, pressure-like, pulling, or burning rather than electric shock-like. The pain often begins in a limited area, such as the maxillary, mandibular, infraorbital, or perioral region, but it may later extend within the same side of the face.

A key diagnostic feature is that the pain does not follow the distribution of a peripheral nerve and is not restricted to a single trigeminal branch. In contrast to trigeminal neuralgia, pain-free intervals are uncommon, the attacks are not brief and stereotyped, and tactile trigger zones are usually absent.

Many patients report fluctuation in intensity during the day, sometimes with lesser pain on waking, but the overall course is chronic and undulating rather than paroxysmal. Routine clinical examination is typically unrevealing, which partly explains why the disorder is so frequently misunderstood in dental and neurological practice [275].

Etiopathogenesis of PIFP The etiopathogenesis of PIFP is multifactorial and incompletely understood. Current evidence does not support a purely psychogenic model, but neither does it identify a single structural lesion or biomarker. Instead, PIFP is best interpreted as a heterogeneous pain condition in which peripheral triggers, altered nociceptive processing, and psychosocial factors may interact. Some patients report onset after minor dental treatment, extraction, endodontic therapy, or local trauma; however, the temporal association is often ambiguous, and the magnitude of the original tissue event is usually insufficient to explain persistent pain lasting months or years.

This supports the view that, in at least some patients, an initiating event may act as a trigger rather than an ongoing nociceptive source. The persistent and disproportionate nature of symptoms also supports a nociplastic or chronic primary pain component, with altered central pain modulation, sleep disturbance, emotional distress, and functional impairment reinforcing one another over time. Importantly, these factors should be understood as contributors to persistence and disability, not as evidence that the pain is unreal [274, 276].

Clinical Features of PIFP The hallmark clinical picture is daily facial and/or oral pain for more than 3 months, present for more than 2 h per day, with poor localization and a dull, aching, or pressure-like quality. The pain is often unilateral at onset, though bilateral or spreading patterns may occur later. Neurological examination should be normal on routine bedside assessment; if clear hypoesthesia, allodynia, hyperalgesia, or another objective trigeminal sensory deficit is present, painful post-traumatic trigeminal neuropathy or another neuropathic pain disorder should be reconsidered (see Table 8.12.)

PIFP must also be distinguished carefully from persistent idopathic dentoalveolar pain (PIDAP), which is centered specifically in the dentoalveolar tissues, and from trigeminal neuralgia, in which brief electric shock-like attacks, innocuous triggerability, and neuroanatomically plausible trigeminal distribution are the dominant features (see Table 8.13). Importantly, trigeminal neuralgia may include concomitant continuous pain, so the presence of background pain alone does not exclude it. Broader differential diagnosis should include temporomandibular disorders, myofascial pain, sinus-related pain, salivary disease, painful cranial neuropathies, and facial presentations of primary headache disorders [277–279].

Diagnosis is based on positive recognition of the clinical pattern together with structured exclusion of competing causes, not on proving the disorder with a single test. A careful history should document onset, temporal profile, pain descriptors, spread, aggravating and relieving factors, prior dental procedures, trauma history, headache symptoms, sleep disturbance, medication exposure, and psychosocial context. Examination should include cranial nerves, intraoral and extraoral structures, temporomandibular joints, masticatory muscles, cervical musculature, and sensory screening. Investigations should be targeted, not excessive: dental imaging and local examination are used to exclude odontogenic disease, whereas cranial imaging is reserved for cases with red flags, atypical neurological findings, or suspicion of secondary pathology. The goal is to avoid both under-investigation and the opposite problem of repeated, low-yield diagnostic escalation that further medicalizes the pain experience [280].

Management of PIFP Management should be conservative, multidisciplinary, and explicitly non-destructive. Patients need a clear explanation that the pain is real, chronic, and treatable, even though no structural lesion has been identified.

Table 8.12 Diagnostic criteria for persistent idiopathic facial pain

Criterion	Description
A	Facial and/or oral pain recurring daily for **>2 h/day for >3 months**
B	Pain is poorly localized and does not follow the distribution of a peripheral nerve
C	Pain is usually dull, aching, nagging, pressure-like, pulling, or burning
D	Routine clinical neurological examination is normal
E	Appropriate history, examination, and indicated investigations exclude dental or other local causes
F	The pain is not better accounted for by another orofacial pain diagnosis, including trigeminal neuralgia, painful post-traumatic trigeminal neuropathy, PIDAP, temporomandibular disorders, primary headache disorders presenting in the face, sinus disease, or salivary pathology

Adapted from ICOP, 1st edition, and supplemented by recent review literature

Additional irreversible dental procedures—such as repeat endodontics, extraction, implant removal, or exploratory surgery—should be avoided unless there is demonstrable pathology, because they rarely relieve PIFP and may aggravate chronification. Pharmacologic therapy is usually extrapolated from chronic neuropathic and chronic primary pain practice: low-dose tricyclic antidepressants are commonly used first, while serotonin–noradrenaline reuptake inhibitors, gabapentin, or pregabalin may be considered in selected patients. Contemporary clinical series suggest that medication-based care is more helpful than further

Table 8.13 Comparison of PIFP, trigeminal neuralgia, and PIDAP

Feature	PIFP	Trigeminal neuralgia (TN)	PIDAP
Core disorder	Idiopathic facial pain	Cranial neuralgia	Idiopathic dentoalveolar pain
Pain quality	Dull, aching, nagging, pressure-like, pulling, sometimes burning	Electric shock-like, stabbing, shooting; may have background continuous pain	Persistent deep aching, pressure, dull, sometimes burning
Temporal pattern	Persistent or near-continuous, fluctuating	Brief recurrent paroxysms, with or without concomitant continuous pain	Persistent daily pain
Typical episode duration	Hours to continuous	Fraction of a second to 2 minutes per paroxysm	>2 h/day for >3 months
Localization	Poorly localized facial/oral region	One or more trigeminal divisions	Tooth, extraction site, gingiva, alveolus, or localized dentoalveolar region
Distribution	No peripheral nerve pattern	Neuroanatomically plausible trigeminal distribution	Dentoalveolar site rather than nerve-branch pattern
Triggering	Variable and non-stereotyped	Often triggered by light touch, chewing, speaking, brushing	May worsen with function or pressure, but lacks classic trigger zones
Neurological findings	Usually normal on bedside exam	Usually normal in classical TN; may be abnormal in secondary TN	Usually normal; subtle somatosensory changes may occur in some patients
Investigations	Used to exclude other facial/oral causes	MRI important to evaluate secondary causes and neurovascular conflict	Dental exam and imaging are used to exclude odontogenic pathology
Key clinical clue	Chronic diffuse non-neuralgic facial pain	Brief shock-like triggerable attacks	Persistent pain centered in dentoalveolar tissues
Effect of dental procedures	Usually unhelpful and potentially harmful when no pathology exists	Not a dental disorder	Dental procedures often precede presentation, but no explanatory pathology is found

Adapted from ICOP, 1st edition, and supplemented by recent review literature

surgery in appropriately diagnosed patients, although treatment response is often partial rather than curative. Multimodal management should also address sleep, mood, function, illness beliefs, and pain coping, with referral for psychological treatment when anxiety, depression, catastrophizing, or repeated health-seeking behavior contributes to disability. Long-term follow-up should track pain intensity, function, sleep quality, medication tolerability, and avoidance of unnecessary interventions [281, 282].

PIFP is a chronic idiopathic orofacial pain disorder defined by a characteristic clinical phenotype and careful exclusion of better explanations. Its modern interpretation should avoid both lesion-based reductionism and outdated psychogenic labeling. For neurostomatology, PIFP is best approached through diagnostic precision, conservative management, and integration of dental, neurologic, and pain-science perspectives within a biopsychosocial framework.

8.14 Eagle Syndrome

Abstract Eagle syndrome is an uncommon but treatable condition caused by elongation of the styloid process and/or calcification of the stylohyoid ligament complex, resulting in irritation or compression of adjacent neurovascular structures. Two major clinical forms are recognized: the classic neurogenic variant, characterized by cervicopharyngeal pain, dysphagia, foreign-body sensation, and referred otalgia, and the vascular variant, in which carotid or jugular compression may cause positional headache, dizziness, syncope, transient ischemic attack, stroke, or other cerebrovascular manifestations. Because styloid elongation is relatively common on imaging but often asymptomatic, diagnosis requires careful clinicoradiologic correlation. Computed tomography with three-dimensional reconstruction is the preferred imaging modality, while computed tomography angiography is essential when vascular involvement is suspected. Management ranges from conservative therapy, including analgesics, neuropathic pain medications, and local injections, to surgical styloidectomy for refractory or vascular cases. Early recognition is crucial, particularly in vascular Eagle syndrome, because timely diagnosis and targeted treatment can prevent serious complications and provide durable symptom relief.

Keywords Eagle syndrome, Stylohyoid syndrome, Elongated styloid process, Stylohyoid ligament calcification, Cervicopharyngeal pain, Styloidectomy

Definition Eagle syndrome (ES), also known as stylohyoid syndrome or stylocarotid syndrome, is a rare but treatable clinical condition caused by an abnormally elongated styloid process of the temporal bone (typically >30 mm) or ossification/calcification of the stylohyoid ligament. This anatomical abnormality leads to mechanical irritation or compression of adjacent neurovascular structures, including branches of the glossopharyngeal nerve, the internal carotid artery, and occasionally the internal jugular vein [283]. The modern clinical entity was first described in 1937 by the American otorhinolaryngologist Watt Weems Eagle, although elongated styloid processes had been noted anatomically centuries earlier. Contemporary understanding recognizes two distinct phenotypes: the classic neurogenic variant, characterized by neuropathic pain and pharyngeal symptoms, and the vascular variant, in which dynamic or static compression of carotid or jugular vessels produces cerebrovascular or autonomic manifestations [284]. Accurate subtyping is essential because it directly influences diagnostic workup, therapeutic strategy, and prognosis.

Elongated styloid processes are identified on imaging in 4%–30% of the asymptomatic general population, yet only a small proportion develop symptomatic Eagle syndrome. Estimated prevalence of clinically manifest cases ranges from 0.16% to approximately 4%, underscoring the condition's underdiagnosis [285]. Presentation occurs predominantly in middle-aged adults, with a mean age of 47 years. The classic neurogenic form shows a slight female predominance, whereas the vascular form exhibits a clear male preponderance (approximately 70%). Bilateral styloid elongation is common, occurring in over 60% of vascular cases. A prior history of tonsillectomy is a well-documented risk factor, particularly for the classic variant [285].

Etiopathogenesis of Eagle Syndrome The formation of an aberrant stylohyoid complex arises through multiple overlapping mechanisms that collectively contribute to the development of an elongated styloid process or calcification of the stylohyoid ligament complex. Embryologically, the styloid process, the stylohyoid ligament, and the lesser horn of the hyoid bone derive from Reichert's cartilage of the second pharyngeal arch during early embryonic development. Any disruption during this critical developmental stage, such as anomalous segmentation of the cartilage, persistence of embryonic cartilaginous rests, or subsequent abnormal metaplastic ossification, can result in pathological elongation or calcification of the stylohyoid apparatus [286]. Beyond these congenital predispositions, several acquired factors serve as important additional triggers in the pathogenesis. These include progressive age-related degenerative changes within the complex, chronic low-grade inflammation, and reactive ossification that may occur following trauma, local infection, or previous surgical manipulation in the cervical region, especially tonsillectomy [287].

In patients who develop symptoms, the elongated or ossified structure exerts sustained direct mechanical pressure on adjacent neurovascular structures within the parapharyngeal space. This compression and irritation lead to the characteristic manifestations of Eagle syndrome. Specifically, classic neurogenic symptoms arise from repeated irritation of the glossopharyngeal nerve branches or the pharyngeal mucosa. In contrast, vascular manifestations result from repetitive microtrauma to the intima of the internal carotid artery or from dynamic impingement during neck rotation, potentially leading to serious outcomes such as arterial dissection, thrombosis, or hemodynamic compromise [284].

Recent analyses emphasize that the manifestation of symptoms in Eagle syndrome frequently depends more heavily on positional dynamics rather than on static anatomical features alone [285].

The multiple overlapping mechanisms involved in the etiopathogenesis of Eagle syndrome are illustrated in Fig. 8.16.

Clinical Features of Eagle Syndrome Clinical manifestations are highly variable and depend on the subtype. In classic neurogenic Eagle syndrome, patients describe sharp, lancinating, or deep aching pain localized to the tonsillar fossa, base of the tongue, mandibular angle, or submandibular region, often radiating to the ear, temporomandibular joint, or upper neck. Pain is typically unilateral and reproducibly provoked or exacerbated by swallowing, yawning, chewing, head turning, or mandibular protrusion. Associated pharyngeal complaints include persistent foreign-body sensation, dysphagia, odynophagia, and referred otalgia [284].

The vascular variant presents with cerebrovascular or autonomic features, including positional headache, dizziness, vertigo, visual disturbances, syncope, Horner syndrome, transient ischemic attack, or ischemic stroke. Neck rotation or extension frequently precipitates episodes through dynamic carotid compression. Overlap between subtypes occurs, and some individuals exhibit mixed neuropathic and vascular symptoms. Less common but documented features include pulsatile tinnitus, hearing changes, and facial paresthesia [285]. Headache and Horner syndrome predominate in vascular cases, while dysphagia and localized neck pain are more characteristic of the classic form [285].

Diagnosis requires integration of clinical history, targeted physical examination, and confirmatory imaging; isolated radiographic elongation is insufficient. A detailed history should elicit symptom triggers (particularly positional) and any antecedent trauma or pharyngeal procedures. The hallmark physical finding is reproduction of the patient's characteristic pain upon digital palpation of the styloid tip in the tonsillar fossa, often perceived as a hard, elongated bony projection [283].

Multidetector computed tomography (CT) of the neck with three-dimensional reconstruction is the gold-standard imaging modality. It provides precise measurement of styloid length and angulation while demonstrating spatial relationships to the carotid sheath and cranial nerves [286]. When vascular compromise is suspected, CT angiography (CTA) is mandatory to identify luminal narrowing, intimal flaps, pseudoaneurysms, or jugular vein stenosis [288]. Plain radiographs or orthopantomograms serve only as low-cost screening tools and lack sufficient three-dimensional detail. Dynamic color-Doppler ultrasonography performed during provocative neck maneuvers is emerging as a valuable adjunct for real-time documentation of flow-velocity changes caused by styloid compression; it is particularly useful when static imaging is equivocal or for procedural guidance [289].

Differential diagnoses are broad and include glossopharyngeal neuralgia, temporomandibular joint dysfunction, cervical spondylosis with radiculopathy, migraine

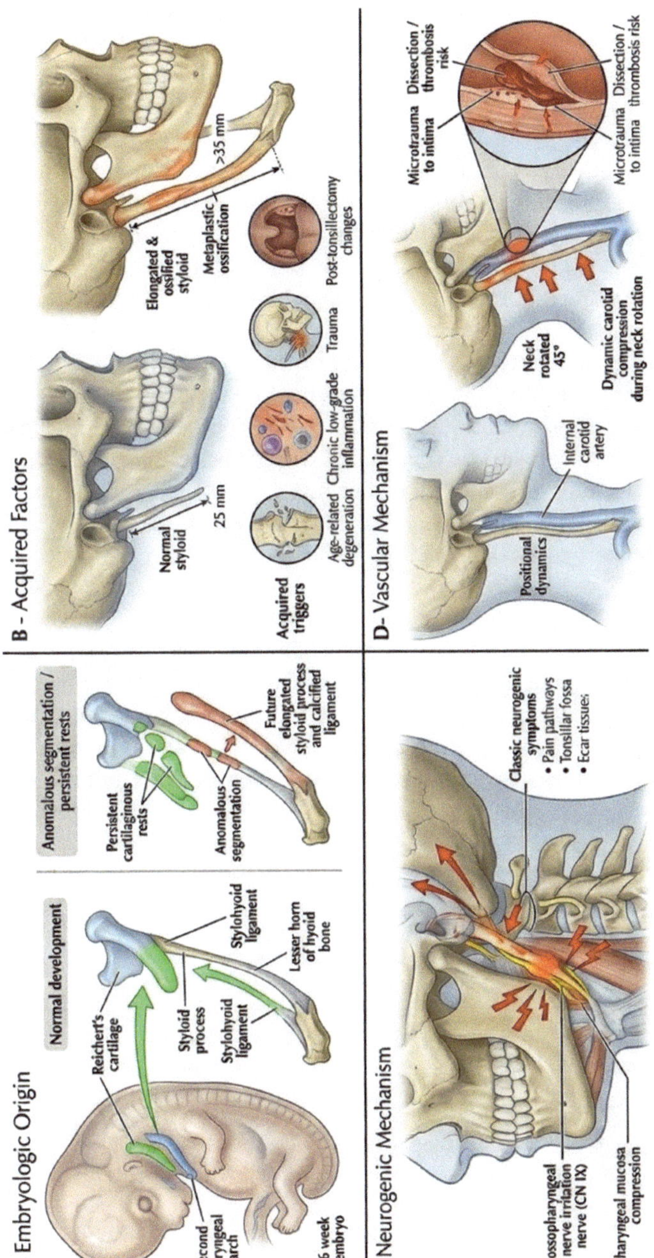

Fig. 8.16 Etiopathogenesis of Eagle syndrome

variants, spontaneous carotid artery dissection, fibromuscular dysplasia, giant-cell arteritis, and parapharyngeal space neoplasms. Multidisciplinary input from otolaryngology, neurology, and vascular surgery often refines the diagnosis [290]. The highly variable clinical manifestations of Eagle syndrome and the characteristic differences between its classic neurogenic and vascular subtypes are illustrated in Fig. 8.17.

Management of Eagle Syndrome Mild or intermittent symptoms are initially managed conservatively. Pharmacotherapy includes non-steroidal anti-inflammatory drugs for inflammatory pain, anticonvulsants (gabapentin, pregabalin, or carbamazepine) or tricyclic antidepressants for neuropathic components, and muscle relaxants when appropriate. Adjunctive physical therapy targeting cervical posture and gentle manual release may provide additive benefit [283].

Diagnostic-therapeutic peristyloid injections offer both confirmation of the diagnosis and temporary symptom control. Performed under ultrasound or landmark guidance, the patient is positioned supine. A reference line is drawn from the mastoid process to the mandibular angle. The needle is advanced toward the expected styloid tip (approximately at the midpoint of this line), negative aspiration for blood or cerebrospinal fluid is confirmed, and incremental injection of 4–5 mL of 0.5% preservative-free lidocaine combined with 40–80 mg methylprednisolone is administered. Subsequent injections, if needed, use reduced steroid dosing. Symptom relief strongly supports the diagnosis and may afford weeks to months of relief [291].

Refractory cases or those with vascular complications warrant surgical intervention, which provides the most durable outcomes. The objective is shortening or complete resection of the symptomatic styloid process (styloidectomy), with or without release of the calcified ligament. Two principal approaches exist: the intraoral (transoral) route, which avoids external scarring and allows faster recovery but offers limited

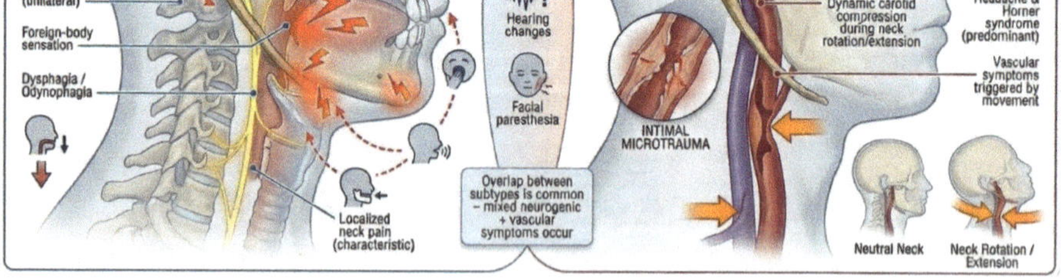

Fig. 8.17 Clinical manifestations of Eagle syndrome

visualization, and the transcervical (external) route, which affords superior exposure and is preferred when vascular reconstruction or complex anatomy is anticipated. Endoscopic-assisted and robotic techniques are increasingly utilized to minimize tissue trauma [285].

High-quality evidence from recent systematic reviews and meta-analyses demonstrates markedly superior results with surgery. In an analysis of 497 patients, 97.8% of those undergoing styloidectomy achieved improvement or complete resolution, compared with 65.9% of medically managed patients [285]. For vascular Eagle syndrome, styloidectomy combined with antiplatelet therapy, endovascular stenting, or open vascular repair dramatically reduces recurrence of ischemic events [288]. Postoperative complications are infrequent (<5%) and include transient dysphagia, hematoma, or minor nerve injury. Long-term recurrence after adequate resection is rare [292].

Prognosis following appropriate management is excellent. Most surgically treated patients experience sustained symptom relief and improved quality of life. Untreated vascular cases, however, carry a substantial risk of recurrent cerebrovascular events; therefore, early recognition is paramount. Future directions include three-dimensional-printed patient-specific models for preoperative planning, refined dynamic imaging protocols, and prospective comparative trials of surgical techniques. Greater inclusion of Eagle syndrome in neurology and vascular curricula will likely shorten diagnostic delays [285].

Eagle syndrome exemplifies a condition in which precise anatomical diagnosis and targeted intervention can dramatically transform patient outcomes. Heightened awareness among clinicians, coupled with the liberal use of modern CT and CT angiography (CTA) together with effective multidisciplinary collaboration, enables timely recognition and highly effective treatment of the syndrome. Surgical styloidectomy remains the definitive therapy for refractory or vascular cases, consistently offering high success rates with low morbidity. Continued research into its underlying etiology, potential imaging biomarkers, and the advancement of minimally invasive techniques will further optimize care and significantly improve long-term outcomes for this frequently underrecognized clinical entity.

References

1. Headache Classification Committee of the International Headache Society (IHS). The International Classification of Headache Disorders, 3rd edition. Cephalalgia. 2018;38(1):1–211. doi:https://doi.org/10.1177/0333102417738202
2. May A, Bahra A, Büchel C, Frackowiak RSJ, Goadsby PJ. Hypothalamic activation in cluster headache attacks. Lancet. 1998;352(9124):275–78. https://doi.org/10.1016/S0140-6736(98)02470-2.
3. Leone M, Franzini A, D'Andrea G, Broggi G, Casucci G, Bussone G. Deep brain stimulation to relieve drug-resistant SUNCT. Ann Neurol. 2005;57(6):924–27. https://doi.org/10.1002/ana.20507.
4. Filho Rocha PAS, Galvão ACR, Teixeira MJ, et al. SUNCT syndrome associated with pituitary tumor. Arq Neuropsiquiatr. 2006;64(2B):507–10. 10.1590/s0004-282×2006000300029.

5. Cohen AS. Short-lasting unilateral neuralgiform headache attacks with conjunctival injection and tearing. Cephalalgia. 2007;27(7):824–32. https://doi.org/10.1111/j.1468-2982.2007.01352.x.
6. Kang MK, Cho SJ. SUNCT, SUNA and short-lasting unilateral neuralgiform headache attacks: debates and an update. Cephalalgia. 2024;44(2):03331024241232256 https://doi.org/10.1177/03331024241232256.
7. Al Khalili Y, Verapaneni KD. Short lasting unilateral neuralgiform headache attacks (SUNCT/SUNA). In: StatPearls. Treasure Island (FL): StatPearls Publishing; 2023. Available from: https://www.ncbi.nlm.nih.gov/books/NBK559088/
8. Cohen AS, Matharu MS, Goadsby PJ. Short-lasting unilateral neuralgiform headache attacks with conjunctival injection and tearing (SUNCT) or cranial autonomic features (SUNA)—a prospective clinical study of SUNCT and SUNA. Brain. 2006;129(Pt 10):2746–60. https://doi.org/10.1093/brain/awl202.
9. Favoni N, Bizzi A, D'Amico D, et al. Epidemiology and clinical features of short-lasting unilateral neuralgiform headache attacks: a systematic review and meta-analysis. Cephalalgia. 2024. doi:https://doi.org/10.1177/03331024241271976
10. Osiowski A, Osiowski M, Taterra D, Klepinowski T. SUNCT and SUNA, distinct disorders or spectrum of a single clinical entity? A meta-analysis. Neurol Sci Neurophysiol. 2024;41(4):193–202. doi:https://doi.org/10.4103/nsn.nsn_122_24.
11. Burish M. Cluster headache, SUNCT, and SUNA. Continuum (Minneap Minn). 2024;30(2):391–410. https://doi.org/10.1212/con.0000000000001411.
12. Wei J, et al. SUNCT/SUNA in pediatric age: a review of pathophysiology and therapeutic options. Brain Sci. 2021;11(9):1252 https://doi.org/10.3390/brainsci11091252.
13. May A, et al. Functional magnetic resonance imaging in spontaneous attacks of SUNCT: Short-lasting neuralgiform headache with conjunctival injection and tearing. Ann Neurol. 1999;46(5):791–4. 10.1002/1531-8249(199911)46:5<791::aid-ana18>3.0.co;2-8.
14. Sprenger T, et al. SUNCT: bilateral hypothalamic activation during headache attacks and resolving of symptoms after trigeminal decompression. Pain. 2005;113(3):422–6. https://doi.org/10.1016/j.pain.2004.09.033.
15. Matharu MS, Levy A. Short-lasting unilateral neuralgiform headache attacks. Ann Indian Acad Neurol. 2018;21(5):S31–S38. https://doi.org/10.4103/aian.aian_356_17.
16. Prakash S, et al. SUNCT and SUNA: an update. Neurol India. 2021;69(1):144–59. https://doi.org/10.4103/0028-3886.314529.
17. Goadsby PJ, et al. Trigeminal autonomic cephalalgias: paroxysmal hemicrania, SUNCT/SUNA, and hemicrania continua. Semin Neurol. 2010;30(2):186–91. https://doi.org/10.1055/s-0030-1248135.
18. Pomeroy JL, Nahas SJ. SUNCT/SUNA: a review. Curr Pain Headache Rep. 2015;19(8):38. https://doi.org/10.1007/s11916-015-0511-2.
19. Miller S, Akram H, Lagrata S, Hariz M, Zrinzo L, Matharu M. Ventral tegmental area deep brain stimulation in refractory short-lasting unilateral neuralgiform headache attacks. Brain. 2016;139(Pt 10):2631–40. https://doi.org/10.1093/brain/aww204. (open access available via the journal or academic repositories).
20. Cosentino G, et al. Trigeminal neurovascular contact in SUNCT and SUNA: a cross-sectional magnetic resonance study. Brain. 2020;143(12):3619–31. https://doi.org/10.1093/brain/awaa307.
21. Williams MH, Broadley SA. SUNCT and SUNA: Clinical features and medical treatment. J Clin Neurosci. 2008;15(5):526–34. https://doi.org/10.1016/j.jocn.2006.09.006.
22. Rodrigues GGR, Bordini CA, Dach F, Eckeli A, Speciali JG. SUNCT syndrome: report of a possible symptomatic case. Arquivos de Neuro-Psiquiatria. 2007;65(3B):852–4. https://doi.org/10.1590/S0004-282×2007000500025.
23. Chitsantikul P, Becker WJ. SUNCT, SUNA and pituitary tumors: clinical characteristics and treatment. Cephalalgia. 2013;33(3):160–70. https://doi.org/10.1177/0333102412468672.

24. Cao Y, et al. Secondary short-lasting unilateral neuralgiform headache with conjunctival injection and tearing: a new case and a literature review. J Clin Neurol. 2018;14(4):433–43. https://doi.org/10.3988/jcn.2018.14.4.433.
25. Makeeva T, Khutorov D, Koreshkina M. Secondary SUNCT syndrome in a patient with Rathke's cleft cyst and hyperprolactinemia: a case report. SN Compr Clin Med. 2020. Advance online publication. https://doi.org/10.1007/s42399-020-00581-w
26. Cavalcante FN, da S, Silva-Néto RP, Bordini CA. SUNCT syndrome associated with pituitary Rathke's cyst and responsive to indomethacin: a case report. Headache Med. 2025;16(3):171–5. https://doi.org/10.48208/HeadacheMed.2025.27.
27. Pareja JA, Shen JM, Kruszewski P, Caballero V, Pamo M, Sjaastad O. SUNCT syndrome: duration, frequency, and temporal distribution of attacks. Headache. 1996;36(3):161–5. https://doi.org/10.1046/j.1526-4610.1996.3603161.x.
28. Pareja JA, Caminero AB, Sjaastad O. SUNCT syndrome: diagnosis and treatment. CNS Drugs. 2002;16(6):373–83. https://doi.org/10.2165/00023210-200216060-00002.
29. Sjaastad O, Saunte C, Salvesen R, Fredriksen TA, Seim A, Røe OD, Fostad K, Løbben OP, Zhao JM. Shortlasting, unilateral neuralgiform headache attacks with conjunctival injection, tearing, sweating, and rhinorrhea. Cephalalgia. 1989;9(2):147–56. https://doi.org/10.1046/j.1468-2982.1989.0902147.x.
30. Duggal AK, Chowdhury D. SUNCT and SUNA: An update. Neurology India. 2021;69(Suppl. 1):S144–59. https://doi.org/10.4103/0028-3886.315990.
31. Weng HY, et al. Similarities and differences between SUNCT and SUNA: a cross-sectional, multicentre study of 76 patients in China. J Headache Pain. 2022;23(1):139. https://doi.org/10.1186/s10194-022-01509-6.
32. Lambru G, Stubberud A, Rantell K, Lagrata S, Tronvik E, Matharu MS. Medical treatment of SUNCT and SUNA: a prospective open-label study including single-arm meta-analysis. J Neurol Neurosurg Psychiatry. 2021;92(3):233–41. https://doi.org/10.1136/jnnp-2020-323999.
33. Marziniak M, Breyer R, Evers S. SUNCT syndrome successfully treated with the combination of oxcarbazepine and gabapentin. Pain Med. 2009;10(8):1497–500. https://doi.org/10.1111/j.1526-4637.2009.00729.x.
34. Trauninger A, Alkonyi B, Kovács N, Komoly S, Pfund Z. Methylprednisolone therapy for short-term prevention of SUNCT syndrome. Cephalalgia. 2010;30(6):735–39. https://doi.org/10.1111/j.1468-2982.2009.01971.x.
35. Arroyo AM, Durán XR, Beldarrain MG, Pinedo A, García-Moncó JC. Response to intravenous lidocaine in a patient with SUNCT syndrome. Cephalalgia. 2010;30(1):110–2. https://doi.org/10.1111/j.1468-2982.2009.01871.x.
36. Porta-Etessam J, Cuadrado ML, Galán L, Sampedro A, Valencia C. Temporal response to bupivacaine bilateral great occipital block in a patient with SUNCT syndrome. J Headache Pain. 2010;11(2):179. https://doi.org/10.1007/s10194-010-0204-5.
37. Zabalza RJ. Sustained response to botulinum toxin in SUNCT syndrome. Cephalalgia. 2012;32(11):869–72. https://doi.org/10.1177/0333102412452045.
38. Lambru G, Nesbitt A, Shanahan P, Matharu MS. Coexistence of hemiplegic migraine with SUNCT or SUNA: a case series. Cephalalgia. 2012;32(3):258–62. https://doi.org/10.1177/0333102411434168.
39. Chakraborty AR, Sunshine K, Miller JP, Sweet JA. Current applications of ablative therapies for trigeminal neuralgia. Neurosurg Clin N Am. 2023;34(2):285–90. https://doi.org/10.1016/j.nec.2022.12.005.
40. Niemeyer CS, Harlander-Locke M, Bubak AN, Rzasa-Lynn R, Birlea M. Trigeminal postherpetic neuralgia: from pathophysiology to treatment. Curr Pain Headache Rep. 2024;28(6):295–306. https://doi.org/10.1007/s11916-023-01209-z.
41. Verjans GMGM, Hintzen RQ, Van Dun JM, et al. Selective retention of herpes simplex virus-specific T cells in latently infected human trigeminal ganglia. Proc Natl Acad Sci U S A. 2007;104(9):3496–501. https://doi.org/10.1073/pnas.0610847104.

42. Kong CL, Thompson RR, Porco TC, Kim E, Acharya NR. Incidence rate of herpes zoster ophthalmicus: a retrospective cohort study from 1994 through 2018. Ophthalmology. 2020;127(3):324–30. https://doi.org/10.1016/j.ophtha.2019.10.001.
43. Bendtsen L, Zakrzewska JM, Abbott J, et al. European Academy of Neurology guideline on trigeminal neuralgia. Eur J Neurol. 2019;26(6):831–49. https://doi.org/10.1111/ene.13950.
44. Headache Classification Committee of the International Headache Society (IHS). The International Classification of Headache Disorders, 3rd edition. Cephalalgia. 2018;38(1):1–211. doi:https://doi.org/10.1177/0333102417738202
45. Cunningham AL, Lal H, Kovac M, et al. Efficacy of the herpes zoster subunit vaccine in adults 70 years of age or older. N Engl J Med. 2016;375(11):1019–32. https://doi.org/10.1056/NEJMoa1603800.
46. Finnerup NB, Attal N, Haroutounian S, et al. Pharmacotherapy for neuropathic pain in adults: a systematic review and meta-analysis. Lancet Neurol. 2015;14(2):162–73. https://doi.org/10.1016/S1474-4422(14)70251-0. (Updated evidence aligns with 2020+ guidelines.).
47. Jajou R, van Puijenbroek E, Overbeek J, Hek K, Mulder E, van Hunsel F, et al. GP consultations for herpes zoster after COVID-19 vaccination: a self-controlled cohort study based on electronic health record data from the Netherlands. Drug Saf. 2026;49(5):529–39. https://doi.org/10.1007/s40264-025-01638-2
48. Song L, He L, Pei Q, et al. Efficacy of Gasserian ganglion high-voltage, long-duration pulsed radiofrequency combined with block on acute/subacute zoster-related trigeminal neuralgia. Pain Res Manag. 2024;2024:1992483 https://doi.org/10.1155/2024/1992483.
49. Sun Z, Liu L, Liu H, Luo F. CT-guided Gasserian ganglion block with local anesthetics and steroids for zoster-related trigeminal neuralgia: a multicenter retrospective study. J Pain Res. 2022;15:2303–13. https://doi.org/10.2147/JPR.S375257.
50. Zheng B, Song L, Liu H. Gasserian ganglion injected with Adriamycin successfully relieves intractable trigeminal nerve postherpetic neuralgia for an elderly patient: a case report. Medicine (Baltimore). 2018;97(38):e12388 https://doi.org/10.1097/MD.0000000000012388.
51. Catapano M, Chiaramonte C, Cavallo C. Outcome comparison of drug-resistant trigeminal neuralgia surgical treatments—An umbrella review of meta-analyses and systematic reviews. Brain Sci. 2023;13(4):530 https://doi.org/10.3390/brainsci13040530.
52. Canfora F. Burning mouth syndrome: Updates on pathogenesis and diagnostic algorithms. J Oral Facial Pain Headache. 2025;39(4):1–30. https://pubmed.ncbi.nlm.nih.gov/41436103/.
53. Pereira JV, Normando AGC, Rodrigues-Fernandes CI, Rivera C, Santos-Silva AR, Lopes MA. The impact on quality of life in patients with burning mouth syndrome: a systematic review and meta-analysis. Oral Surg Oral Med Oral Pathol Oral Radiol. 2021;131(2):186–94. https://doi.org/10.1016/j.oooo.2020.11.019.
54. Scala A, Checchi L, Montevecchi M, Marini I, Giamberardino MA. Update on burning mouth syndrome: overview and patient management. Crit Rev Oral Biol Med. 2003;14(4):275–291. https://doi.org/10.1177/154411130301400402.
55. Tan HL, Renton T. Burning mouth syndrome: an update. Cephalalgia Rep. 2020;3:Article 2515816320970143 https://doi.org/10.1177/2515816320970143.
56. Teruel A, Patel S. Burning mouth syndrome: A review of etiology, diagnosis, and management. Gen Dent. 2019;67(2):24–29. https://pubmed.ncbi.nlm.nih.gov/30875303/.
57. Headache Classification Committee of the International Headache Society. (2018). The International Classification of Headache Disorders, 3rd edition. Cephalalgia 38(1):1–211. https://doi.org/10.1177/0333102417738202
58. International Classification of Orofacial Pain. International classification of orofacial pain, 1st edition. Cephalalgia, 2020;40(2):129–221. https://doi.org/10.1177/0333102419893823
59. International Association for the Study of Pain. 2021. IASP terminology. https://www.iasp-pain.org/resources/terminology/
60. Adamo D, Canfora F, Calabria E, Coppola N, Pecoraro G, Mignogna MD. Burning mouth syndrome: Unraveling multifactorial etiologies and advancing management strategies. Oral Dis. 2023. Advance online publication. https://doi.org/10.1111/odi.14716

61. Chmieliauskaite M, Stelson EA, Epstein JB, Klasser GD, Farag A, Carey B, Heir GM. Consensus agreement to rename burning mouth syndrome and improve International Classification of Diseases-11 disease criteria: an international Delphi study. Pain. 2021;162(10):2548–57. https://doi.org/10.1097/j.pain.0000000000002243.
62. Adamo D, Canfora F, Calabria E, Coppola N, Leuci S, Mignogna MD. Burning mouth syndrome: Unlocking the future of pharmacological treatment. J Oral Pathol Med. 2025;54(5):251–262. https://doi.org/10.1111/jop.13641.
63. Kishore J, Shaikh F, Mirza S, Raffat M A, Ikram S, Akram Z. Cytokine levels and their role in the etiopathogenesis of Burning Mouth Syndrome: a systematic review. Cephalalgia. 2019;39(12):1586–94. https://doi.org/10.1177/0333102419854052.
64. Tan HL, Smith JG, Hoffmann J, Renton T. A systematic review of treatment for patients with burning mouth syndrome. Cephalalgia Rep. 2022;5:Article 25158163221079475 https://doi.org/10.1177/25158163221079475.
65. Jääskeläinen SK. Pathophysiology of primary burning mouth syndrome with special focus on taste dysfunction: a review. Oral Dis. 2022;28(3):543–551. https://doi.org/10.1111/odi.13763.
66. Kim MJ, Kho HS. Understanding of burning mouth syndrome based on psychological aspects. Chin J Dent Res. 2023;26(2):67–73. https://doi.org/10.3290/j.cjdr.b4133385.
67. Tan HL, Smith JG, Hoffmann J, Renton T. Managing burning mouth syndrome: current and future directions. Oral Diseases. 2023;29(8):3050–60. https://doi.org/10.1111/odi.14325.
68. Madariaga VI, Tanaka H, Ernberg M. Psychophysical characterisation of burning mouth syndrome—a systematic review and meta-analysis. J Oral Rehabil. 2020;47(12):1590–605. https://doi.org/10.1111/joor.13028.
69. Wu S, Zhang W, Yan J, Noma N, Young A, Yan Z. Worldwide prevalence estimates of burning mouth syndrome: a systematic review and meta-analysis. Oral Dis. 2022;28(6):1431–1440. https://doi.org/10.1111/odi.13868.
70. De Lima-Souza RA, Mourão CF, Rados PV, Figueiredo MAZ, Martins MD. Clinical and epidemiological profile of burning mouth syndrome patients following the International Headache Society classification: a systematic review and meta-analysis. Oral Surg Oral Med Oral Pathol Oral Radiol. 2024;137(1):12–24. https://doi.org/10.1016/j.oooo.2023.08.013.
71. Acharya S, Carlsson AP, Raza M, Berglundh T, Andersson P. Burning mouth syndrome: An epidemiological study of etiology, risk factors, and clinical management in Swedish adults. J Oral Facial Pain Headache. 2023;37(2):107–115. https://doi.org/10.11607/ofph.3226.
72. Suga T, Watanabe T, Aota Y, Nagamine T, Toyofuku A. Burning mouth syndrome: the challenge of an aging population. Geriatrics. 2021;6(1):Article 18 https://doi.org/10.3390/geriatrics6010018.
73. El Toum S, El Toum S. Prevalence profile of burning mouth syndrome in a sample of Egyptian population: a cross-sectional clinical based study. Explor Med. 2024;5(1):1–12. https://doi.org/10.37349/emed.2024.00204.
74. Souza FT, Santos TP, Bernardes VF, Teixeira AL, Kummer AM, Silva TA. The impact of burning mouth syndrome on quality of life. Health Qual Life Outcomes. 2020;18(1):Article 312. https://doi.org/10.1186/s12955-020-01536-5.
75. Galli F, Lodi G, Sardella A, Vegni E. Role of psychological factors in burning mouth syndrome: a systematic review and meta-analysis. Cephalalgia. 2021;41(1):110–21. https://doi.org/10.1177/0333102420964414.
76. Sardella A, Lodi G, Demarosi F, Bez C, Cassano S, Carrassi A. Burning mouth syndrome: a retrospective study investigating spontaneous remission and response to treatments. Br Dent J. 2006;200(5):267–70. https://doi.org/10.1038/sj.bdj.4813305.
77. Lamey PJ, Lewis MA. Oral medicine in practice: burning mouth syndrome. Br Dent J. 2021;230(7):435–40. https://doi.org/10.1038/s41415-021-2807-4.
78. Kouri M, Adamo D, Vardas E, Georgaki M, Canfora F, Mignogna MD, Nikitakis N. Small fiber neuropathy in burning mouth syndrome: a systematic review. Int J Mol Sci. 2024;25(21):Article 11442 https://doi.org/10.3390/ijms252111442.

79. de Pedro M, et al. Burning mouth syndrome underlying factors: a roadmap from a network perspective. Oral Dis. 2025. https://doi.org/10.1111/odi.15219
80. De Tommaso M, Arendt-Nielsen L, Mastroluca A, Serpino C, Vecchio E, Franco G, Sardaro M, de Venere D. Small fiber neuropathy in burning mouth syndrome: a systematic review. Int J Mol Sci. 2024;25(21):Article 11650 https://doi.org/10.3390/ijms252111650.
81. Kim YJ, Kim YK, Myung SC, Kim EJ, Yoon DY. Burning mouth syndrome in association with genetic polymorphisms: a case-control study. Pain Medicine. 2020;21(11):2990–2997. https://doi.org/10.1093/pm/pnaa199.
82. Vecchio E, Bove F, Ricci K, Di Iorio R, De Tommaso M, Delussi M. Acupuncture and osteopathic manipulative treatment for burning mouth syndrome: a pilot randomized crossover study. Healthcare. 2024;12(5):Article 558 https://doi.org/10.3390/healthcare12050558.
83. Adamo D, Sardella A, Varoni E, Lajolo C, Biasotto M, Ottaviani G, Vescovi P, Simonazzi T, Lodi G, Mignogna MD. The association between burning mouth syndrome and urologic chronic pelvic pain syndrome: a case-control study. J Oral Pathol Med. 2021;50(5):490–496. https://doi.org/10.1111/jop.13166.
84. Suga T, Tu T T H, Watanabe M, Inoue T, Toyofuku A. Burning mouth syndrome in younger populations: gender disparities and bipolar tendencies. Psychiatry Clin Neurosci Rep. 2025;4(2):e70110 https://doi.org/10.1002/pcn5.70110.
85. Argiuolo, L., et al. (2025). Inside the fire: Exploring the impact of anxiety, depression, and sleep disturbances on pain perception in burning mouth syndrome: a cross-sectional study of 200 patients. J Oral Pathol Med. Advance online publication. (PMC12521075)
86. Coculescu EC, Radu A, Coculescu BI. Burning mouth syndrome: a review on diagnosis and treatment. J Med Life. 2014;7(4):512–5. https://www.ncbi.nlm.nih.gov/pmc/articles/PMC4316128
87. Lebel A, et al. The role of stress in burning mouth syndrome triggered by dental treatments: a two-step hypothesis. J Oral Rehabil. 2025;52:1001–14. https://doi.org/10.1111/joor.13964.
88. Wang O, Mei L, Polonowita A, Guan G. Exploring the link between burning mouth syndrome and human herpesvirus-6B: a narrative review on pathogenesis. J R Soc N Z. 2025;55:1384–99. https://doi.org/10.1080/03036758.2024.2406818.
89. Pellegrini L, Canfora F, Ottaviani G, et al. Obsessive-compulsive symptoms and traits in patients with burning mouth syndrome: a cross-sectional multicentric analysis. Clin Oral Investig. 2025;25(4):223 https://doi.org/10.1007/s00784-025-06293-6.
90. Chang MC, Lee S, Kim HS, Lee YC, Lee S. Clinical features and variations of pain expressions in 834 burning mouth syndrome patients with or without psychiatric comorbidities. Cureus. 2023;15(12):Article e51120 https://doi.org/10.7759/cureus.51120.
91. Çinar S, et al. Efficacy of alpha-lipoic acid in patients with burning mouth syndrome compared to that of placebo or other interventions: a systematic review with meta-analyses. J Dent Anesth Pain Med. 2022;22(5):323–40. https://doi.org/10.17245/jdapm.2022.22.5.323.
92. Coculescu EC, Tovaru S, Radu A, Coculescu BI. Burning mouth syndrome – known and unknown. Folia Med. 2024;66(6):770–775. https://doi.org/10.3897/folmed.66.e139442.
93. de Pedro M, et al. Effectiveness of nonpharmacologic treatments of burning mouth syndrome: a systematic review. J Oral Facial Pain Headache. 2021;35(3):175–98. https://doi.org/10.11607/ofph.2868.
94. de Pedro M, et al. Burning mouth syndrome treated with low-level laser and clonazepam: a randomized, single-blind clinical trial. Biomedicines. 2024;12(5):Article 1048 https://doi.org/10.3390/biomedicines12051048.
95. Farag AM, Kuten-Shorrer M, Natto Z, Ariyawardana A, Mejia LM, Albuquerque R, Patton LL. WWOM VII: effectiveness of systemic pharmacotherapeutic interventions in the management of BMS: a systematic review and meta-analysis. Oral Dis. 2023;29(2):343–368. https://doi.org/10.1111/odi.13817.
96. Treede R-D, et al. Chronic pain as a symptom or a disease: the IASP Classification of Chronic Pain for ICD-11. Pain. 2019;160(1):19–27. https://doi.org/10.1097/j.pain.0000000000001384

97. Tirupathi SP, et al. Burning measure for burning mouth syndrome: a systematic review. J Korean Assoc Oral Maxillofac Surg. 2024;50(2):63–69. https://doi.org/10.5125/jkaoms.2024.50.2.63.
98. Kim MJ, Kho HS. Understanding of burning mouth syndrome based on psychological aspects. Chin J Dent Res. 2022;25(1):9–19. https://doi.org/10.3290/j.cjdr.b2788641.
99. Liu YF, et al. Effects of low-level laser therapy on burning pain and quality of life in patients with burning mouth syndrome: a systematic review and meta-analysis. BMC Oral Health. 2023;23:Article 734. https://doi.org/10.1186/s12903-023-03441-w.
100. Monteserín-Matesanz M, Esparza-Gómez GC, López-López J, López-Jornet P. Central sensitization in burning mouth syndrome: a cross-sectional study. J Clin Med. 2022;11(24):Article 7477 https://doi.org/10.3390/jcm11247477.
101. Nouioua S, et al. The efficacy of alpha-lipoic acid in the management of burning mouth syndrome: An updated systematic review of randomized controlled clinical trials. Health Sci Rep. 2023;6(4):Article e1186 https://doi.org/10.1002/hsr2.1186.
102. Orliaguet M, Misery L. Neuropathic and psychogenic components of burning mouth syndrome: the new frontiers of two old concepts. Biology. 2023;12(4):Article 623 https://doi.org/10.3390/biology12040623.
103. Ritchie A, Kramer JM. Recent advances in the etiology, diagnosis, and treatment of burning mouth syndrome: a narrative review. J Dent Anesth Pain Med. 2021;21(6):509–21. https://doi.org/10.17245/jdapm.2021.21.6.509.
104. Sinding C, Gransjøen AM, Schlumberger G, Valence F, Saint-Eve A, Ballester J, Thomas-Danguin T. Oral burning without altered taste perception characterises patients with burning mouth syndrome: a case-control study with new descriptors of dysgeusia. Clin Otolaryngol. 2021;46(6):1270–79. https://doi.org/10.1111/coa.13809.
105. Škrinjar I, Vučićević Boras V, Bakale I, et al. Stratification of patients with burning mouth syndrome in the Croatian population: a single-center cross-sectional study. Neurosciences (Riyadh). 2024;29(2):113–119. https://doi.org/10.17712/nsj.2024.2.20230088.
106. Tan HL, Smith JG, Hoffmann J, Renton T. A systematic review of treatment for patients with burning mouth syndrome. Cephalalgia. 2022;42(2):128–161. https://doi.org/10.1177/03331024211036152.
107. Bašić K, Peroš K, Bošnjak Z, Šutej I. Subgingival microbiota profile in association with cigarette smoking in young adults: a cross-sectional study update. Dent J. 2021;9(12):150 https://doi.org/10.3390/dj9120150.
108. Wu Y, Chang JYF, Ghazi P, Cheng B. Is burning mouth syndrome associated with extraoral dryness? A systematic review. J Clin Med. 2023;12(20):Article 6525 https://doi.org/10.3390/jcm12206525.
109. Lin J, Wan J, Zha T, et al. Burning mouth syndrome: Unraveling multifactorial etiologies and advancing management strategies. Oral Dis. 2026;32(3):602–16. https://doi.org/10.1111/odi.70116
110. de Pedro M, López-Pintor RM, Casañas E, Hernández G. General health status of a sample of patients with burning mouth syndrome: a case-control study. Oral Dis. 2020;26(7):1490–98. https://doi.org/10.1111/odi.13369.
111. Maniaci A, Merlino F, Cocuzza S, et al. Endoscopic surgical treatment for rhinogenic contact point headache: systematic review and meta-analysis. Eur Arch Otorhinolaryngol. 2021;278(6):1743–53. https://doi.org/10.1007/s00405-021-06724-6. PMID: 33677741
112. Swanson D, Lewis K, Thornhill K, Fabian I, Kaye AD, Viswanath O, Urits I. An update on the diagnosis, treatment, and management of sphenopalatine neuralgia. Int J Otorhinolaryngol Head Neck Surg. 2022;8(3):286–291. https://doi.org/10.18203/issn.2454-5929.ijohns20220486.
113. Lavalle S, Pace A, Magliulo G, Lentini M, Lechien JR, Calvo-Henriquez C, Parisi FM, Iannella G, Maniaci A, Messineo D. Impact of nasal anatomical variation subtype on surgical outcomes for rhinogenic contact point headache. Diagnostics. 2025;15(2):121 https://doi.org/10.3390/diagnostics15020121.

114. Maniaci A, Lechien JR, Calvo-Henriquez C, et al. Long-term stability of outcomes of endoscopic surgery for rhinogenic contact point headache (Sluder's neuralgia). Am J Otolaryngol. 2022;43(2):103368. https://doi.org/10.1016/j.amjoto.2021.103368.
115. Bernichi JV, et al. Rhinogenic and sinus headache – literature review. Am J Otolaryngol. 2021;42(6):103113 https://doi.org/10.1016/j.amjoto.2021.103113.
116. Bautista AL, Coyne K, Bautista A, Abd-Elsayed A. Sphenopalatine ganglion blocks in headache management: a review. Brain Sci. 2025;15(7):672 https://doi.org/10.3390/brainsci15070672.
117. Yadav R, Chakraborty P, Saini R, Kumar R, Saroj SK, Yadav S. Anterior ethmoidal nerve syndrome – an under-diagnosed cause of headache. Bangladesh J Otorhinolaryngol. 2022;30(1):1–7. https://doi.org/10.47210/bjohns.2022.v30i1.648.
118. Cho W, Eom TW, Kim JH. A case of mucosal contact point headache caused by septal spur. Korean J Otorhinolaryngol Head Neck Surg. 2014;57(6):407–10 https://doi.org/10.3342/kjorl-hns.2014.57.6.407
119. Kwon KW, Jang YJ. Management of chronic nasal pain developing after rhinoseptoplasty or nasal trauma. Facial Plast Surg Aesthet Med. 2023. Advance online publication. https://doi.org/10.1089/fpsam.2022.0229
120. Orfanou C, et al. Sphenopalatine ganglion block using the Tx360 nasal applicator for trigeminal neuralgia: a prospective observational pilot study. Cureus. 2025;17(1):e78727 https://doi.org/10.7759/cureus.78727.
121. Ertilav E, et al. Successful treatment of sphenopalatine ganglion radiofrequency thermocoagulation in trigeminal autonomic cephalalgia. Turk J Neurol. 2021;27(3):1–5. https://doi.org/10.4274/tnd.2021.34356.
122. Fakoya AO, et al. Anatomy, head and neck, nasal concha. In: StatPearls. StatPearls Publishing; 2024. https://www.ncbi.nlm.nih.gov/books/NBK546636/
123. Mahajan SK, Paliwal VK. Sphenopalatine neuralgia as a differential diagnosis of complex regional pain syndrome of face. Cephalalgia Rep 2025;8. https://doi.org/10.1177/25158163251327173
124. Swain SK. Rhinogenic contact point headache – a review. Matrix Sci Med. 2022;6(3):65–69.
125. Pohl H. History of cluster headache. Cephalalgia Rep 2022;5. https://doi.org/10.1177/25158163221128183
126. Lana Shahabaddin, Emin A, Kasho H. Effect of septoplasty on relieving contact point headache. Int Tinnitus J. 2024;28(1):157–62. https://doi.org/10.5935/0946-5448.20240023.
127. Alexander CE. Sphenopalatine ganglion block. In: StatPearls. StatPearls Publishing; 2022. https://www.ncbi.nlm.nih.gov/books/NBK557751/
128. Shah RJ, et al. Sphenopalatine ganglion radiofrequency thermocoagulation. In: StatPearls. StatPearls Publishing; 2023. https://www.ncbi.nlm.nih.gov/books/NBK536944/
129. He W, Zhang Y, Long T, Pan Q, Zhang S and Zhou J. Sphenopalatine neuralgia: an independent neuralgia entity. Pooled analysis of a case series and literature review. Headache. 2019;59(3): 358–370 https://doi.org/10.1111/head.13469
130. Vila-Pueyo M, Campoy S, Fernández-Fernández S. Galcanezumab reduces trigeminal nociception and is effective in preclinical models of migraine and trigeminal autonomic cephalalgias. Headache. 2025;65(1):23–34. https://doi.org/10.1111/head.15006.
131. Lamas Pérez R, et al. Efficacy and safety of galcanezumab as chronic cluster headache preventive treatment under real world conditions: observational prospective study. Cephalalgia. 2024;44(1):03331024231226181 https://doi.org/10.1177/03331024231226181.
132. Lasso Andrade FA, et al. Thermal radiofrequency of the sphenopalatine ganglion in persistent idiopathic facial pain using a suprazygomatic approach: case report. Interv Pain Med. 2025;4(4):100716. https://doi.org/10.1016/j.inpm.2025.100716
133. Xin B, Wang Y, Zhang L. Long-term follow-up safety and effectiveness of CT-guided radiofrequency thermocoagulation of sphenopalatine ganglion in refractory headache treatment. Pain Ther. 2022;11(3):789–802. https://doi.org/10.1007/s40122-022-00401-0.

134. Vadim Tashlykov, Genov P, Sheinfeld R, et al. Two-stage sphenopalatine ganglion neurostimulation for refractory craniofacial pain: a retrospective study. Neuromodulation. 2025 https://doi.org/10.1016/j.neurom.2025.09.317
135. Andreani JCM, Bruera O, Lisicki M, et al. Long-term effectiveness of combined unilateral sphenopalatine and occipital nerve stimulation in patients with refractory chronic cluster headache. Ther Adv Neurol Disord. 2025;18:1–12. https://doi.org/10.1177/25158163251355554.
136. Michał But, Wernicki K, Król A, Kosson D. A novel method of noninvasive sphenopalatine ganglion stimulation in craniofacial pain: case series and literature review. Neuromodulation. 2026 https://doi.org/10.1016/j.neurom.2025.11.010
137. Ng CC, Yuan H. 2025 highlights in non-invasive neurostimulation and neuromodulation for headache management. Cephalalgia. 2026;46(1):12–25. https://doi.org/10.1177/03331024251405935.
138. Sokal P, Harat M, Litwin B. Sphenopalatine ganglion stimulation for the treatment of trigeminal neuropathic pain. Front Neurol. 2025;16:1535102 https://doi.org/10.3389/fneur.2025.1535102.
139. Chakraborty P, Yadav R, Saini R, Kumar R, Saroj SK, Yadav S. Anterior ethmoidal nerve syndrome – an under-diagnosed cause of headache. Beng J Otolaryngol Head Neck Surg. 2022;30(1):45–50. https://doi.org/10.47210/bjohns.2022.v30i1.648.
140. Bora KK, Sharma K, Sukai A. A prospective study of association of anterior ethmoidal nerve syndrome with nasal septal spur and its management. Glob J Res Anal 2023;12(3). https://doi.org/10.36106/gjra/3200127
141. Basandrai A, Bhatia P, Saharan D. Sluder's neuralgia: revisiting a rare cause of craniofacial pain. Eur J Pharm Med Res. 2015;12(5):516–517.
142. Adjei P, Asasu SA, Ayeebo ML, Donkor EA, Amoakohene-Yeboah A. Maxillary herpes zoster with nasociliary nerve involvement and Klebsiella superinfection: a rare occurrence. Clin Case Rep. 2024;12(9):e9356. https://doi.org/10.1002/ccr3.9356.
143. Janjua MS, Singh P, Sajja K. Nummular headache. In: StatPearls. StatPearls Publishing; 2023 (Bookshelf ID: NBK540991).
144. Veerapaneni KD, Kapoor N, Veerapaneni P, Lui F, Nalleballe K. Trigeminal neuropathy. In: StatPearls. StatPearls Publishing; 2024 (Bookshelf ID: NBK554518).
145. De Corso E, et al. Facial pain: sinus or not? Acta Otorhinolaryngol Ital. 2018;38(6):485–96. https://doi.org/10.14639/0392-100X-1721.
146. Bendtsen L, Zakrzewska JM, Heinskou TB, Hodaie M, Leal PRL, Nurmikko T, Obermann M, Cruccu G, Maarbjerg S. Guidelines for the management of trigeminal neuralgia. Cleve Clin J Med. 2023;90(6):355–62. https://doi.org/10.3949/ccjm.90a.22091.
147. Oomen KPQ, et al. Sluder's neuralgia: a trigeminal autonomic cephalalgia? Cephalalgia. 2010;30(11):1364–1368. https://doi.org/10.1111/j.1468-2982.2009.01919.x.
148. Singh MP, Mehrotra V, Singh A. A rare case of Charlin's syndrome. Arch Otorhinol Head Neck Surg. 2017;29(2):123–25.
149. Nierenburg H, Morris MS. External nasal neuralgia: an update. Curr Pain Headache Rep. 2017;21(11):44 https://doi.org/10.1007/s11916-017-0645-5. PMID: 28900824.
150. Shilash OB, Chen C, Patel VM, Cohen JA. Trigeminal neuralgia management in patients with multiple sclerosis: a systematic review of approaches and outcomes. J Clin Neurosci. 2025;123:1–10. https://doi.org/10.1016/j.jocn.2025.111220.
151. De Corso E, Kar M, Cantone E, Lucidi D, Settimi S, Mele D, Salvati A, Muluk NB, Paludetti G, Cingi C. Facial pain: sinus or not?. Acta Otorhinolaryngol Ital. 2018;38(6):485–496. https://doi.org/10.14639/0392-100X-1721.
152. Puma M, Cruccu G. A case of post-traumatic persistent nasal pain. Front Neurol. 2020;10:1409 https://doi.org/10.3389/fneur.2019.01409.
153. Khadilkar SV, Patil VA. Medical management of trigeminal neuralgia. Neurol India. 2021;69(Suppl 1):S159–S166. https://doi.org/10.4103/0028-3886.315992.
154. Zhang Y, Li X, Wang Z, Liu H. Safety and efficacy of carbamazepine in the treatment of trigeminal neuralgia: a meta-analysis. Math Biosci Eng. 2024;21(4):5123–5140. https://doi.org/10.3934/mbe.2024235.

155. Pinto Moreira M, Santos AR, Oliveira JF, Costa AM. Anticonvulsant therapy in trigeminal neuralgia: a class-oriented systematic review. Medicines. 2026;13(1):3 https://doi.org/10.3390/medicines13010003.
156. Xu R, Xie ME, Jackson CM, Lim M. Trigeminal neuralgia: current approaches and emerging interventions. J Pain Res. 2021;14:3437–63. https://doi.org/10.2147/JPR.S331036.
157. Alshihri A, Alqahtani M, Alqahtani A, Alqahtani S, Alqahtani A. What is new and effective in treating refractory trigeminal neuralgia?. Cureus. 2024;16(11):e73012 https://doi.org/10.7759/cureus.73012.
158. Zhu G, Sun C, Ting H, Zhang M, Jiang Y, Mao H, Huang M, Feng W. Global trends and hotspots in trigeminal neuralgia research from 2001 to 2021: a bibliometric analysis. Front Neurol. 2022;13:894006 https://doi.org/10.3389/fneur.2022.894006.
159. Swain SK, Behera IC, Mohanty S, Sahu MC. Rhinogenic contact point headache – frequently missed clinical entity. Apollo Med. 2016;13(3):169–173. https://doi.org/10.1016/j.apme.2016.08.001.
160. Clark MV, Zakrzewska JM, Renton T. Pharmacological management of orofacial pain – a clinician's guide. Oral Surg. 2020;13(4):422–49. https://doi.org/10.1111/ors.12537.
161. Han A, Rampersaud E, Hidas G, Sternbach JM. Glossopharyngeal neuralgia: epidemiology, risk factors, pathophysiology, differential diagnosis, and treatment options. Health Psychol Res. 2022;10(3):36044 https://doi.org/10.52965/001c.36044.
162. Cruccu G, et al. Trigeminal neuralgia: new classification and diagnostic grading for practice and research. Neurology. 2016;87(2):220–8. https://doi.org/10.1212/WNL.0000000000002840
163. Alshammari AF, et al. Systematic review of pharmacotherapy for atypical facial pain: evaluation of pain reduction, depression, anxiety and quality of life. Ann Med. 2025;57(1):2476050 https://doi.org/10.1080/07853890.2025.2476050.
164. Rubis A, Juodzbalys G. The use of botulinum toxin A in the management of trigeminal neuralgia: a systematic literature review. J Oral Maxillofac Res. 2020;11(2):e2 https://doi.org/10.5037/jomr.2020.11202.
165. Lin CS, et al. Interventional treatments for postherpetic neuralgia: a systematic review. Pain Phys. 2019;22(3):209–28.
166. Sdrulla AD, Xu Q, He SQ, Tiwari V, Yang F, Zhang C, Shu B, Shechter R, Raja SN, Wang J, Dong X, Guan Y. Minimally invasive procedures for neuropathic pain. Pain Med. 2016;17(4):771–781. https://doi.org/10.1093/pm/pnw011.
167. Matejowsky HG, et al. Interventional procedures for refractory neuropathic pain. Explor Neurosci. 2023;2(1):1–15. https://doi.org/10.37349/en.2023.00028.
168. Pearl C, et al. Management of trigeminal neuralgia with botulinum toxin type A: report of two cases. Dent J. 2022;10(11):207 https://doi.org/10.3390/dj10110207.
169. Kandari A, et al. Botulinum toxin-A as a treatment option for refractory idiopathic trigeminal neuralgia of the ophthalmic branch: a case report and literature review. Oral Surg Oral Med Oral Pathol Oral Radiol. 2024;138(1):1–7. https://doi.org/10.1016/j.oooo.2024.02.015.
170. Wu S, Lian Y, Zhang H, Chen Y, Wu C, Li S, Zheng Y, Chen Y, Wang H, Huang Z. Botulinum toxin type A for refractory trigeminal neuralgia in older patients: a better alternative to invasive procedures. J Pain Res. 2019;12:941–949. https://doi.org/10.2147/JPR.S194653.
171. Hu X, et al. Efficacy and safety of botulinum toxin type A in the treatment of trigeminal neuralgia: an update on systematic review with meta-analyses. Clin J Pain. 2024;40(6):383–92. https://doi.org/10.1097/AJP.0000000000001207.
172. Van der Cruyssen F, et al. Consensus guidelines on training, diagnosis, treatment and follow-up care of trigeminal nerve injuries. Int J Oral Maxillofac Surg. 2024;53(1):1–10. https://doi.org/10.1016/j.ijom.2023.03.013.
173. Kim ME. A review of recent evidence on trigeminal neuralgia. J Oral Med Pain. 2023;48(1):3–10. https://doi.org/10.14476/jomp.2023.48.1.3.
174. Ovsepian SV, Waxman SG. Gene therapy for chronic pain: emerging opportunities in target-rich peripheral nociceptors. Nat Rev Neurosci. 2023;24(4):252–265. https://doi.org/10.1038/s41583-022-00673-7

175. Curran D, Doherty TM, Lecrenier N, Breuer T. Healthy ageing: Herpes zoster infection and the role of zoster vaccination. npj Vaccines. 2023;8:184. https://doi.org/10.1038/s41541-023-00757-0.
176. Fujiwara T, Uraguchi K. Antiviral treatment for Ramsay Hunt syndrome: a systematic review and meta-analysis. Auris Nasus Larynx. 2024;51(3):488–91. https://doi.org/10.1016/j.anl.2024.01.001.
177. Sukphulloprat P. Ramsay Hunt syndrome treatment & management. Medscape 2025. https://emedicine.medscape.com/article/1166804-treatment
178. Yu C, Lee H-Y, Chen Y-C. Early diagnosis and treatment of Ramsay Hunt syndrome: a case report. International Journal of Emergency Medicine, 2025;18(1). https://doi.org/10.1186/s12245-024-00807-x
179. Osman HA, Deifa AA, Ali M. Clinical insights into Ramsay Hunt syndrome: a case report highlighting diagnostic challenges and management considerations. Int Med Case Rep J. 2025;18:1709–13. https://doi.org/10.2147/IMCRJ.S558967.
180. Jayaram PR, Choradia SC, Agarwal R. (2026). Ramsay-Hunt syndrome in childhood: a systematic review and treatment update. Indian J Otolaryngol Head Neck Surg. 2026. https://doi.org/10.1007/s12070-026-06355-5
181. Khandelwal A, Ahmed KA, Sarma K, Shunyu NB. When pain relief backfires: Ramsay Hunt syndrome after intra-articular steroid injection – a rare complication with 6-month follow-up. J Orthop Case Rep. 2026;16(2):55–57. https://doi.org/10.13107/jocr.2026.v16.i02.6742.
182. Goswami Y, Bayazit N, Liu Y, Faraz A, Cerqueira AV. Ramsay Hunt syndrome: an introduction, signs and symptoms, and treatment. Cureus. 2023;15(1):e33678 https://doi.org/10.7759/cureus.33678.
183. Crouch AE, Hohman MH, Moody MP. Ramsay Hunt syndrome. In: StatPearls. StatPearls Publishing; 2023. https://www.ncbi.nlm.nih.gov/books/NBK557409
184. Malhotra R, Kannan SR, Nespola C. Clinical features and prognosis of facial palsy and hearing loss in patients with Ramsay Hunt syndrome. Cureus. 2022;14(11):e31266 https://doi.org/10.7759/cureus.31266.
185. Mayo Clinic. Ramsay Hunt syndrome: Diagnosis & treatment. 2025. https://www.mayoclinic.org/diseases-conditions/ramsay-hunt-syndrome/diagnosis-treatment/drc-20351789
186. Cleveland Clinic. Ramsay Hunt syndrome (herpes zoster oticus): Causes & treatment. 2022. https://my.clevelandclinic.org/health/diseases/6093-ramsay-hunt-syndrome
187. Monsanto RDC, Bittencourt AG, Bobato Neto NJ, Beilke SCA, Klagenberg KF, de la Hoz AF. Treatment and prognosis of facial palsy on Ramsay Hunt syndrome: results based on a review of the literature. Int Arch Otorhinolaryngol. 2016;20(4):394–400. https://doi.org/10.1055/s-0036-1586485.
188. Jeon Y, Lee H. Ramsay Hunt syndrome. J Dent Anesth Pain Med. 2018;18(6):333–337. https://doi.org/10.17245/jdapm.2018.18.6.333.
189. BMJ Best Practice. Ramsay Hunt syndrome. 2026. https://bestpractice.bmj.com/topics/en-us/3000322
190. Mayo Clinic. Ramsay Hunt syndrome: Symptoms & causes. 2025. https://www.mayoclinic.org/diseases-conditions/ramsay-hunt-syndrome/symptoms-causes/syc-20351783
191. Callen E. Ramsay Hunt syndrome and the pharmacist. U.S. Pharmacist. 2021;46(9):22–26. https://www.uspharmacist.com/article/ramsay-hunt-syndrome-and-the-pharmacist.
192. Amrutkar CV, Burton, EV. Tolosa-Hunt syndrome. In: StatPearls. StatPearls Publishing; 2025. https://www.ncbi.nlm.nih.gov/books/NBK459225/
193. Chądzyński PR, Pomianowski A, Brydak-Godowska J. Tolosa-Hunt syndrome: a review of diagnostic criteria based on a case series. Postępy Psychiatrii i Neurologii. 2024;33(1):26–34. https://doi.org/10.5114/ppn.2023.135176.
194. Kim H-J, Lee S-U, Lee E-S, Choi J-Y, Kim J-S. Recurrence and long-term outcomes of Tolosa-Hunt syndrome. J Neurol. 2024;271(2):935–943. https://doi.org/10.1007/s00415-023-12044-y.

195. Ahmed HS, Shivananda DB, Pulkurthi SR, Dias AF, Sahoo PP. Clinical profile and outcomes in Tolosa-Hunt Syndrome; a systematic review. J Clin Neurosci. 2024;129:110858 https://doi.org/10.1016/j.jocn.2024.110858.
196. Park S-H, Jang S-I, Lee E-J, Kim N-H. Case report: Tolosa-Hunt syndrome—expanding the neuromyelitis optica spectrum disorder phenotype. Front Neurol. 2024;15:1326867 https://doi.org/10.3389/fneur.2024.1326867.
197. Headache Classification Committee of the International Headache Society (IHS). The International Classification of Headache Disorders, 3rd edition. Cephalalgia 2018;38(1):1–211. https://doi.org/10.1177/0333102417738202
198. Yu S, Chen T. Pearls & Oy-sters: idiopathic orbital inflammation and Tolosa-Hunt syndrome with intracranial extension. Neurology. 2023;101(8):371–374. https://doi.org/10.1212/WNL.0000000000207368.
199. Mullen E, Rutland JW, Green MW, Bederson J, Shrivastava R. Reappraising the Tolosa-Hunt syndrome diagnostic criteria: a case series. Headache. 2020;60(1):259–64. https://doi.org/10.1111/head.13692.
200. Ahmed HS, et al. Tolosa–Hunt syndrome in children and adolescents: a systematic review. Headache J Head Face Pain. 2025;65(6):1027–1040. https://doi.org/10.1111/head.14890.
201. Etheridge T, Jones J, Caskey EL, Digre KB, Warner JEA, Seay MD. Tolosa-Hunt syndrome after COVID-19 infection. Cephalalgia Rep 2022;5. https://doi.org/10.1177/25158163221078977
202. Nilofar F, Khan A, Singh R. Tolosa-Hunt syndrome as the initial presentation of systemic lupus erythematosus. Cureus. 2024;16(6):e61692 https://doi.org/10.7759/cureus.61692.
203. He W, Li Y, Zhang X. A case report on recurrent alternating Tolosa-Hunt syndrome due to bacterial sphenoid sinusitis: Rediscussing the diagnostic terminology and classification. BMC Neurol. 2023;23(1):67. https://doi.org/10.1186/s12883-023-03067-z.
204. Alshamlan S, Almukhaimar N, Bukhamseen L, Alnaimi Y. Tolosa-Hunt syndrome: a case report. Cureus. 2025;17(5):e84606 https://doi.org/10.7759/cureus.84606.
205. Chowdhury SI, Mifty SK, Khan TAR, et al. Tolosa–Hunt syndrome presenting as painful unilateral ophthalmoplegia in a 65-year-old woman: a case report. Clin Case Rep. 2026;14(2):e72012 https://doi.org/10.1002/ccr3.72012.
206. Bommala S, et al. The hidden cavernous mystery behind a headache: a case report of Tolosa-Hunt syndrome. Cureus. 2025;17(10):e95669 https://doi.org/10.7759/cureus.95669.
207. Dinaki K, Tanaka T, Saito N. Intracavernous aneurysm mimicking Tolosa-Hunt syndrome. Neurol Med Chir. 2024;64(3):120–25. https://doi.org/10.2176/nmc.cr.2023-0156.
208. Young A, Okuyemi OT. Frey syndrome. In: StatPearls. StatPearls Publishing; 2023. https://www.ncbi.nlm.nih.gov/books/NBK562247/
209. Mantelakis A, Lafford G, Lee CW, Spencer H, Deval J-L, Joshi A. Frey's syndrome: a review of aetiology and treatment. Cureus. 2021;13(12):Article e20107 https://doi.org/10.7759/cureus.20107.
210. Motz KM, Kim YJ. Auriculotemporal syndrome (Frey syndrome. Otolaryngol Clin N Am. 2016;49(2):501–509. https://doi.org/10.1016/j.otc.2015.10.010.
211. Neumann A, Rosenberger D, Vorsprach O, Dazert S. The incidence of Frey syndrome following parotidectomy: results of a survey and follow-up. HNO. 2011;59(2):173–78. https://doi.org/10.1007/s00106-010-2223-6.
212. Golding CN, et al. The incidence of Frey syndrome and treatment with botulinum toxin in the Central Denmark Region. Laryngoscope Investig Otolaryngol. 2022;7(6):1840–46. https://doi.org/10.1002/lio2.926.
213. Pinkston JA, Cole P. Incidence rates of salivary gland tumors: Results from a population-based study. Otolaryngol Head Neck Surg. 1999;120(6):834–40. https://doi.org/10.1016/S0194-5998(99)70323-2.
214. de Santana BWJ, et al. Incidence and profile of benign epithelial tumors of salivary glands from a single center in Northeast of Brazil. Braz J Otorhinolaryngol. 2021;87(3):298–303. https://doi.org/10.1016/j.bjorl.2019.10.013.

215. Iwanaga J, Fisahn C, Watanabe K, et al. Parotid branches of the auriculotemporal nerve: An anatomical study with implications for Frey syndrome. J Craniofac Surg. 2017;28(1):262–64. https://doi.org/10.1097/SCS.0000000000003260.
216. Minor V. Ein neues Verfahren zu der klinischen Untersuchung der Schweissabsonderung. Deutsche Zeitschrift für Nervenheilkunde. 1928;101:302–308.
217. Choi HG, et al. Comparisons of three indicators for Frey's syndrome: subjective symptoms, Minor's starch iodine test, and infrared thermography. Acta Oto-Laryngologica. 2013;133(11):1193–97. https://doi.org/10.3109/00016489.2013.807940.
218. Mashrah MA, Aldhohrah T, Abdelrehem A, Koraitim M, Wang L. What is the best method for prevention of postparotidectomy Frey syndrome? Network meta-analysis. Head Neck. 2021;43(4):1345–58. https://doi.org/10.1002/hed.26597.
219. Clayman MA, Clayman SM, Seagle MB. A review of the surgical and medical treatment of Frey syndrome. Ann Plast Surg. 2006;57(5):581–4. https://doi.org/10.1097/01.sap.0000237085.59782.65.
220. Li C, Wu F, Zhang Q, Li C, Wang L. Interventions for the treatment of Frey's syndrome. Cochrane Database Syst Rev. 2015;2015(3):Article CD009959 https://doi.org/10.1002/14651858.CD009959.pub2.
221. Drobik C, Laskawi R. Frey's syndrome: Treatment with botulinum toxin. Acta Oto-Laryngol. 1995;115(2):459–61. https://doi.org/10.3109/00016489509139348.
222. Shan X-F, et al. Botulinum toxin type A for Frey's syndrome. Cancer Med. 2015;4(11):1639–50. https://doi.org/10.1002/cam4.518
223. De Virgilio A, et al. Different surgical strategies in the prevention of Frey syndrome. Laryngoscope. 2021;131(8):1761–68 https://doi.org/10.1002/lary.29414.
224. Barbera G, Ottaviano C, Lobbia G, Rizzo T, Zatta E, Nocini R. Frey's syndrome surgical treatment after parotidectomy: a scoping review. J Clin Med. 2025;14(2):Article 415 https://doi.org/10.3390/jcm14020415.
225. Dulguerov P, et al. Frey syndrome treatment with botulinum toxin. Otolaryngol Head Neck Surg. 2000;122(6):821–7. https://doi.org/10.1067/mhn.2000.99738.
226. Headache Classification Committee of the International Headache Society (IHS). (2018). The International Classification of Headache Disorders, 3rd edition. Cephalalgia 2018;38(1):1–211. https://doi.org/10.1177/0333102417738202
227. International Classification of Orofacial Pain, 1st edition (ICOP). Cephalalgia 2020;40(2):129–221. https://doi.org/10.1177/0333102419893823
228. Pigg M, Nixdorf DR, Law AS, Renton T, Sharav Y, Baad-Hansen L, List T. New International Classification of Orofacial Pain: what is in it for endodontists. J Endod. 2021;47(3):345–57. https://doi.org/10.1016/j.joen.2020.12.002.
229. Han A, Montgomery C, Zamora A, Winder E, Kaye AD, Carroll C, Cornett EM. Glossopharyngeal neuralgia: Epidemiology, risk factors, pathophysiology, differential diagnosis, and treatment options. Health Psychol Res. 2022;10:36042 https://doi.org/10.52965/001c.36042.
230. Park JS, Ahn YH. Glossopharyngeal neuralgia. J Korean Neurosurg Soc. 2023;66(1):12–23. https://doi.org/10.3340/jkns.2022.0178.
231. Hurth H, Roder C, Tatagiba M, Ebner FH. Diagnostic and treatment pitfalls in glossopharyngeal neuralgia: evidence from a case series. Acta Neurochir. 2024;166(1):415 https://doi.org/10.1007/s00701-024-06313-6.
232. Wu L, Xiong J, Huang Y, Han K, Cai K, Fu X. Case report: trigeminal neuralgia misdiagnosed as glossopharyngeal neuralgia. Front Neurol. 2023;14:1079914 https://doi.org/10.3389/fneur.2023.1079914.
233. Berckemeyer MA, Suarez-Meade P, Villamonte Carcelen MF, Dyer Ricci M, Cheshire WP, Trifiletti DM, Middlebrooks EH, Quinones-Hinojosa A, Grewal SS. Current advances in the surgical treatment of glossopharyngeal neuralgia. Neurosurg Rev. 2023;46(1):47 https://doi.org/10.1007/s10143-023-01948-y.

234. Zheng W, Zhao P, Song H, Liu B, Zhou J, Fan C, Wang D, Liu R. Prognostic factors for long-term outcomes of microvascular decompression in the treatment of glossopharyngeal neuralgia: a retrospective analysis of 97 patients. J Neurosurg. 2022;137(3):820–827. https://doi.org/10.3171/2021.9.JNS21877.
235. Hajikarimloo B, Mohammadzadeh I, Tos SM, Mortezaei A, Habibi MA. Microvascular decompression in glossopharyngeal neuralgia: a systematic review and meta-analysis. Acta Neurol Belgica. 2025. Advance online publication. https://doi.org/10.1007/s13760-025-02978-7
236. You S, Qin X, Tong L, Feng Z. Long-term follow-up of ultrasound-guided glossopharyngeal nerve block treatment for glossopharyngeal neuralgia: a retrospective clinical study of 43 cases. J Pain Res. 2024;17:913–21. https://doi.org/10.2147/JPR.S437609.
237. Jia Y, Shrestha N, Wang X, Wang T, Luo F. The long-term outcome of CT-guided pulsed radiofrequency in the treatment of idiopathic glossopharyngeal neuralgia: a retrospective multi-center case series. J Pain Res. 2020;13:2093–2102. https://doi.org/10.2147/JPR.S259994.
238. Li F, Lu H, Gong G, Ou C, Zhang Y. The efficacy of ultrasound-guided pulsed radiofrequency in the treatment of primary glossopharyngeal neuralgia. Front Neurol. 2024;15:1453598 https://doi.org/10.3389/fneur.2024.1453598.
239. Wu Z, Zhao Y, Wu F, Fan Y, Yang Y. Percutaneous radiofrequency thermocoagulation and microvascular decompression for treating glossopharyngeal neuralgia: a retrospective clinical study. BMC Neurol. 2023;23(1):384. https://doi.org/10.1186/s12883-023-03415-z.
240. Siempis T, Rehder R, Voulgaris S, Alexiou GA. Stereotactic radiosurgery for idiopathic glossopharyngeal neuralgia: a systematic review. World Neurosurg X. 2024;22:100325 https://doi.org/10.1016/j.wnsx.2024.100325.
241. Hajikarimloo B, Mohammadzadeh I, Mortezaei A, Ebrahimi A, Habibi MA. Radiosurgery for glossopharyngeal neuralgia: A systematic review and meta-analysis. World Neurosurg. 2025;202:124389 https://doi.org/10.1016/j.wneu.2025.124389.
242. Verhoeff MC, et al. Updating the bruxism definitions: Report of an international consensus meeting. J Oral Rehabil 2025;52(9):1335–42. https://doi.org/10.1111/joor.13985.
243. Manfredini D, et al. Standardised tool for the assessment of bruxism. J Oral Rehabil. 2024;51(1):29–58. https://doi.org/10.1111/joor.13411.
244. Cid-Verdejo R, Chávez Farías C, Martínez-Pozas O, Meléndez Oliva E, Cuenca-Zaldívar JN, Ardizone García I, Martínez Orozco FJ, Sánchez Romero EA. Instrumental assessment of sleep bruxism: a systematic review and meta-analysis. Sleep Med Rev. 2024;74:101906 https://doi.org/10.1016/j.smrv.2024.101906.
245. Zieliński G, Pająk A, Wójcicki M. Global prevalence of sleep bruxism and awake bruxism in pediatric and adult populations: a systematic review and meta-analysis. J Clin Med. 2024;13(14):4259 https://doi.org/10.3390/jcm13144259.
246. Stanisic N, Saracutu OI, Colonna A, Wu W, Manfredini D, Häggman-Henrikson B. Awake bruxism prevalence across populations: A systematic review and meta-analysis. J Evid Based Dent Pract. 2025;25(3):102171 https://doi.org/10.1016/j.jebdp.2025.102171.
247. Uchima Koecklin KH, Aliaga-Del Castillo A, Li P. The neural substrates of bruxism: current knowledge and clinical implications. Front Neurol. 2024;15:1451183 https://doi.org/10.3389/fneur.2024.1451183.
248. Polmann H, Réus JC, Massignan C, Serra-Negra JM, Dick BD, Flores-Mir C, Lavigne GJ, De Luca Canto G. Association between sleep bruxism and stress symptoms in adults: a systematic review and meta-analysis. J Oral Rehabil. 2021;48(5):621–631. https://doi.org/10.1111/joor.13142.
249. Kuang B, Li D, Lobbezoo F, de Vries R, Hilgevoord A, de Vries N, Huynh N, Lavigne G, Aarab G. Associations between sleep bruxism and other sleep-related disorders in adults: a systematic review. Sleep Med. 2022;89:31–47. https://doi.org/10.1016/j.sleep.2021.11.008.
250. Mortazavi N, Tabatabaei AH, Mohammadi M, Rajabi A. Is bruxism associated with temporomandibular joint disorders? a systematic review and meta-analysis. Evid Based Dent. 2023;24(3):144 https://doi.org/10.1038/s41432-023-00911-6.

251. Bronkhorst H, Kalaykova S, Huysmans M-C, Loomans B, Pereira-Cenci T. Tooth wear and bruxism: a scoping review. J Dent. 2024;145:104983 https://doi.org/10.1016/j.jdent.2024.104983.
252. Nykänen L, Manfredini D, Lobbezoo F, Kämppi A, Bracci A, Ahlberg J. Assessment of awake bruxism by a novel bruxism screener and ecological momentary assessment among patients with masticatory muscle myalgia and healthy controls. J Oral Rehabil. 2024;51(1):162–69. https://doi.org/10.1111/joor.13462.
253. Colonna A, Bracci A, Ahlberg J, Câmara-Souza MB, Bucci R, Conti PCR, Dias R, Emodi-Perlman A, Favero R, Häggman-Henrikson B, Lobbezoo F, Michelotti A, Raphael KG, Wetselaar P, Yoshizawa S, Manfredini D. Ecological momentary assessment of awake bruxism behaviors: a scoping review of findings from smartphone-based studies in healthy young adults. J Clin Med. 2023;12(5):1904 https://doi.org/10.3390/jcm12051904.
254. Ainoosah S, Farghal AE, Alzemei MS, Saini RS, Gurumurthy V, Quadri SA, Okshah A, Mosaddad SA, Heboyan A. Comparative analysis of different types of occlusal splints for the management of sleep bruxism: a systematic review. BMC Oral Health. 2024;24(1):29. https://doi.org/10.1186/s12903-023-03782-6.
255. Buzatu R, Luca MM, Castiglione L, Sinescu C. Efficacy and safety of botulinum toxin in the management of temporomandibular symptoms associated with sleep bruxism: a systematic review. Dent J. 2024;12(6):156 https://doi.org/10.3390/dj12060156.
256. Häggman-Henrikson B, Ali D, Aljamal M, Chrcanovic BR. Bruxism and dental implants: a systematic review and meta-analysis. J Oral Rehabil. 2024;51(1):202–217. https://doi.org/10.1111/joor.13567.
257. Herrero Babiloni A, Lavigne GJ, Svensson P, Dal Fabbro C, Carra MC. Navigating the assessment of sleep-related bruxism: Clinical guidance for applying the 2023 ICSD-3-TR criteria. Sleep Med. 2026;137:106891 https://doi.org/10.1016/j.sleep.2025.106891.
258. Headache Classification Committee of the International Headache Society (IHS). (2018). The International Classification of Headache Disorders, 3rd edition. Cephalalgia 2018;38(1):1–211. https://doi.org/10.1177/0333102417738202
259. International Classification of Orofacial Pain (ICOP). Cephalalgia, 2020;40(2):129–221. https://doi.org/10.1177/0333102419893823
260. Park H-J, Ahn J-M, Ryu J-W. Post-traumatic trigeminal neuropathic pain: a narrative review of understanding, management, and prognosis. Biomedicines. 2024;12(9):Article 2058 https://doi.org/10.3390/biomedicines12092058.
261. Eliav T, Benoliel R, Korczeniewska OA. Post-traumatic trigeminal neuropathy: neurobiology and pathophysiology. Biology. 2024;13(3):Article 167 https://doi.org/10.3390/biology13030167.
262. Neal TW, Zuniga JR. Post-traumatic trigeminal neuropathic pain: factors affecting surgical treatment outcomes. Front Oral Health. 2022;3:Article 904785 https://doi.org/10.3389/froh.2022.904785.
263. Van der Cruyssen F, Peeters F, De Laat A, Jacobs R, Politis C, Renton T. Prognostic factors, symptom evolution, and quality of life of posttraumatic trigeminal neuropathy. Pain. 2022;163(4):e557–71. https://doi.org/10.1097/j.pain.0000000000002408.
264. International Classification of Orofacial Pain, 1st edition (ICOP). Cephalalgia 2020;40(2):129–221. https://doi.org/10.1177/0333102419893823
265. Sanner F, Sonntag D, Hambrock N, Zehnder M. Patients with persistent idiopathic dentoalveolar pain in dental practice. Int Endod J. 2022;55(3):231–9. https://doi.org/10.1111/iej.13664.
266. Heir GM, Ananthan S, Kalladka M, Kuchukulla M, Renton T. Persistent idiopathic dentoalveolar pain: is it a central pain disorder. Dent Clin N Am. 2023;67(1):71–83. https://doi.org/10.1016/j.cden.2022.07.006.
267. Malacarne A, Spierings ELH, Lu C, Maloney GE. Persistent dentoalveolar pain disorder: a comprehensive review. J Endod. 2018;44(2):206–11. https://doi.org/10.1016/j.joen.2017.09.009.

268. Tamaki K, Saito N, Tomita H. Serum 3-hydroxybutyrate is expected to serve as one of the supportive diagnostic markers of persistent idiopathic dentoalveolar pain (PDAP. J Pain Res. 2023;16:4005–4013. https://doi.org/10.2147/JPR.S436034.
269. Park H-J, Jeong S. Comprehensive management approach to persistent idiopathic dentoalveolar pain: a narrative review. Oral Biol Res. 2024;48(3):69–74. https://doi.org/10.21851/obr.48.03.202409.69.
270. Rupel K, Martina F, Ottaviani G, Di Lenarda R, Manganotti P. Efficacy of botulinum neurotoxin A in persistent idiopathic dentoalveolar pain: a case series. Clin Oral Investigations. 2024;28(10):548 https://doi.org/10.1007/s00784-024-05935-5.
271. Jang JH, Chung JW. Clinical features of the persistent idiopathic dentoalveolar pain compared with inflammatory dental pain. J Oral Med Pain. 2022;47(2):87–94. https://doi.org/10.14476/jomp.2022.47.2.87.
272. International Classification of Orofacial Pain, 1st edition (ICOP). Cephalalgia 2020;40(2):129–221. https://doi.org/10.1177/0333102419893823
273. Pigg M, Nixdorf DR, Law AS, Renton T, Sharav Y, Baad-Hansen L, List T. New International Classification of Orofacial Pain: what is in it for endodontists?. J Endod. 2021;47(3):345–7. https://doi.org/10.1016/j.joen.2020.12.002.
274. Barke A, Korwisi B, Rief W. Chronic pain in the ICD-11: New diagnoses that clinical psychologists should know about. Clin Psychol Europe. 2022;4(Spec Issue):e9933 https://doi.org/10.32872/cpe.9933.
275. Ziegeler C, Beikler T, Gosau M, May A. Idiopathic facial pain syndromes—an overview and clinical implications. Deutsches Ärzteblatt Int. 2021;118(6):81–87. https://doi.org/10.3238/arztebl.m2021.0006.
276. Ziegeler C, Brauns G, May A. Characteristics and natural disease history of persistent idiopathic facial pain, trigeminal neuralgia, and neuropathic facial pain. Headache. 2021;61(9):1441–51. https://doi.org/10.1111/head.14212.
277. Coulter J, Nixdorf DR. A review of persistent idiopathic dentoalveolar pain (formerly PDAP/atypical odontalgia. Oral Surg. 2020;13(4):371–8. https://doi.org/10.1111/ors.12472.
278. Chong MS, Bahra A, Zakrzewska JM. Guidelines for the management of trigeminal neuralgia. Cleve Clin J Med. 2023;90(6):355–62. https://doi.org/10.3949/ccjm.90a.22052.
279. May A, Benoliel R, Imamura Y, Pigg M, Baad-Hansen L, Svensson P, Hoffmann J. Orofacial pain for clinicians: a review of constant and attack-like facial pain syndromes. Cephalalgia. 2023;43(8):3331024231187160 https://doi.org/10.1177/03331024231187160.
280. Montisano DA, Raggi A, De Tommaso M, Grazzi L. Persistent idiopathic facial pain, a challenging diagnostic entity: clinical series from a third-level headache center. Confin Cephalalg 2025;34(2). https://doi.org/10.4081/cc.2024.15773
281. Bae S-H, Shim G-J, Kwon T-G. Diagnosis and management of persistent idiopathic facial pain following dental procedures: a retrospective study. J Korean Assoc Oral Maxillofac Surg. 2025;51(5):269–77. https://doi.org/10.5125/jkaoms.2025.51.5.269.
282. Guan G, Polonowita AK, Mei L, Polonowita DA, Polonowita AD. Chronic orofacial pain and pharmacological management: a clinical guide. Oral Surg Oral Med Oral Pathol Oral Radiol. 2025;140(1):e1–e21. https://doi.org/10.1016/j.oooo.2025.02.005.
283. Pagano S, Ricciuti V, Mancini F, Barbieri FR, Chegai F, Marini A, Marruzzo D, Paracino R, Ricciuti RA. Eagle syndrome: an updated review. Surg Neurol Int. 2023;14:389 https://doi.org/10.25259/SNI_666_2023.
284. Tadjer J, Béjot Y. Vascular variant of Eagle syndrome: a review. Front Neurol. 2024;15:Article 1463275 https://doi.org/10.3389/fneur.2024.1463275.
285. Hassani M, Grønlund EW, Albrechtsen SS, Kondziella D. Neurological phenotypes and treatment outcomes in Eagle syndrome: Systematic review and meta-analysis. PeerJ. 2024;12:Article e17423 https://doi.org/10.7717/peerj.17423.
286. Akkaşoğlu Taş A, Çetin H, Çalışkan S. Computed tomography evaluation of elongated styloid process and calcified stylohyoid ligament: Revisiting Eagle's syndrome. J Surg Med. 2025;9(11):240–244. https://doi.org/10.28982/josam.8456.

287. Thielen A, Brizzi V, Majoufre C, Nicot R, Schlund M. Eagle syndrome and vascular complications—a systematic review. Int J Oral Maxillofac Surg. 2025;54(1):31–42. https://doi.org/10.1016/j.ijom.2024.09.011. MID: 39406633.
288. Held ME, et al. Eagle syndrome presentation and outcomes in a large surgical case series. Laryngoscope Investig Otolaryngol. 2024;9(4):Article e1289 https://doi.org/10.1002/lio2.1289.
289. Williams DM, et al. Eagle syndrome: case report, literature review, proposed classification, and role of ultrasound in its diagnosis and management. J Diagn Med Sonogr 2024. https://doi.org/10.1177/15443167241281927
290. Sharifi A, Kouhi A. Management of Eagle syndrome. Oral Maxillofac Surg Clin N Am. 2023;35(3):345–352. https://doi.org/10.1016/j.coms.2023.01.005.
291. Swanson, et al. (2023). Eagle syndrome: pathophysiology, differential diagnosis and treatment options. Health Psychol Res. 2023;10(5):67851. https://doi.org/10.52965/001c.67851
292. Ferreira MS, et al. Complications in intraoral versus external approach for surgical treatment of Eagle syndrome: a systematic review and meta-analysis. Cranio. 2024;42(4):570–82. https://doi.org/10.1080/08869634.2021.2020995.

Myofascial Pain Syndrome in Neurostomatology

9

Abstract

Myofascial pain syndrome (MPS) is a common cause of chronic regional pain in the head, neck, and orofacial regions and is highly relevant to neurostomatology. It is characterized by localized muscle pain and tenderness associated with myofascial trigger points within taut bands of skeletal muscle, producing local and referred pain, restricted movement, and functional impairment. In neurostomatologic practice, MPS frequently overlaps with temporomandibular disorders, tension-type headache, cervicogenic pain, and treatment-related pain conditions. Because referred pain from masticatory and cervical muscles can mimic odontogenic, temporomandibular joint, otologic, and neuropathic disorders, MPS is an important source of diagnostic confusion. Its pathophysiology is multifactorial and involves muscle overload, biomechanical strain, local ischemia, peripheral sensitization, and, in chronic cases, central sensitization. Diagnosis remains primarily clinical, based on history and palpation of taut bands and hypersensitive trigger points that reproduce familiar pain. Management is multimodal and includes education, correction of perpetuating factors, exercise, physical therapy, dry needling or trigger-point injections, selected pharmacotherapy, and psychosocial support when indicated. Early recognition and multidisciplinary care are essential to prevent chronicity and improve outcomes in neurostomatology.

Keywords

Myofascial pain syndrome · Trigger points · Temporomandibular disorders · Referred pain

A. Kalanov, *Neurostomatology*, https://doi.org/10.1007/978-981-92-2552-1_9

9.1 Definition

Building on the interdisciplinary framework of neurostomatology, myofascial pain syndrome (MPS), also known as myofascial pain (MFP), represents one of the most common causes of chronic regional musculoskeletal pain, particularly in the head, neck, and orofacial regions [1]. It manifests as localized muscle tenderness and pain originating from myofascial trigger points (TrPs)—hyperirritable spots within taut bands of skeletal muscle fibers. These TrPs produce both local and referred pain patterns, often accompanied by restricted motion, muscle weakness, and autonomic phenomena (see Table 9.1.). In neurostomatology, MPS frequently overlaps with temporomandibular disorders (TMD), chronic tension-type headaches, cervicogenic pain, and post-treatment complications in head and neck oncology, making accurate recognition essential for effective multidisciplinary care [2].

Recent epidemiological data confirm the high burden of MPS in clinical populations. In patients with chronic nonspecific neck pain, prevalence reaches 88.9%, with multiple TrPs commonly identified in muscles such as the infraspinatus and scalenes [3]. Among middle-aged adults, estimates range from 37% in men to 65% in women, rising to 85% in those over 65 years in selected cohorts [4]. In head and neck cancer survivors 3 months post-treatment, point prevalence is 68.3%, with the sternocleidomastoid most frequently involved (55%), followed by masseter (29.2%), trapezius (25.8%), and temporalis (15%) [5]. These figures underscore MPS as a frequent primary or contributing source of pain in neurostomatologic practice, especially in patients with TMD, whiplash-associated disorders, and postural strain from prolonged dental or computer work. Prevalence varies due to diagnostic heterogeneity, but consistent findings highlight underdiagnosis in routine orofacial evaluations [1, 4].

9.2 Etiopathogenesis of MPS

The etiopathogenesis of MPS involves complex peripheral and central mechanisms. Peripheral TrPs arise from sustained low-level muscle contraction, overuse, trauma, or biomechanical overload, leading to localized ischemia, metabolic stress, and release of

Table 9.1 Clinical characteristics of myofascial trigger points

Category	Key findings
Essential diagnostic features	• Taut band (palpable with cross-fiber or pincer palpation) • Hypersensitive spot within the taut band • Referred pain or symptom reproduction on stimulation
Confirmatory signs	• Local twitch response on snapping palpation or needling • Restricted range of motion and muscle weakness • Reproduction of patient's familiar pain pattern
Supporting phenomena	• Autonomic changes (sweating, lacrimation in orofacial region) • Pain relief after TrP inactivation • Associated sensory changes (*paresthesia, numbness*)

Adapted from International Consensus [1, 2]

sensitizing substances (bradykinin, substance P, calcitonin gene-related peptide) [1, 6]. Recent multimodal studies combining shear-wave elastography, ultrasound, and histology reveal key tissue changes: affected muscles exhibit increased thickness and stiffness (Young's modulus significantly elevated), deep fascia thickening, reduced perfusion, and hypoechoic areas. Histologically, biopsies demonstrate muscle fiber atrophy, fibrofatty replacement, connective-tissue hyperplasia, microvascular proliferation, and perivascular inflammatory infiltration (CD3+ T-cells), without classic contraction knots in many cases. These findings support a revised model of metabolic exhaustion, fibrotic remodeling, and neuroimmune crosstalk, expanding beyond the original energy-crisis hypothesis [7].

Centrally, persistent nociceptive input from active myofascial trigger points (TrPs) drives central sensitization in the spinal cord and supraspinal centers, leading to heightened hyperalgesia and the development of referred pain patterns. Although MPS is primarily considered a nociceptive pain condition resulting from peripheral nociceptor activation within the muscle and fascia, it frequently overlaps with nociplastic or neuropathic pain components, especially in chronic and widespread cases [6]. In orofacial MPS, involvement of the key masticatory muscles (masseter, temporalis, and pterygoids) significantly amplifies symptoms of temporomandibular disorders through convergence in the trigeminal system, thereby contributing to associated complaints such as headaches, otalgia, and dizziness [8]. Genetic predisposition, hormonal fluctuations, metabolic disturbances (e.g., vitamin D deficiency), and psychosocial stressors all modulate an individual's susceptibility to MPS, ultimately forming a self-sustaining vicious cycle characterized by persistent pain, protective muscle guarding, and symptom perpetuation [1, 2].

The peripheral trigger points in masticatory muscles and the development of central sensitization in neurostomatology are depicted in Fig. 9.1.

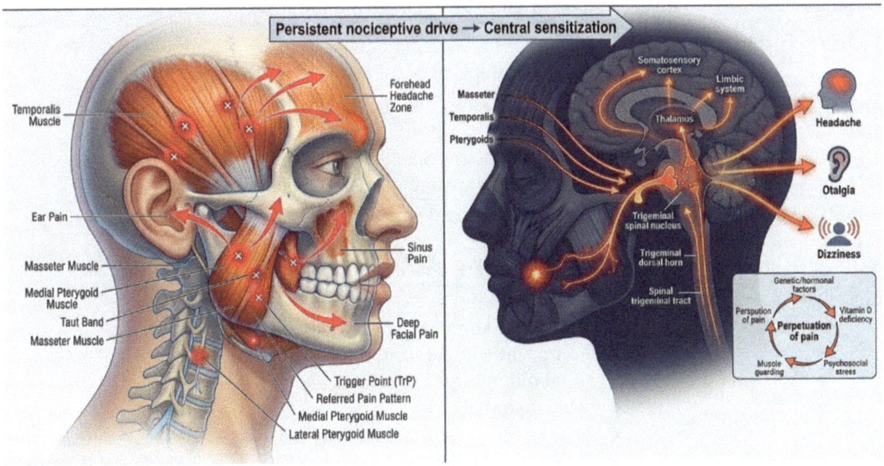

Fig. 9.1 Schematic illustration of myofascial pain syndrome pathophysiology in neurostomatology

9.3 Clinical Features of MPS

MPS typically presents with regional dull, aching, or burning pain that is often exacerbated by muscle movement, exposure to cold, emotional stress, or sustained postures, while it is usually relieved by the application of heat, gentle stretching, or light activity.

Patients commonly report associated symptoms including muscle stiffness, restricted range of motion, localized weakness, and generalized fatigue. Autonomic phenomena such as localized sweating, piloerection (goosebumps), or vasomotor changes (flushing or pallor) may also accompany the pain; proprioceptive disturbances, including dizziness, imbalance, and tinnitus, are particularly prominent in cases involving cervical and masticatory muscles [2, 4] (see Table 9.2).

Referred pain from myofascial trigger points follows highly predictable and characteristic patterns—for instance, trigger points in the masseter muscle often refer pain to the teeth and ear, those in the temporalis muscle refer to the temple and forehead, while medial pterygoid trigger points frequently refer to the temporomandibular joint (TMJ) and ear region. These patterns are critical in neurostomatology for accurate differential diagnosis, helping clinicians distinguish MPS from primary dental pathology, trigeminal neuralgia, or other orofacial pain conditions [8].

In the field of neurostomatology, thorough intraoral palpation of the medial and lateral pterygoid muscles is essential because these deep masticatory muscles are frequently affected following prolonged mouth opening during dental procedures, inadvertent trauma from local anesthetic injections, or occlusal imbalances. Furthermore, iatrogenic factors such as excessive static loading on the masticatory muscles during lengthy restorative dentistry or orthodontic treatments can significantly exacerbate symptoms, affecting both patients and dental professionals alike [8].

Table 9.2 Clinical characteristics of myofascial pain syndrome

Category	Details
Trigger points in muscle	• Localized tenderness on palpation • Pain reproduction (local or referred)
Associated symptoms	• Otologic (tinnitus, ear fullness) • Paresthesias • Visual disturbances or dizziness (cervical/orofacial) • Dermatographism or skin hypersensitivity
Pain in zone of reference	• Consistent referral patterns (e.g., masseter → teeth, trapezius → occiput) • Relief upon TrP deactivation
Contributing factors	• Trauma or whiplash • Occupational/repetitive strain (dentists, office workers) • Poor posture or parafunctional habits (bruxism, clenching) • Metabolic/nutritional deficits • Sleep disturbances • Psychosocial stress and anxiety

Adapted from International Consensus [1, 2]

Active TrPs reproduce the patient's spontaneous pain; latent TrPs cause pain only on palpation [2, 6]. Diagnosis remains clinical, relying on history and physical examination. The 2021 Chinese expert consensus and international Delphi-derived criteria propose the triad of (1) a taut band, (2) a hypersensitive spot within the taut band, and (3) referred pain or symptom reproduction as essential for definite diagnosis, although in clinical practice high confidence is often achieved with two of the three features plus confirmatory signs (e.g., local twitch response) [2, 9]. While manual palpation by trained examiners is reliable, inter-rater agreement remains only moderate ($\kappa \approx 0.4$–0.6), highlighting the utility of ultrasound elastography for improved objectivity [1, 7]. Ancillary tools include ultrasound for TrP localization and stiffness measurement, though not mandatory [1, 7].

Differential diagnosis must carefully exclude TMD joint pathology (e.g., disc displacement without reduction), trigeminal neuralgia, atypical odontalgia, sinusitis, dental infection, cervical disc disease, and central neuropathic pain following radiotherapy or stroke [8]. Imaging (MRI, ultrasound) rules out structural lesions, while psychological screening identifies comorbid depression or anxiety, present in up to 75% of head/neck cancer-related MPS cases [5]. Standardized regional protocols, including intraoral pterygoid examination, improve construct validity and prevent misdiagnosis as purely neurological or odontogenic pain [1, 8]. Forward head posture from prolonged dental chair time or screen use, occlusal discrepancies after restorations, and post-radiotherapy fibrosis are unique perpetuators in this field.

These features distinguish MPS from systemic conditions like fibromyalgia (widespread pain without focal TrPs) and emphasize the need for regional expertise in neurostomatology [2, 4].

Trauma initiates many cases, but persistence arises from multiple interrelated factors: postural strain, repetitive occupational loading (common in dentistry), parafunctional habits (bruxism, tongue thrusting), metabolic issues (hypothyroidism, iron/vitamin deficiency), sleep disorders, and psychosocial stress. In neurostomatology, forward head posture from screen use or dental procedures, occlusal imbalances, and post-radiotherapy fibrosis are key perpetuators [2, 4, 5]. Failure to address these leads to progression from single-muscle to widespread chronic pain syndromes [6].

9.4 Management of MPS

Effective MPS management is multimodal, individualized, and addresses all affected muscles, comorbidities, and perpetuating factors. First-line approaches include patient education, posture correction, and home stretching/exercise programs. Physical therapy modalities—manual release, dry needling (substantial evidence for short- to medium-term pain reduction, particularly when combined with physical therapy), extracorporeal shockwave, and kinesiology taping—show good outcomes in cervical and orofacial MPS [1, 2, 10].

Trigger-point injections (local anesthetic ± corticosteroid or botulinum toxin) provide rapid relief, especially when ultrasound-guided. Pharmacotherapy is adjunctive: NSAIDs or COX-2 inhibitors for acute flares, muscle relaxants (tizanidine), SNRIs (duloxetine) for chronic pain with mood disturbance, and gabapentinoids for neuropathic overlap [9].

In refractory orofacial cases, intraoral splints combined with TrP therapy improve TMD symptoms [8]. Psychological interventions (cognitive-behavioral therapy, biofeedback) are crucial given high comorbidity with anxiety/depression [5]. Emerging options include low-level laser, ozone injection, and radiofrequency ablation [9].

Simple cases (single muscle, few perpetuators) respond to basic interventions; complex cases (multiple muscles, cancer survivorship, psychosocial burden) require coordinated care involving orofacial pain specialists, physiotherapists, psychologists, and oncologists. Comprehensive treatment prevents chronicity and improves quality of life [1, 6].

Myofascial pain syndrome is a prevalent, treatable contributor to chronic head, neck, and orofacial pain in neurostomatology. Updated understanding of its pathophysiology—emphasizing fibrotic remodeling and neuroimmune changes—combined with standardized clinical criteria, objective imaging adjuncts, and multimodal, factor-directed management offers excellent outcomes. Future research should focus on objective biomarkers (ultrasound elastography) and region-specific protocols, including dental iatrogenic prevention strategies, to further advance care. Early recognition and holistic intervention remain the cornerstone of successful neurostomatologic practice.

References

1. Steen JP, Jaiswal KS, Kumbhare D. Myofascial pain syndrome: an update on clinical characteristics, etiopathogenesis, diagnosis, and treatment. Muscle Nerve. 2025 May; 71(5):889–910. https://doi.org/10.1002/mus.28377
2. Cao Q-W, Peng B-G, Wang L, Huang Y-Q, Jia D-L, Jiang H, Lv Y, Liu X-G, Liu R-G, Li Y, Song T, Shen W, Yu L-Z, Zheng Y-J, Liu Y-Q, Huang D. Expert consensus on the diagnosis and treatment of myofascial pain syndrome. World J Clin Cases. 2021;9(9):2077–89. https://doi.org/10.12998/wjcc.v9.i9.2077.
3. Ezzati K, Ravari M, Alizadeh F. Prevalence of cervical myofascial pain syndrome and its correlation with the severity of pain and disability in patients with chronic non-specific neck pain. Arch Bone Jt Surg. 2021;9(2):230–4. https://doi.org/10.22038/abjs.2020.48697.2415.
4. Lam C, Francio VT, Gustafson K, Carroll M, York A, Chadwick AL. Myofascial pain—A major player in musculoskeletal pain. Best Pract Res Clin Rheumatol. 2024;38(1):101944. https://doi.org/10.1016/j.berh.2024.101944.
5. Praveen A, Rao K, Gayatri S, Damani AD, Ghoshal A, Nair S, Kumar NAN, Salins SL, Pai A, Singh A, Udupa KS, Usman N, Salins NS. Prevalence, patterns, and impact of myofascial pain in patients with head and neck cancer after cancer treatment—a single-center cross-sectional study in India. BMC Palliat Care. 2025;24:109. https://doi.org/10.1186/s12904-025-01745-y.
6. Fernández-de-las-Peñas C, Nijs J, Cagnie B, Gerwin RD, Plaza-Manzano G, Valera-Calero JA, Arendt-Nielsen L. Myofascial pain syndrome: a nociceptive condition comorbid with neuropathic or nociplastic pain. Life. 2023;13(3):694. https://doi.org/10.3390/life13030694.
7. Cao W, Yin L, Sun H, Ge H, Guo R, Lyu X. Ultrasound features of myofascial trigger points: a multimodal study integrating preliminary histological findings from the upper trapezius. Sci Rep. 2025;15:20510. https://doi.org/10.1038/s41598-025-05869-2.
8. Golanska P, Saczuk K, Domitrz I. Temporomandibular myofascial pain syndrome—aetiology and biopsychosocial modulation. A narrative review. Int J Environ Res Public Health. 2021;18(15):7807. https://doi.org/10.3390/ijerph18157807.
9. Chen C, Liao D-M. Myofascial pain disorder: a narrative review. Tungs' Medical J. 2024;18(2):68–73. https://doi.org/10.4103/ETMJ.ETMJ-D-24-00029.
10. Kalladka M, et al. Musculoskeletal disorders and orofacial pain: a narrative review. Front Oral Maxillofac Med. 2024;6:17. https://doi.org/10.21037/fomm-21-103.

Complex Regional Pain Syndrome in the Head and Neck

10

Abstract

Complex regional pain syndrome (CRPS) is a chronic regional pain disorder marked by pain disproportionate to any inciting event and accompanied by sensory, vasomotor, sudomotor/edema, and motor/trophic changes. Although usually described in the limbs, it may rarely involve the head and neck, including the trigeminal region. In neurostomatology, craniofacial CRPS should be viewed as a regional manifestation of a broader pain disorder rather than a primary orofacial entity. Reported triggers include dental and maxillofacial procedures, facial trauma, sinonasal surgery, and trigeminal nerve injury. Diagnosis is clinical, based on Budapest criteria adapted to craniofacial assessment and exclusion of alternative causes. Important differential diagnoses include painful post-traumatic trigeminal neuropathy, trigeminal neuralgia, persistent idiopathic facial pain, temporomandibular disorders, and trigeminal autonomic cephalalgias. Management requires early, multimodal, function-oriented care. Further research is needed to clarify epidemiology, validate diagnostic criteria in the trigeminal territory, and optimize treatment strategies.

Keywords

Complex regional pain syndrome · Craniofacial pain · Head and neck pain · Neurostomatology · Trigeminal pain · Budapest criteria · Neuropathic pain · Autonomic dysfunction

Definition Complex Regional Pain Syndrome (CRPS) is a chronic, regional pain disorder characterized by continuing pain that is disproportionate to any inciting event and accompanied by sensory, vasomotor, sudomotor/edema, and motor/trophic abnormalities. Contemporary consensus retains the subdivision into CRPS Type I, occurring without confirmed major nerve injury, and CRPS Type II, occurring in the

A. Kalanov, *Neurostomatology*, https://doi.org/10.1007/978-981-92-2552-1_10

presence of a defined nerve lesion [1]. Although originally described in the extremities, CRPS is now understood as a systemic pain disorder with potential involvement of any body region, including the craniofacial territory.

In the context of neurostomatology, the concept of craniofacial CRPS requires careful nosological positioning. Orofacial pain classifications, including the International Classification of Orofacial Pain (ICOP), recognize multiple neuropathic and idiopathic facial pain conditions, yet CRPS is not defined as a primary orofacial entity [2]. Therefore, when CRPS involves the trigeminal system, it represents regional expression of a systemic disorder rather than a distinct dental or maxillofacial diagnosis. The defining characteristic of CRPS is not neuropathic pain alone, but multisystem regional dysfunction. Persistent pain must be accompanied by objective and subjective disturbances across sensory, vasomotor, sudomotor, or motor/trophic domains. This requirement distinguishes CRPS from isolated post-traumatic trigeminal neuropathy and other chronic facial pain disorders.

Population-based data estimate the overall incidence of CRPS at approximately 5–26 cases per 100,000 person-years, with a clear female predominance and peak incidence in midlife [3]. Robust epidemiological data for craniofacial CRPS remain unavailable. Existing evidence derives predominantly from case reports and small retrospective series following dental extractions, orthognathic procedures, implant placement, sinus surgery, and trigeminal nerve injury [4]. Within neurostomatology, CRPS should be considered when patients present with persistent facial pain that exceeds expected postoperative recovery, particularly when regional autonomic features accompany pain. Delayed recognition may lead to repeated dental interventions, reinforcement of peripheral nociceptive signaling, and progressive central sensitization. Contemporary reviews emphasize that early identification of CRPS in any anatomical region is associated with improved functional outcomes [3].

Etiopathogenesis in the Craniofacial Region. Craniofacial CRPS has been described following a variety of dental and maxillofacial interventions, most commonly dental extractions and endodontic procedures. Cases have also been reported after implant placement, orthognathic surgery, and temporomandibular joint operations, where tissue trauma and neural irritation may act as precipitating factors. Additional triggers include facial trauma and sinonasal surgical procedures involving structures within the trigeminal territory. In some instances, particularly in CRPS Type II, the condition develops after documented trigeminal nerve injury, whereas in other cases no clearly defined nerve lesion can be identified. CRPS Type II may occur after identifiable trigeminal nerve damage, whereas Type I may follow minor trauma or inflammatory insult without clear nerve transection [1].

Risk factors identified in limb CRPS cohorts include severe acute pain, prolonged immobilization, and psychological vulnerability [3]. Although these associations are extrapolated to the craniofacial region, direct region-specific validation is lacking. Importantly, not all persistent postoperative facial pain represents CRPS; careful clinical assessment is required to avoid diagnostic overextension.

CRPS is widely regarded as a multifactorial disorder involving peripheral neuroinflammation, central sensitization, autonomic dysregulation, and cortical reorganization [3].

Tissue injury initiates a cascade of peripheral inflammatory events that play a central role in the early pathogenesis of Complex Regional Pain Syndrome. Following trauma or surgical intervention, damaged tissues release proinflammatory cytokines, neuropeptides such as substance P and calcitonin gene-related peptide, and mast cell–derived mediators, which together promote neurogenic inflammation and microvascular dysfunction [5]. This inflammatory milieu contributes to vasodilation, increased vascular permeability, and local edema, while simultaneously sensitizing peripheral nociceptors. Sustained peripheral sensitization lowers activation thresholds and amplifies nociceptive signaling, thereby perpetuating pain beyond the expected period of tissue healing and facilitating transition to chronicity. At the central level, persistent peripheral input may induce enhanced excitability of nociceptive neurons within the dorsal horn and, in craniofacial presentations, within the trigeminal nucleus caudalis [6]. This phenomenon of central sensitization is characterized by increased synaptic efficacy, reduced inhibitory control, and expansion of receptive fields. Clinically, it manifests as hyperalgesia and allodynia, including pain provoked by normally innocuous stimuli. Central amplification mechanisms are strongly implicated in chronic CRPS and are supported by experimental pain models demonstrating long-term potentiation–like changes within nociceptive pathways.

Autonomic dysregulation further contributes to the complex phenotype of CRPS. Abnormal coupling between sympathetic efferents and nociceptive afferents may produce temperature asymmetry, skin color change, and sudomotor alterations [3]. In the craniofacial region, these signs may be subtle and require careful bilateral comparison, yet they remain clinically significant indicators of regional autonomic imbalance. Functional neuroimaging studies have demonstrated altered somatosensory cortical representation in CRPS patients, reflecting maladaptive neuroplasticity [7]. Although such findings are not diagnostic, they support the concept that cortical reorganization contributes to symptom persistence and altered body perception. The trigeminal system's dense integration with cranial autonomic pathways may explain the persistence and regional distribution of autonomic features in craniofacial CRPS. Unlike trigeminal autonomic cephalalgias, in which autonomic signs are episodic and attack-related, CRPS-related autonomic disturbances tend to be continuous and regionally maintained.

Clinical Phenotype Tissue injury initiates a cascade of peripheral inflammatory events that play a central role in the early pathogenesis of Complex Regional Pain Syndrome. Following trauma or surgical intervention, damaged tissues release proinflammatory cytokines, neuropeptides such as substance P and calcitonin gene-related peptide, and mast cell–derived mediators, which together promote neurogenic inflammation and microvascular dysfunction [5]. This inflammatory milieu contributes to vasodilation, increased vascular permeability, and local edema, while simultaneously sensitizing peripheral nociceptors. Sustained peripheral sensitization lowers activation thresholds and amplifies nociceptive signaling, thereby perpetuating pain beyond the expected period of tissue healing and facilitating transition to chronicity.

At the central level, persistent peripheral input may induce enhanced excitability of nociceptive neurons within the dorsal horn and, in craniofacial presentations, within the trigeminal nucleus caudalis [6]. This phenomenon of central sensitization is characterized by increased synaptic efficacy, reduced inhibitory control, and expansion of receptive fields. Clinically, it manifests as hyperalgesia and allodynia, including pain provoked by normally innocuous stimuli. Central amplification mechanisms are strongly implicated in chronic CRPS and are supported by experimental pain models demonstrating long-term potentiation–like changes within nociceptive pathways.

Autonomic dysregulation further contributes to the complex phenotype of CRPS. Abnormal coupling between sympathetic efferents and nociceptive afferents may produce temperature asymmetry, skin color change, and sudomotor alterations [3]. In the craniofacial region, these signs may be subtle and require careful bilateral comparison, yet they remain clinically significant indicators of regional autonomic imbalance.

Diagnosis of CRPS remains clinical and is based on the Budapest criteria, which improve specificity while maintaining sensitivity [1] (see Table 10.1).

Functional neuroimaging studies have demonstrated altered somatosensory cortical representation in CRPS patients, reflecting maladaptive neuroplasticity [7]. Although such findings are not diagnostic, they support the concept that cortical reorganization contributes to symptom persistence and altered body perception.

The trigeminal system's dense integration with cranial autonomic pathways may explain the persistence and regional distribution of autonomic features in craniofacial CRPS. Unlike trigeminal autonomic cephalalgias, in which autonomic signs are episodic and attack-related, CRPS-related autonomic disturbances tend to be continuous and regionally maintained.

Both symptoms and signs must be present, and objective findings should be documented at the time of examination. In craniofacial assessment, bilateral comparison is essential. Surface temperature measurement after environmental acclimatization may aid detection of vasomotor asymmetry. Imaging studies are primarily employed to exclude structural pathology rather than confirm CRPS.

The principal differential diagnoses that must be systematically considered when evaluating suspected craniofacial CRPS are summarized in Table 10.2, highlighting the key distinguishing clinical features that aid in accurate diagnostic discrimination.

Table 10.1 Budapest diagnostic criteria for CRPS

Criterion	Clinical requirement	Craniofacial operationalization
Continuing pain	Disproportionate to inciting event	Pain exceeding expected healing trajectory after dental/surgical trauma
Symptoms (≥1 in ≥3 categories)	Sensory; Vasomotor; Sudomotor/edema; Motor/trophic	Patient-reported facial allodynia; temperature change; facial swelling; jaw stiffness
Signs (≥1 in ≥2 categories)	Objective findings on examination	Measured temperature asymmetry; observed edema; evoked allodynia; color change
Exclusion	No alternative diagnosis better explains symptoms	Exclusion of odontogenic, neuropathic, sinus, TMJ, or primary headache disorders

Adapted for Craniofacial Application

Table 10.2 Differential diagnosis of craniofacial CRPS

Condition	Pain pattern	Autonomic features	Distinguishing characteristics
Painful post-traumatic trigeminal neuropathy	Localized neuropathic pain	Typically absent or minimal	Sensory deficit in defined nerve distribution
Trigeminal neuralgia	Paroxysmal electric shock pain	Episodic tearing possible	Attack-based; no persistent regional edema
Persistent idiopathic facial pain	Diffuse dull pain	Absent	No objective autonomic/ trophic signs
Temporomandibular disorder	Activity-related jaw pain	Absent	Mechanical dysfunction; imaging correlation
Trigeminal autonomic cephalalgias	Short severe attacks	Prominent but episodic	Strictly attack-related autonomic signs
Postherpetic neuralgia	Dermatomal neuropathic pain	Absent	History of herpes zoster

Painful post-traumatic trigeminal neuropathy lacks persistent vasomotor or sudomotor disturbances. Trigeminal neuralgia is characterized by paroxysmal electric-shock pain without continuous regional autonomic dysfunction [8]. Persistent idiopathic facial pain does not exhibit objective autonomic signs. Temporomandibular disorders are associated with mechanical dysfunction and localized musculoskeletal findings. Trigeminal autonomic cephalalgias demonstrate prominent but episodic cranial autonomic symptoms [9]. The presence of continuous regional autonomic and trophic features strongly favors CRPS.

Management of craniofacial CRPS. Management of craniofacial CRPS should be early, multimodal, and function-oriented [3]. Education and reassurance constitute foundational steps. Functional rehabilitation, including graded jaw mobilization and desensitization therapy, aims to prevent disuse and central amplification. Pharmacologic therapy follows contemporary neuropathic pain recommendations [10]. Gabapentinoids, tricyclic antidepressants, and serotonin–norepinephrine reuptake inhibitors may provide partial relief.

Short-course corticosteroids may be considered in the early inflammatory phase, based on extrapolated limb CRPS evidence [3]. Sympathetic blocks, such as stellate ganglion block, may be reserved for selected refractory cases. Neuromodulation techniques are considered only in specialized centers when conservative measures fail.

Psychological interventions, particularly cognitive behavioral therapy, address pain-related fear and maladaptive coping.

The stepwise therapeutic framework for craniofacial CRPS, outlining interventions and corresponding levels of evidence, is systematically presented in Table 10.3 to guide structured clinical decision-making.

Early diagnosis of craniofacial complex regional pain syndrome (CRPS) is consistently associated with improved clinical outcomes and reduced risk of chronic disability [3]. Delayed recognition may permit progression to persistent central sensitization, autonomic dysregulation, and significant psychosocial impairment. Chronic cases often

Table 10.3 Management ladder for craniofacial CRPS

Step	Intervention	Evidence strength
1.	**Patient education** and reassurance	**Strong** (general CRPS)
2.	**Early functional rehabilitation** (jaw mobilization, desensitization)	**Moderate** (extrapolated)
3.	**Neuropathic pain pharmacotherapy** (gabapentin, pregabalin, SNRIs, TCAs)	**Moderate**
4.	**Short-course corticosteroids** in early inflammatory phase	**Moderate** (early CRPS data)
5.	**Sympathetic blocks** (e.g., stellate ganglion block)	**Limited**; specialist indication
6.	**Neuromodulation** (peripheral nerve stimulation)	**Limited**; refractory cases
7.	**Psychological therapy** (CBT, coping skills)	**Strong** adjunctive evidence

demonstrate entrenched pain behaviors, functional limitation, and reduced quality of life. Structured follow-up should incorporate standardized pain intensity scales, detailed mapping of allodynic regions, and serial assessment of jaw range of motion.

Evaluation of functional oral activities—including chewing, speaking, and oral hygiene tolerance—is also essential to monitor rehabilitation progress. Given the multidimensional nature of the disorder, multidisciplinary management involving pain specialists, dentists, psychologists, and physiotherapists is frequently required.

Despite growing recognition, major research gaps remain, including the true incidence of craniofacial CRPS and the validation of Budapest diagnostic criteria within the trigeminal territory. Additional priorities include clarifying the role of quantitative sensory testing, defining long-term neuromodulation outcomes, and establishing standardized diagnostic pathways in dental practice through well-designed multicenter prospective studies.

References

1. Birklein F, Dimova V. Complex regional pain syndrome–up-to-date. PAIN Rep. 2017;2(6):e624. https://doi.org/10.1097/PR9.0000000000000624.
2. International Headache Society. International classification of orofacial pain (ICOP), 1st edition. Cephalalgia. 2020;40(2):129–221. https://doi.org/10.1177/0333102419893823.
3. Harden RN, McCabe CS, Goebel A, et al. Complex regional pain syndrome: Practical diagnostic and treatment guidelines, 5th edition. Pain Med. 2022;23(Suppl. 1):S1–S53. https://doi.org/10.1093/pm/pnac046.
4. Ferraro MC, O'Connell NE, Sommer C, et al. Complex regional pain syndrome: Advances in epidemiology, pathophysiology, diagnosis, and treatment. Lancet Neurol. 2024;23(5):522–33. https://doi.org/10.1016/S1474-4422(24)00076-0.
5. Urits I, Shen AH, Jones MR, Viswanath O, Kaye AD. Complex regional pain syndrome: current concepts and treatment options. Curr Pain Headache Rep. 2018;22(2):10. https://doi.org/10.1007/s11916-018-0667-7.

6. Baron R, Binder A, Wasner G. Neuropathic pain: diagnosis, pathophysiological mechanisms, and treatment. Lancet Neurol. 2010;9(8):807–19. https://doi.org/10.1016/S1474-4422(10)70143-5.
7. Maihöfner C, Handwerker HO, Neundörfer B, Birklein F. Patterns of cortical reorganization in complex regional pain syndrome. Neurology. 2003;61(12):1707–15. https://doi.org/10.1212/01.WNL.0000098939.02752.8E.
8. Bendtsen L, Zakrzewska JM, Heinskou TB, et al. Advances in diagnosis, classification, pathophysiology, and management of trigeminal neuralgia. Lancet Neurol. 2020;19(9):784–96. https://doi.org/10.1016/S1474-4422(20)30233-7.
9. Wang Z, Yang X, Zhao B, Li W. Primary headache disorders: from pathophysiology to neurostimulation therapies. Heliyon. 2023;9(4):e14786. https://doi.org/10.1016/j.heliyon.2023.e14786.
10. Finnerup NB, Kuner R, Jensen TS. Neuropathic pain: from mechanisms to treatment. Physiol Rev. 2021;101(1):259–301. https://doi.org/10.1152/physrev.00045.2019.

Headaches in Neurostomatology

11

Abstract

Headache disorders are a major concern in neurostomatology because pain arising within the trigeminovascular and trigeminocervical systems may be referred to the face, jaws, oral mucosa, and teeth, thereby mimicking odontogenic or temporomandibular disease. This overlap is clinically important, as patients with migraine, trigeminal autonomic cephalalgias, and related neurovascular pain syndromes often first present to dental practitioners with complaints of toothache, jaw pain, or sinus pain, leading to delayed diagnosis and unnecessary dental procedures. Contemporary neurostomatologic practice requires integration of headache and orofacial pain classifications, together with recognition of trigeminal convergence, central sensitization, and calcitonin gene-related peptide–mediated mechanisms underlying referred dental pain and allodynia. Diagnosis depends on exclusion of true dental pathology, identification of headache-associated sensory and autonomic features, awareness of secondary headache red flags, and recognition of overlap with temporomandibular disorders. Management is based on accurate classification, interdisciplinary care, and headache-directed therapy while avoiding unwarranted irreversible dental treatment. A modern neurostomatologic approach is essential to improve diagnostic accuracy and patient outcomes in headache-related facial and dental pain.

Keywords

Headache · Neurostomatology · Orofacial pain · Migraine · Trigeminal autonomic cephalalgias · Referred dental pain · Trigeminovascular system · Temporomandibular disorders · Central sensitization · Differential diagnosis

A. Kalanov, *Neurostomatology*, https://doi.org/10.1007/978-981-92-2552-1_11

11.1 Introduction

Headache disorders occupy a central place in neurostomatology because pain generated within the trigeminovascular and trigeminocervical systems often extends beyond the cranium and is perceived in the orbit, maxilla, mandible, oral mucosa, or teeth. In daily practice, patients with migraine, cluster headache, or other neurovascular pain syndromes may first seek dental care because the dominant complaint is "toothache," "jaw pain," or "sinus pain" rather than "headache." This overlap creates a major diagnostic challenge: when facial or dental pain is interpreted exclusively through an odontogenic lens, patients may undergo unnecessary endodontic therapy, extraction, or occlusal treatment without relief. For this reason, the neurologist must understand headache biology, recognize the facial presentations of primary headache disorders, and distinguish them from true dental disease and from painful cranial neuropathies [3].

11.2 Classification of Headache Disorders in Neurostomatologic Context

Contemporary diagnostic thinking in neurostomatology requires integration of general headache taxonomy with orofacial pain taxonomy. Within modern headache classification, disorders are broadly separated into primary headaches, secondary headaches, and painful cranial neuropathies, whereas the International Classification of Orofacial Pain (ICOP) provides a parallel framework for pain affecting dentoalveolar tissues, masticatory muscles, temporomandibular joints, cranial nerves, idiopathic facial pain, and orofacial pains resembling primary headaches. This latter category is especially important for dentists because it acknowledges that migraine-like and trigeminal autonomic cephalalgia-like pains may occur predominantly in the face and oral region even when a conventional headache complaint is absent. Accordingly, classification is not merely academic: it determines whether the patient should receive dental treatment, neurologic treatment, or interdisciplinary care [2].

Headache disorders remain among the most prevalent and disabling conditions worldwide. A major epidemiologic update estimated that approximately 52% of the global population experience an active headache disorder, with migraine affecting about 14% and tension-type headache about 26%. Migraine is not only common but also highly disabling, particularly in people of working age, and it frequently coexists with sensory hypersensitivity, nausea, cognitive inefficiency, sleep disturbance, and impaired quality of life. In neurostomatology, this burden is amplified by diagnostic misdirection: facially expressed headache may be mistaken for pulpitis, apical disease, sinus-related toothache, temporomandibular disorder, or atypical odontalgia. Thus, the true clinical burden extends beyond neurologic disability to include delayed diagnosis, repeated consultations, irreversible dental procedures, and persistent patient frustration [1, 4].

The major diagnostic categories of the *International Classification of Headache Disorders, 3rd edition (ICHD-3)* are summarized in Table 1, with emphasis on those

Table 11.1 Major diagnostic categories in the International Classification of Headache Disorders, 3rd edition (ICHD-3), and their relevance to neurostomatology

Major category	Principal subgroups/examples	Relevance to neurostomatology
Primary headache disorders	Migraine; tension-type headache; trigeminal autonomic cephalalgias; other primary headache disorders	Major source of referred facial, jaw, and tooth pain without odontogenic pathology; migraine and TACs are especially important in dental misdiagnosis.
Secondary headache disorders	Headache attributed to trauma or injury to the head and/or neck	May mimic post-traumatic orofacial pain and requires careful history-based differentiation.
	Headache attributed to cranial or cervical vascular disorder	Important because some vascular disorders are urgent and may present with facial pain or atypical headache.
	Headache attributed to non-vascular intracranial disorder	Structural intracranial causes must be excluded in atypical or progressive craniofacial pain.
	Headache attributed to a substance or its withdrawal	Medication-overuse and drug-related headache may complicate chronic facial pain presentations.
	Headache attributed to infection	Critical differential diagnosis when facial pain coexists with fever, systemic illness, or rapid deterioration.
	Headache attributed to disorder of homeostasis	Metabolic and systemic disturbances may produce diffuse craniofacial pain.
	Headache or facial pain attributed to disorders of the skull, neck, eyes, ears, nose, paranasal sinuses, teeth, mouth, or other facial/cervical structures	Particularly important in dental medicine because it includes pain arising from teeth, mouth, and adjacent craniofacial structures.
	Headache attributed to psychiatric disorder	Relevant in selected chronic pain cases, but only after adequate exclusion of neurologic and odontogenic causes.
Painful cranial neuropathies and other facial pain	Pain attributed to a lesion or disease of the trigeminal nerve, including trigeminal neuralgia and painful trigeminal neuropathies	Highly relevant because these disorders frequently mimic pulpal, periodontal, or post-procedural dental pain.
	Other cranial neuralgias/ neuropathies, including occipital neuralgia, nervus intermedius neuralgia, glossopharyngeal neuralgia, and related syndromes	Important in differential diagnosis of regional head, ear, throat, and facial pain.
	Burning mouth syndrome (BMS)	Directly relevant to neurostomatology and oral medicine.
	Persistent idiopathic facial pain (PIFP)	A major nonodontogenic facial pain diagnosis often encountered after unnecessary dental treatment.
	Central neuropathic pain	Relevant in selected neurologic cases with persistent facial pain.
	Other headache disorders	Residual category for headache disorders not classified elsewhere.

most relevant to neurostomatologic practice. The International Headache Society continues to list ICHD-3 as the current edition

Headache Phenotypes Most Relevant to Dental and Orofacial Practice

The headache phenotypes most relevant to neurostomatology are migraine, trigeminal autonomic cephalalgias, and neurovascular orofacial pain presentations. Migraine may involve throbbing, pressure-like, or aching pain and may be accompanied by photophobia, phonophobia, nausea, vestibular symptoms, and cutaneous or mucosal allodynia. Importantly, migraine pain is not confined to the cranium; facial and intraoral presentations have been increasingly recognized, particularly in the maxillary and mandibular divisions of the trigeminal nerve. Neurovascular orofacial pain and orofacial migraine represent clinically important entities in which attacks mimic pulpal or periodontal pain, often showing episodic recurrence, unilateral distribution, and aggravation by ordinary activities or sensory stimuli. Trigeminal autonomic cephalalgias, especially cluster headache, are likewise crucial for dentists to recognize because they can present as excruciating unilateral orbital, supraorbital, temporal, maxillary, or dental-region pain accompanied by ipsilateral lacrimation, conjunctival injection, rhinorrhea, nasal congestion, eyelid edema, facial sweating, ptosis, or miosis. A patient with cluster headache may therefore describe "tooth pain" or "sinus pain" while the true syndrome is neurovascular and autonomic rather than odontogenic [6, 8].

Within neurostomatology, the most clinically important ICHD-3 categories are primary headaches with facial referral, secondary headaches involving craniofacial structures, and painful cranial neuropathies/other facial pains, all of which may present as apparent dental or jaw pain in the absence of true odontogenic disease.

Contemporary Pathophysiology

Modern headache science supports a multidimensional model in which peripheral trigeminal activation, neuropeptide signaling, central sensitization, and altered brain-network modulation interact dynamically. Among these mechanisms, calcitonin gene-related peptide (CGRP) has assumed major importance. CGRP is released from trigeminal afferents and contributes to vasodilatory and nociceptive signaling, thereby helping sustain migraine attacks and related hypersensitivity states. The therapeutic success of CGRP-targeting monoclonal antibodies and gepants has strongly reinforced the biologic relevance of this pathway. In parallel, repeated nociceptive input can sensitize second-order neurons in the trigeminal nucleus caudalis and higher pain-processing networks, lowering the threshold for pain and widening the perceived field of symptoms. For the neurologist, the practical implication is that a structurally healthy tooth may become the perceived site of pain because the nervous system is amplifying and mislocalizing trigeminal input rather than because the tooth itself is diseased [5].

Trigeminal Convergence, Referred Dental Pain, and Allodynia

The neuroanatomical basis for headache-related dental pain lies in convergence. Afferent inputs from intracranial vessels, dura mater, cervical structures, facial tissues, and dentoalveolar regions converge within the trigeminocervical complex and related central pathways. Once these neurons are sensitized, pain may be referred from one trigeminal territory to another, producing maxillary or mandibular pain during a primary headache disorder. Recent oral-health research has shown that migraine is associated with pain and sensitivity in teeth and gums, particularly in patients with allodynia, supporting the clinical impression that dental structures can become symptom targets during migraine attacks.

This mechanism explains why percussion sensitivity, thermal intolerance, or diffuse "toothache" may occur in the absence of caries, pulpal necrosis, fracture, or periapical disease. It also explains why repeated dental procedures often fail in these patients: the source is neurovascular and central, whereas the treatment is directed only at the tooth [7].

Clinical Differentiation in Neurostomatologic Practice

Diagnostic differentiation rests on pattern recognition, exclusion of dental disease, and attention to neurologic warning signs. Several features should raise suspicion of headache-related or nonodontogenic pain: recurrent attacks with symptom-free intervals; pain spreading beyond a single tooth; associated photophobia, phonophobia, nausea, lacrimation, rhinorrhea, or restlessness; allodynia to light touch; and poor correlation between pain intensity and dental findings. Conversely, odontogenic pain is more likely when there is clear structural pathology, localized provocation, and a distribution matching the diseased tooth or supporting tissues.

The clinician must also screen for red flags suggesting a secondary headache disorder, including abrupt onset, progressive pattern change, neurologic deficit, systemic illness, older age at onset, painful eye symptoms, post-traumatic onset, immunologic compromise, or other SNNOOP10 warning features. In atypical cases, local anesthetic testing can be helpful but should never be interpreted in isolation. Failure of complete pulpal or peridental anesthesia to abolish the familiar pain should prompt reconsideration of the diagnosis and possible neurologic referral [10, 11, 14].

Temporomandibular Disorders and Headache Overlap

Temporomandibular disorders (TMD) frequently coexist with primary headache disorders, particularly migraine. This association is clinically important in neurostomatology because patients with combined headache and jaw pain often present first to dental or orofacial pain clinics rather than to neurology services [9]. Muscle hyperactivity, bruxism, and cervical dysfunction may increase trigeminal nociceptive input and thereby exacerbate migraine frequency or intensity in susceptible individuals [16]. However, the relationship is neither simple nor unidirectional; rather, it appears to be complex and bidirectional, with headache-related central sensitization potentially amplifying masticatory pain, while TMD-related peripheral input may worsen headache burden.

Recent controlled studies assessing occlusal adjustment and disclusion time reduction therapies have reported potential benefit in selected patients with TMD accompanied by headache or migraine [15, 16]. These findings suggest that, in carefully chosen cases, reduction of abnormal occlusal loading and masticatory muscle hyperactivity may contribute to symptom improvement. Nevertheless, the overall quality of evidence remains moderate, the studied populations are relatively specific, and the generalizability of these results to broader headache populations is limited. Earlier reports also proposed that disclusion time-reduction strategies might reduce chronic myofascial pain-dysfunction symptoms and headache frequency, but these studies should be interpreted with caution because of methodological limitations and the evolving standards of evidence-based headache medicine [17].

For this reason, neurologists must avoid attributing primary migraine exclusively to occlusal imbalance in the absence of supportive diagnostic criteria. Migraine is a neurobiological disorder with established central and trigeminovascular mechanisms, and although occlusal or musculoskeletal factors may act as aggravating or perpetuating influences in some individuals, they should not be regarded as universal or primary causes. The prevailing clinical view is that TMD and headache may coexist, interact, and amplify each other through shared pathways such as central sensitization, parafunction, sleep disruption, stress-related muscle overactivity, and cervical-trigeminal convergence. Accordingly, management should be individualized, conservative, and interdisciplinary, with irreversible occlusal interventions considered only after a careful diagnostic workup and clear evidence that the patient's symptoms are meaningfully linked to temporomandibular dysfunction.

11.3 Management Principles

Management begins with correct diagnosis. When the clinical picture suggests primary headache with facial referral, the priority is not dental intervention but headache-directed care and interdisciplinary coordination. Dentists should treat confirmed odontogenic pathology, but they should defer irreversible procedures when examination and imaging do not support a dental source.

For migraine, acute treatment commonly includes nonsteroidal anti-inflammatory drugs, triptans, gepants, or other migraine-specific rescue options, while preventive therapy is considered when attacks are frequent, prolonged, or disabling. Current guidance now recognizes CGRP-targeting therapies as a first-line preventive option alongside earlier established preventive approaches. In cluster headache and related trigeminal autonomic cephalalgias, urgent neurologic management is particularly important because the attacks are severe, stereotyped, and often highly responsive to disorder-specific therapy rather than dental treatment. Within the dental office, adjunctive counseling on sleep, stress, trigger recognition, oral parafunction, and avoidance of procedure cascades is often crucial. The neurologist should function as a diagnostic gatekeeper who identifies when the pain belongs to headache medicine, when it belongs to dentistry, and when it belongs to both [12, 13].

Future work in neurostomatology should aim to refine diagnostic algorithms for headache-related facial pain, validate practical chairside indicators of referred tooth pain, and integrate headache medicine more closely with oral medicine, endodontics, and TMD care. Particular research priorities include biomarkers of trigeminal sensitization, imaging correlates of referred dental pain, and prospective studies on how often headache disorders lead to unnecessary dental treatment. In conclusion, headache disorders are not peripheral to neurostomatology; they are one of its core diagnostic challenges.

A modern neurostomatologic approach must therefore recognize that facial and dental pain may represent the clinical expression of primary headache biology rather than simply local tissue disease confined to the teeth, periodontium, or surrounding oral structures. Accurate classification, careful exclusion of odontogenic pathology, awareness of neurologic and systemic red flags, and timely interdisciplinary referral are all essential to prevent misdiagnosis, avoid unnecessary dental procedures, ensure appropriate treatment, and ultimately improve patient outcomes and quality of care.

References

1. Stovner LJ, Hagen K, Linde M, Steiner TJ. The global prevalence of headache: an update, with analysis of the influences of methodological factors on prevalence estimates. J Headache Pain. 2022;23(1):34. https://doi.org/10.1186/s10194-022-01402-2.
2. International Classification of Orofacial Pain, 1st edition (ICOP). Cephalalgia, 2020;40(2):129–221. https://doi.org/10.1177/0333102419893823
3. Pigg M, Nixdorf DR, Law AS, Renton T, Sharav Y, Baad-Hansen L, List T. New International Classification of Orofacial Pain: what is in it for endodontists? J Endod. 2021;47(3):345–57. https://doi.org/10.1016/j.joen.2020.12.002.
4. Ashina M, Terwindt GM, Al-Karagholi MAM, de Boer I, Lee MJ, Hay DL, Schulte LH, Hadjikhani N, Sinclair AJ, Ashina H, Schwedt TJ, Goadsby PJ. Migraine: disease characterisation, biomarkers, and precision medicine. The Lancet. 2021;397(10283):1496–504. https://doi.org/10.1016/S0140-6736(20)32162-0.
5. Edvinsson L. CGRP and migraine: from bench to bedside. Revue Neurologique. 2021;177(7):785–90. https://doi.org/10.1016/j.neurol.2021.06.003.
6. Sharav Y, Haviv Y, Benoliel R. Orofacial migraine or neurovascular orofacial pain from pathogenesis to treatment. Int J Mol Sci. 2023;24(3):2456. https://doi.org/10.3390/ijms24032456.
7. Yilmaz M, Kale E, Ozge A. The link between cutaneous allodynia and pain/sensitivity in teeth and gums during migraine episodes. BMC Oral Health. 2024;24(1):1043. https://doi.org/10.1186/s12903-024-04705-9.
8. Petersen AS, Lund N, Goadsby PJ, Belin AC, Wang S-J, Fronczek R, Burish M, Cho S-J, Peres MFP, Jensen RH. Recent advances in diagnosing, managing, and understanding the pathophysiology of cluster headache. Lancet Neurol. 2024;23(7):712–24. https://doi.org/10.1016/S1474-4422(24)00143-1.
9. Réus JC, Polmann H, Souza BDM, Flores-Mir C, Gonçalves DAG, de Queiroz LP, Okeson J, De Luca Canto G. Association between primary headaches and temporomandibular disorders: a systematic review and meta-analysis. J Am Dent Assoc. 2022;153(2):120–131.e6. https://doi.org/10.1016/j.adaj.2021.07.021.

10. Dodick DW. Diagnosing secondary and primary headache disorders. Continuum. 2021;27(3):572–85. https://doi.org/10.1212/CON.0000000000000980.
11. García-Azorín D, Abelaira-Freire J, González-García N, Rodriguez-Adrada E, Schytz HW, Barloese M, Guerrero ÁL, Porta-Etessam J, Martín-Sánchez FJ. Sensitivity of the SNNOOP10 list in the high-risk secondary headache detection. Cephalalgia. 2022;42(14):1521–31. https://doi.org/10.1177/03331024221120249.
12. Charles AC, Digre KB, Goadsby PJ, Robbins MS, Hershey A. Calcitonin gene-related peptide-targeting therapies are a first-line option for the prevention of migraine: an American Headache Society position statement update. Headache. 2024;64(4):333–41. https://doi.org/10.1111/head.14692.
14. Beecroft EV, Edwards D, Allison JR. Other secondary headaches: odontogenic pain and other painful orofacial conditions. Neurol Clin. 2024;42(2):615–32. https://doi.org/10.1016/j.ncl.2023.12.013.
13. Burch R. Acute treatment of migraine. Continuum. 2024;30(2):344–63. https://doi.org/10.1212/CON.0000000000001402.
15. Thumati P, Thumati RP, Poovani S, et al. A multi-center disclusion time reduction (DTR) randomized controlled occlusal adjustment study using occlusal force and timing sensors synchronized with muscle physiology sensors. Sensors. 2021;21(23):7804. https://doi.org/10.3390/s21237804.
16. Thumati PP, Thumati RP, Thumati P, Kerstein RB, Radke J. Efficacy of disclusion time reduction occlusal therapy for headache and migraine. Adv Dent Tech. Published online July 2021;31:2021.
17. Sutter BA. Incidence of headaches related to occlusion and bite force imbalance: a case study. Cranio May 2016;34(3):195–207. https://doi.org/10.1179/2151090315Y.0000000006. PMID: 25885423.

General Principles of Differential Diagnosis in Neurostomatology

12

Abstract

Differential diagnosis in neurostomatology requires structured evaluation of oral, facial, and cranial pain because odontogenic and non-odontogenic disorders often overlap. This chapter emphasizes pain phenotype, temporal profile, mechanism, anatomical distribution, and key associated features. Focused history-taking, examination, and selective investigations improve diagnostic accuracy and support appropriate interdisciplinary management of orofacial pain.

Keywords

Neurostomatology · Differential diagnosis · Orofacial pain · Trigeminal neuralgia

Differential diagnosis in neurostomatology is the structured process of distinguishing among dentoalveolar, musculoskeletal, neuropathic, neurovascular, idiopathic, and referred sources of pain arising in the oral, facial, and cranial regions. In contemporary practice, this process should be anchored in the *International Classification of Orofacial Pain* (ICOP), which organizes orofacial pain according to diagnostic group, temporal profile, distribution, and presumed mechanism [1]. This framework is especially important because orofacial pain is common but highly heterogeneous: prevalence varies widely across diagnostic subgroups, and many patients first seek care from dental practitioners even when the dominant disorder is neurologic, headache-related, or musculoskeletal rather than odontogenic [2, 3].

While specific differential distinctions between individual pathologies are addressed in the corresponding chapters, this chapter outlines the overarching principles, diagnostic logic, and clinical mechanisms underlying the differential diagnosis of orofacial pain in neurostomatology.

As illustrated in Fig. 12.1, classification by temporal profile and dominant diagnostic mechanism helps organize neurostomatologic differential diagnosis and highlights the core clinical questions that separate major pain categories.

A. Kalanov, *Neurostomatology*, https://doi.org/10.1007/978-981-92-2552-1_12

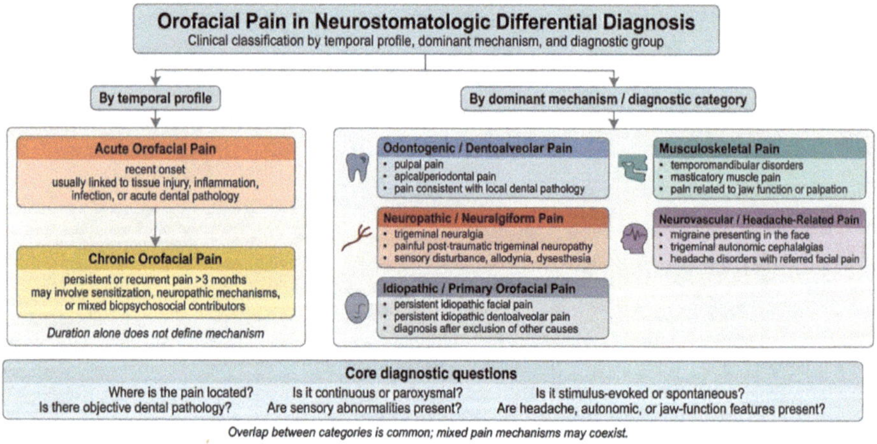

Fig. 12.1 Mechanism-informed framework for differential diagnosis in neurostomatology

Duration alone does not define mechanism, and mixed pain presentations are common, particularly in persistent disorders shaped by peripheral input, altered somatosensory processing, sleep disturbance, psychosocial burden, and altered pain modulation. Accordingly, mechanism-based reasoning should guide diagnostic thinking without forcing every patient into a rigid single-category framework [5, 12].

12.1 Core Diagnostic Principles

A practical diagnostic approach begins with the recognition that site alone is not diagnosis. Patients may describe pain arising from teeth, periodontal tissues, masticatory muscles, temporomandibular joints, cranial nerves, paranasal structures, cervical tissues, or primary headache disorders simply as "tooth pain," "jaw pain," "facial pain," or "head pain." Therefore, the clinician should first classify the pain phenotype before assigning a disease label. The most useful early distinctions are whether the pain is acute tissue-related, persistent inflammatory, neuralgiform, post-traumatic neuropathic, musculoskeletal, headache-related, or idiopathic. These distinctions are clinically relevant because pain in the same anatomic region may reflect very different temporal patterns, mechanisms, and diagnostic implications [4, 5].

12.2 History-Taking and Pain Phenotyping

The history should be systematic and explicitly mechanism-oriented. Essential elements include location, laterality, onset, duration, temporal pattern, quality, radiation, provoking and relieving factors, associated symptoms, previous dental treatment, and the relationship of pain to chewing, jaw movement, touch, temperature, swallowing, speaking, sleep, and posture. The clinician should determine whether the pain is unilateral or bilateral, intraoral or extraoral, well localized or diffuse, continuous or paroxysmal, spontaneous or stimulus-evoked.

Acute and chronic pain differ not only in duration but also in clinical context, dominant mechanisms, and diagnostic implications. The categories shown are intended as practical clinical guides; overlap may occur, and persistent pain does not by itself establish a specific mechanism or diagnosis.

Paroxysmal electric shock-like pain lasting seconds and triggered by innocuous stimuli should raise strong suspicion for a neuralgiform disorder, whereas continuous aching, pressure-like, or burning pain broadens the differential to include temporomandibular disorders, painful post-traumatic trigeminal neuropathy, persistent idiopathic facial pain, persistent idiopathic dentoalveolar pain, burning mouth syndrome, and facial presentations of primary headache disorders [1, 5, 6]. A careful history of prior endodontic treatment, extraction, implant placement, local anesthetic injury, or orthognathic surgery is especially important because procedural history may redirect the differential toward post-traumatic trigeminal pain rather than primary dental disease [10].

12.3 Pain Quality and Mechanism-Based Interpretation

Pain quality remains one of the most clinically useful discriminators in differential diagnosis. Nociceptive and inflammatory pain is commonly aching, throbbing, tender, or pressure-like and is often provoked by local mechanical or inflammatory factors. Neuropathic pain is more likely to be burning, shooting, electric, stabbing, tingling, or associated with numbness, dysesthesia, or allodynia. Musculoskeletal pain related to temporomandibular disorders often varies with mandibular function, palpation, jaw loading, or parafunction. Neurovascular pain may present with episodic attacks, autonomic symptoms, photophobia, phonophobia, nausea, or a history of migraine or other primary headache disorders [8, 11].

Table 12.1 provides a practical side-by-side comparison of acute and chronic orofacial pain patterns relevant to neurostomatologic assessment.

Table 12.2 summarizes the clinical features that help distinguish nociceptive, inflammatory, and neuropathic pain mechanisms in orofacial practice.

In clinical reality, overlap is common, and persistent pain states may show mixed or nociplastic features rather than a single pure mechanism. For this reason, mechanistic classification should support clinical reasoning rather than replace it [8, 12].

12.4 Distinguishing Odontogenic from Non-Odontogenic Pain

One of the central diagnostic tasks in neurostomatology is to determine whether the pain is odontogenic or non-odontogenic. Odontogenic pain usually correlates with a tooth or supporting structure and is interpreted together with clinical examination, percussion, palpation, pulp sensibility or vitality findings, periodontal assessment,

Table 12.1 . Practical comparison of acute and chronic orofacial pain in neurostomatologic differential diagnosis.

Characteristic	Acute orofacial pain	Chronic orofacial pain
Duration	Usually recent onset; commonly up to 3 months	Persists or recurs for more than 3 months
Typical cause	Tissue injury, inflammation, infection, or recent dental pathology	Ongoing peripheral pathology, neuropathic mechanisms, central sensitization, or mixed biopsychosocial contributors
Biologic role	Usually protective and related to a current noxious process	Often loses clear protective value and becomes disease-associated
Pain quality	Frequently sharp, throbbing, localized, or stimulus-linked	Commonly persistent, aching, burning, diffuse, fluctuating, or amplified
Temporal pattern	Often linked closely to an identifiable provoking event	May persist beyond healing or occur with recurrent exacerbations
Associated findings	Local inflammatory signs or identifiable dental/tissue pathology are often present	Sensory amplification, sleep disturbance, distress, disability, or pain-related behavioral change may coexist
Examples	Acute pulpitis, acute apical periodontitis, oral ulcer pain, acute temporomandibular strain	Painful post-traumatic trigeminal neuropathy, persistent idiopathic dentoalveolar pain, persistent idiopathic facial pain, chronic temporomandibular pain

and targeted imaging. By contrast, non-odontogenic pain becomes more likely when the pain distribution is anatomically inconsistent with a single tooth, when repeated dental procedures have failed, when the quality is burning or electric, when somatosensory abnormalities are reported, or when symptoms are linked to non-dental triggers such as light touch, jaw function, or headache-associated features [4, 6]. A common diagnostic error is to equate radiographic abnormality with pain causation. Another is to interpret transient response to local anesthesia as definitive proof of a dental source. Differential diagnosis should therefore proceed by clinicoradiologic correlation rather than by procedural trial and error [3, 4].

12.5 Clinical Examination

Clinical examination should include both stomatognathic and neurologic components. Beyond routine dental and mucosal assessment, the examination should evaluate masticatory muscles, temporomandibular joints, mandibular range of motion, jaw loading, joint sounds, and cervical contribution when indicated. Focused sensory examination in the trigeminal distribution is particularly important when neuropathic pain is suspected. Qualitative sensory testing may identify hypoesthesia, hyperalgesia, allodynia, dysesthesia, or painful numbness, and more detailed psychophysical testing can refine phenotyping when available [7].

Table 12.2 Clinical features that help distinguish nociceptive, inflammatory, and neuropathic pain in orofacial practice

Feature	Nociceptive pain	Inflammatory pain	Neuropathic pain
Primary mechanism	Activation of intact nociceptors by noxious mechanical, thermal, or chemical stimuli	Nociceptor activation and sensitization driven by inflammatory mediators	Pain caused by a lesion or disease of the somatosensory nervous system
Usual tissue/nerve status	No structural nerve lesion required	Inflamed peripheral tissues; nerve structure may remain intact	Somatosensory pathway abnormality is present or strongly suspected
Typical pain quality	Aching, sharp, throbbing, pressure-like	Tender, throbbing, aching, pressure-like; often exaggerated by inflammation	Burning, electric shock-like, shooting, stabbing, tingling, painful numbness
Response to stimuli	Usually provoked by clearly noxious stimulation	Increased sensitivity to local mechanical or thermal provocation	May occur spontaneously or be evoked by both noxious and non-noxious stimuli
Common clinical clues	Pain corresponds to local tissue injury and protective withdrawal	Redness, swelling, heat, tenderness, inflammatory aggravation	Allodynia, hyperalgesia, dysesthesia, paresthesia, sensory loss, neuralgiform attacks
Examples in orofacial practice	Mechanical mucosal trauma; acute dentin hypersensitivity from exposed dentin	Irreversible pulpitis; acute apical inflammation; inflammatory temporomandibular pain	Trigeminal neuralgia; painful post-traumatic trigeminal neuropathy

The examination should also assess whether pain is reproduced by function or palpation, because reproducibility with mandibular movement or muscle palpation favors a musculoskeletal source, although musculoskeletal pain and headache disorders frequently coexist rather than exclude one another [3, 11].

12.6 Key Diagnostic Groups and Common Mimics

Certain diagnostic groups deserve special emphasis because they are frequent sources of misdiagnosis in dental settings. Trigeminal neuralgia classically presents with recurrent unilateral, brief, severe, electric shock-like pain in one or more trigeminal divisions, often provoked by light touch, speaking, chewing, tooth brushing, or facial movement. The diagnosis is primarily clinical, whereas magnetic resonance imaging is used to assess for neurovascular conflict and to exclude secondary causes [9]. Because symptoms may overlap with dental pain, temporomandibular disorders, migraine, or painful trigeminal neuropathy, delayed diagnosis and inappropriate dental treatment remain common [5, 6]. Painful post-traumatic trigeminal neuropathy should be considered when persistent pain, numbness, dysesthesia, paresthesia, or mechanically evoked symptoms follow extraction, implant placement, endodontic treatment, local anesthetic injury, or facial trauma [10]. Persistent idiopathic dentoalveolar pain should be suspected when daily dentoalveolar pain persists without sufficient local pathology and when repeated irreversible treatment has failed or worsened the course [10]. Temporomandibular disorders remain another major mimic because they may present as tooth, ear, temple, or facial pain and frequently overlap with headache disorders [11].

12.7 Red Flags and Diagnostic Pitfalls

A safe neurostomatologic approach must explicitly identify red flags. Urgent reassessment or referral is warranted when facial pain is accompanied by progressive sensory loss, objective cranial nerve deficit, new motor weakness, facial or oral swelling, fever, trismus, unexplained weight loss, suspected malignancy, immunosuppression, systemic inflammatory disease, or a new severe headache pattern. Persistent unilateral numbness, pain out of proportion to local findings, or clustered autonomic symptoms with atypical progression should also prompt broader investigation [3, 4, 6].

Common diagnostic pitfalls include repeated endodontic or surgical procedures without objective pathology, premature labeling of persistent pain as psychogenic, failure to ask about trauma or prior dental procedures, under-recognition of headache presenting in the face, and underestimation of mixed pain mechanisms. These errors are often avoidable when the clinician classifies pain phenotype before treatment and resists interpreting every facial pain complaint through a purely odontogenic lens [4, 5, 12].

12.8 Investigations and Interdisciplinary Referral

Investigations are supportive and should be hypothesis-driven. Periapical radiographs, panoramic imaging, cone-beam computed tomography, magnetic resonance imaging, and medical neuroimaging each have selective roles, but none substitutes for careful history and examination. Dental imaging is most useful when the working diagnosis remains local inflammatory or structural disease; MRI is more relevant when neuralgia, secondary neuropathy, intracranial pathology, or other non-dental causes are suspected [3, 4, 9].

Diagnostic local anesthesia may occasionally help localize pain, but results should be interpreted cautiously and in context. Interdisciplinary referral is appropriate when the phenotype is inconsistent with dental disease, when red flags are present, when somatosensory abnormalities suggest neuropathic pain, or when migraine, trigeminal autonomic cephalalgia, temporomandibular disorder, and persistent idiopathic pain syndromes overlap. The practical goal is not merely to name the pain, but to identify the dominant mechanism well enough to avoid unnecessary irreversible treatment and direct the patient toward the correct management pathway [3, 4, 7].

12.9 Summary

In neurostomatology, accurate differential diagnosis depends on five linked principles: classify the pain phenotype before naming the disorder; distinguish odontogenic from non-odontogenic sources early; combine stomatognathic and neurologic examination; actively search for red flags and sensory abnormalities; and use investigations as targeted adjuncts rather than substitutes for diagnostic reasoning. When interpreted in this way, Fig. 1 and Tables 1–2 function as practical aids to mechanism-informed diagnosis rather than rigid taxonomies. The clinician's task is not to force every patient into a single category, but to build a defensible working diagnosis that recognizes overlap, prioritizes safety, and prevents unnecessary dental intervention [1, 3, 5].

The future of neurostomatology lies in the close convergence of neurology and dentistry, whose rapid scientific and clinical development will progressively enhance diagnostic precision and ultimately improve therapeutic outcomes for patients.

References

1. International Classification of Orofacial Pain, 1st edition (ICOP). Cephalalgia. 2020;40(2):129–221. https://doi.org/10.1177/0333102419893823
2. Porporatti AL, Schroder Â, G D, Lebel A, Moreau N, Guillouet C, Stechman-Neto J, Boucher Y. Prevalence of orofacial and head pain: an umbrella review of systematic reviews. J Oral Facial Pain Headache. 2024 Sep; 38(3):1-14. https://doi.org/10.22514/jofph.2024.022.
3. Barad M, Romero-Reyes M. Orofacial pain. Continuum. 2024;30(5):1397–426. https://doi.org/10.1212/CON.0000000000001488.
4. Labanca M, Gianò M, Franco C, Rezzani R. Orofacial pain and dentistry management: guidelines for a more comprehensive evidence-based approach. Diagnostics. 2023;13(17):2854. https://doi.org/10.3390/diagnostics13172854.
5. May A, Benoliel R, Imamura Y, Pigg M, Baad-Hansen L, Svensson P, Hoffmann J. Orofacial pain for clinicians: a review of constant and attack-like facial pain syndromes. Cephalalgia. 2023 Aug;43(8):3331024231187160. https://doi.org/10.1177/03331024231187160
6. Beecroft EV, Edwards D, Allison JR. Other secondary headaches: odontogenic pain and other painful orofacial conditions. Neurol Clin 2024;42(2):615–32. https://doi.org/10.1016/j.ncl.2023.12.013.
7. Costa YM, Bonjardim LR, Conti PCR, Svensson P. Psychophysical evaluation of somatosensory function in oro-facial pain: achievements and challenges. J Oral Rehabil. 2021;48(9):1066–76. https://doi.org/10.1111/joor.13202.
8. Shraim MA, Sluka KA, Sterling M, Arendt-Nielsen L, Argoff C, Bagraith KS, Baron R, Brisby H, Carr DB, Chimenti RL, Courtney CA, Curatolo M, Darnall BD, Ford JJ, Graven-Nielsen T, Kolski MC, Kosek E, Liebano RE, Merkle SL, Hodges PW. Features and methods to discriminate between mechanism-based categories of pain experienced in the musculoskeletal system: a Delphi expert consensus study. Pain. 2022;163(9):1812–28. https://doi.org/10.1097/j.pain.0000000000002577.
9. Pergolizzi JV, LeQuang JAK, El-Tallawy SN, Ahmed RS, Wagner M, Varrassi G. The challenges in clinical diagnosis of trigeminal neuralgia: a review. Cureus. 2024;16(6):e61898. https://doi.org/10.7759/cureus.61898.

10. Park H-J, Ahn J-M, Ryu J-W. Post-traumatic trigeminal neuropathic pain: a narrative review of understanding, management, and prognosis. Biomedicines. 2024;12(9):2058. https://doi.org/10.3390/biomedicines12092058.
11. Romero-Reyes M, Klasser G, Akerman S. An update on temporomandibular disorders and headache. Curr Neurol Neurosci Rep. 2023;23(10):561–70. https://doi.org/10.1007/s11910-023-01291-1.
12. Canfora F, et al. Advancements in understanding and classifying chronic orofacial pain: key insights from biopsychosocial models and international classifications. Biomedicines. 2023;11(12):3266. https://doi.org/10.3390/biomedicines11123266.

Behavioral and Psychological Aspects of Neurostomatology 13

Abstract

Behavioral and psychological factors are fundamental to neurostomatology because chronic orofacial pain arises through interactions among sensory, cognitive, emotional, and social processes rather than from structural pathology alone. In the trigeminal and stomatognathic regions, these influences affect symptom severity, disability, health-care utilization, and quality of life. Anxiety, depression, catastrophizing, hypervigilance, maladaptive illness behavior, and sleep disturbance are particularly relevant in conditions such as burning mouth syndrome, persistent idiopathic facial pain, persistent idiopathic dentoalveolar pain, trigeminal neuralgia, and painful temporomandibular disorders. This chapter reviews their importance in assessment, patient communication, and multidisciplinary management, and outlines psychologically informed interventions aimed at improving coping, function, and overall outcomes.

Keywords

Neurostomatology · Chronic orofacial pain · Biopsychosocial model · Pain catastrophizing · Anxiety and depression · Cognitive-behavioral therapy · Sleep disturbance · Multidisciplinary care ·

Neurostomatology concerns disorders of oral and maxillofacial sensation, pain, and neurosensory function, including neuropathic, idiopathic, and complex chronic orofacial pain conditions. In this field, behavioral and psychological factors are not secondary or optional considerations; they are central determinants of symptom expression, illness behavior, health-care utilization, treatment adherence, disability, and quality of life. Contemporary classification frameworks recognize that chronic orofacial pain cannot be understood solely through tissue injury or structural pathology, because pain emerges from the interaction of sensory input with cognitive, emotional, autonomic,

A. Kalanov, *Neurostomatology*, https://doi.org/10.1007/978-981-92-2552-1_13

and social processes. For this reason, a modern neurostomatological perspective must be explicitly biopsychosocial rather than narrowly biomedical [1, 2].

13.1 Biopsychosocial Relevance to Neurostomatology

Pain arising in the trigeminal and stomatognathic regions is particularly vulnerable to behavioral modulation because these regions are intimately linked to speech, eating, facial expression, sleep, social interaction, and self-image. Chronic pain in the mouth and face therefore carries consequences that extend well beyond nociception. Patients may become preoccupied with oral sensations, fear serious undetected disease, avoid chewing or speaking, repeatedly inspect oral tissues, or seek repeated dental procedures in the hope of eliminating symptoms that are no longer driven by an active peripheral source. Emotional distress may amplify salience, increase vigilance, lower coping capacity, and intensify pain-related disability, while persistent pain can in turn worsen anxiety and depressed mood. Recent literature consistently shows that clinically significant anxiety and depression are common in chronic orofacial pain populations, and recent systematic reviews further support positive associations between orofacial pain and both anxiety and depression, although causality and effect size vary across diagnoses and study designs [3–5].

A second major domain is maladaptive pain cognition. Pain catastrophizing, anticipatory fear, low self-efficacy, and hypervigilance to bodily symptoms can magnify the affective and evaluative dimensions of pain even when peripheral findings are limited. In specialty orofacial pain populations, higher catastrophizing is associated with higher pain intensity and poorer adjustment. More recent work in older adults with chronic orofacial pain also suggests that anxiety and catastrophizing are linked to greater pain interference, reinforcing the importance of assessing these cognitive-emotional processes during routine neurostomatological care. Such findings are clinically relevant because many difficult consultations do not arise from pain alone, but from the interaction between pain, fear, symptom monitoring, and repeated help-seeking behavior [6, 7].

13.2 Quality of Life, Sleep, and Functional Consequences

Behavioral and psychological influences are also reflected in broader function. Chronic orofacial pain affects eating, speaking, oral hygiene, social confidence, occupational performance, interpersonal relationships, and overall health-related quality of life. Although comparative data suggest that some chronic orofacial pain cohorts may show slightly less global psychosocial impairment than broader non-cancer chronic pain populations, the burden remains substantial, especially in domains related to discomfort, symptoms, sleep, and role limitation. Sleep is particularly important in neurostomatology because insomnia, fragmented sleep, and fatigue may lower resilience, worsen emotional regulation, increase pain sensitivity, and reduce treatment engagement. A recent study in treatment-seeking chronic orofacial pain patients found that insomnia symptoms were associated with higher pain intensity, greater pain interference, higher anxiety/depression scores, and lower life satisfaction, supporting routine sleep-focused inquiry in neurostomatological assessment [8, 9].

13.3 Disorder-Specific Considerations

Burning mouth syndrome (BMS) is one of the clearest examples of why a simplistic separation between "neuropathic" and "psychological" explanations is inadequate. Current literature describes BMS as a chronic intraoral burning or dysesthetic disorder in which peripheral small-fiber dysfunction, altered central pain processing, hormonal influences, and psychosocial factors may all contribute. Recent reviews emphasize that the presence of anxiety, stress, sleep disturbance, or depressive symptoms does not justify reducing BMS to a psychiatric diagnosis. Rather, such factors should be understood as contributors to symptom severity, persistence, and suffering in at least a substantial subgroup of patients. This is especially important in clinical practice, where patients with BMS often arrive after repeated inconclusive evaluations and may already feel invalidated. Proper care therefore requires psychological assessment without psychologizing the disorder itself [10–12].

Persistent idiopathic dentoalveolar pain (PIDAP) and persistent idiopathic facial pain (PIFP) raise related but distinct behavioral issues. These conditions are often diagnostically delayed, easily misunderstood, and sometimes aggravated by unnecessary irreversible dental interventions. In PIDAP, pain may continue after a dental procedure even when no ongoing odontogenic cause can be identified, and central or nociplastic mechanisms are thought to contribute in at least some patients. In PIFP, the pain is often persistent, poorly localized, and medically perplexing, which may encourage repeated consultations, diagnostic uncertainty, and escalating procedural care. Both disorders illustrate how maladaptive illness behavior can emerge when chronic pain is interpreted solely through a structural dental model. Trigeminal neuralgia presents a different pattern: here, the extraordinary severity and unpredictability of attacks often generate anticipatory fear, food avoidance, sleep disruption, and depressive symptoms. Importantly, recent prospective evidence shows that anxiety and depression scores may improve when pain is effectively relieved after microvascular decompression, highlighting the close interdependence of pain biology and psychological suffering [13–15].

13.4 Clinical Assessment in Neurostomatology

Assessment should begin with standard diagnostic domains—pain location, temporal profile, provoking and relieving factors, associated sensory changes, previous treatments, medication exposure, and neurological or dental findings—but should extend to sleep quality, emotional distress, pain beliefs, avoidance patterns, oral parafunctions, functional limitations, and the patient's understanding of the disorder. Multidimensional pain assessment is particularly useful in persistent pain that exceeds expected tissue healing, shows disproportionate disability, follows repeated unsuccessful procedures, or is accompanied by marked distress or diagnostic uncertainty. Reviews of pain assessment in oral medicine emphasize the value of combining clinical examination with rating scales, questionnaires, and where appropriate daily symptom recording in order to capture the sensory, emotional, and functional dimensions of pain more accurately [16].

13.5 Behavioral and Psychological Management

Behavioral management should be presented to patients as a mechanism-informed component of comprehensive care, not as an alternative to diagnosis. Cognitive-behavioral therapy (CBT) is the best established psychological approach in this context because it addresses catastrophic misinterpretation, fear-based avoidance, excessive symptom monitoring, and ineffective coping patterns. Reviews focused specifically on orofacial pain suggest that CBT, alone or integrated with other conservative therapies, can be beneficial in conditions such as temporomandibular disorders, myofascial pain, and burning mouth syndrome. Broader chronic pain literature also indicates that CBT-based interventions may improve depression, anxiety, and quality of life more consistently than they reduce pain intensity itself. This distinction is clinically important: in neurostomatology, the realistic aims of behavioral therapy are improved coping, reduced interference, better self-management, and restoration of function rather than complete pain elimination in every patient [17–19].

Other interventions may also be incorporated within a multidisciplinary plan. Relaxation training, stress regulation, pacing, sleep-focused behavioral care, and biofeedback-informed approaches may be helpful in selected patients, particularly those with painful temporomandibular disorders, parafunctional activity, insomnia, or high autonomic arousal. Hypnosis has recently re-emerged as a possible adjunctive treatment. A 2025 systematic review suggested that hypnosis may reduce pain perception in some chronic orofacial pain settings; however, the evidence base remains limited by small samples, heterogeneous protocols, and variable methodological quality. Consequently, hypnosis may be considered for selected patients, but it should remain adjunctive and should not displace first-line conservative, education-based, and psychologically informed care [20].

13.6 Communication and Multidisciplinary Care

Communication style is itself a therapeutic tool in neurostomatology. Many patients with chronic oral or facial pain have already undergone multiple consultations, imaging studies, and procedures before they reach an appropriate specialist. Some have been told that nothing is wrong; others have been subjected to repeated interventions despite little evidence of ongoing tissue pathology. The clinician must therefore communicate three points clearly and repeatedly: first, the pain is real; second, the absence of visible pathology does not invalidate the symptom; and third, chronic recovery usually depends on a multidimensional treatment plan rather than a purely procedural one. When central sensitization, stress-pain interaction, or behavioral treatment is explained carefully, patients are more likely to understand that psychological care is being used to treat pain-related suffering and disability, not to deny the biological reality of the condition.

The most effective model of care is interdisciplinary. Depending on diagnosis and severity, the team may include oral medicine specialists, dentists with expertise in orofacial pain, neurologists, pain physicians, psychologists, psychiatrists, physical therapists, and when indicated sleep specialists. Such collaboration is especially valuable in patients with overlapping phenotypes—for example, burning mouth syndrome with insomnia and health anxiety, painful temporomandibular disorders with catastrophizing and parafunctional behavior, or persistent dentoalveolar pain after multiple unsuccessful dental procedures. The goal is not to replace biological diagnosis with psychological interpretation, but to align biological, behavioral, and functional management so that treatment becomes more rational, less invasive, and more patient-centered.

Important gaps remain in the evidence base. More longitudinal studies are needed to determine whether anxiety, depression, catastrophizing, and sleep disturbance act primarily as risk factors, consequences, or bidirectional amplifiers across different neurostomatological diagnoses. Better subgrouping of patients is also needed, because the behavioral profile of trigeminal neuralgia is not identical to that of BMS, PIDAP, PIFP, or painful temporomandibular disorders. Further research should also clarify which psychological interventions work best for which diagnostic phenotypes, and whether integrated models combining education, pharmacotherapy, rehabilitation, and targeted behavioral care can reduce unnecessary dental treatment and improve long-term outcomes.

Behavioral and psychological factors are fundamental to the expression, maintenance, and management of chronic disorders encountered in neurostomatology. Anxiety, depression, catastrophizing, sleep disturbance, fear, and maladaptive illness behavior do not make pain less real; rather, they shape how pain is experienced, communicated, and functionally disabling. A contemporary neurostomatology approach should therefore move beyond lesion-based thinking and adopt a biopsychosocial framework that integrates careful diagnosis with structured psychosocial assessment, conservative multimodal treatment, and interdisciplinary collaboration. Such an approach improves diagnostic precision, strengthens the therapeutic alliance, reduces the risk of unnecessary irreversible procedures, and supports better patient-centered outcomes.

References

1. International Classification of Orofacial Pain, 1st edition (ICOP). Cephalalgia. 2020;40(2):129–221. https://doi.org/10.1177/0333102419893823
2. Ettlin DA, Napimoga MH, Meira e Cruz M, Clemente-Napimoga JT. Orofacial musculoskeletal pain: an evidence-based bio-psycho-social matrix model. Neurosci Biobehav Rev. 2021;128:12–20. https://doi.org/10.1016/j.neubiorev.2021.06.008.

3. Karamat A, Smith JG, Melek LNF, Renton T. Psychologic impact of chronic orofacial pain: a critical review. J Oral Facial Pain Headache. 2022;36(2):103–40. https://doi.org/10.11607/ofph.3010.
4. Rahardian MK, Putri FA, Maulina T. Association between orofacial pain and anxiety: a systematic review. J Pain Res. 2024;17:1–10. https://doi.org/10.2147/JPR.S432031
5. Anita H, Putri FA, Maulina T. The association between orofacial pain and depression: a systematic review. J Pain Res. 2024;17:785–96. https://doi.org/10.2147/JPR.S435219
6. Dinan JE, Hargitai IA, Watson N, Smith A, Schmidt JE. Pain catastrophising in the oro-facial pain population. J Oral Rehabil. 2021;48(6):643–53. https://doi.org/10.1111/joor.13166.
7. LaRowe LR, Bakhshaie J, Vranceanu AM, Greenberg J. Anxiety, pain catastrophizing, and pain outcomes among older adults with chronic orofacial pain. J Behav Med. 2024;47(3):537–43. https://doi.org/10.1007/s10865-024-00473-7.
8. Tanner J, Teerijoki-Oksa T, Kautiainen H, Vartiainen P, Kalso E, Forssell H. Health-related quality of life in patients with chronic orofacial pain compared with other chronic pain patients. Clin Exp Dent Res. 2022;8(3):742–9. https://doi.org/10.1002/cre2.560.
9. Alessandri-Bonetti A, Sangalli L, Boggero IA. Relationship between insomnia and pain in patients with chronic orofacial pain. Pain Med. 2024;25(5):319–26. https://doi.org/10.1093/pm/pnae003.
10. Khawaja SN, Alaswaiti OF, Scrivani SJ. Burning mouth syndrome. Dent Clin North Am. 2023;67(1):49–60. https://doi.org/10.1016/j.cden.2022.07.004.
11. Orliaguet M, Misery L. Neuropathic and psychogenic components of burning mouth syndrome: a systematic review. Biomolecules. 2021;11(8):1237. https://doi.org/10.3390/biom11081237.
12. Rezazadeh F, Farahmand F, Hosseinpour H, Shahriarirad R, Sabet Eghlidi A. The association between emotional stress, sleep disturbance, depression, and burning mouth syndrome. BioMed Res Int. 2021;Article 5555316. https://doi.org/10.1155/2021/5555316
13. Heir GM, Ananthan S, Kalladka M, Kuchukulla M, Renton T. Persistent idiopathic dentoalveolar pain: is it a central pain disorder. Dent Clin North Am. 2023;67(1):71–83. https://doi.org/10.1016/j.cden.2022.07.006.
14. Ziegeler C, Brauns G, Beikler T, Gosau M, May A. Idiopathic facial pain syndromes—An overview and clinical implications. Dtsch Arztebl Int. 2021;118:81–7. https://doi.org/10.3238/arztebl.m2021.0006
15. Bauer M, Krigers A, Schoen V, Thomé C, Freyschlag CF. Clinical depression and anxiety are relieved by microvascular decompression in patients with trigeminal neuralgia: a prospective patient-reported outcome study. J Clin Med. 2024;13(8):2329. https://doi.org/10.3390/jcm13082329.
16. Scribante A, Pellegrini M, Pulicari F, Ghizzoni M, Modugno FP, Spadari F. Pain assessment in oral medicine through its different dimensions: a comprehensive review. Dent J. 2023;11(11):246. https://doi.org/10.3390/dj11110246.
17. Noma N, Watanabe Y, Shimada A, Usuda S, Iida T, Shimada A, Tanaka Y, Oono Y, Sasaki K. Effects of cognitive behavioral therapy on orofacial pain conditions. J Oral Sci. 2021;63(1):4–7. https://doi.org/10.2334/josnusd.20-0437.
18. Sanabria-Mazo JP, Colomer-Carbonell A, Fernández-Vázquez Ó, Noboa-Rocamora G, Cardona-Ros G, McCracken LM, Montes-Pérez A, Castaño-Asins JR, Edo S, Borràs X, Sanz A, Feliu-Soler A, Luciano JV. A systematic review of cognitive behavioral therapy-based interventions for comorbid chronic pain and clinically relevant psychological distress. Front Psychol. 2023;14:1200685. https://doi.org/10.3389/fpsyg.2023.1200685
19. Busse JW, Casassus R, Carrasco-Labra A, Durham J, Mock D, Zakrzewska JM, Palmer C, Samer CF, Coen M, Guevremont B, Hoppe T, Guyatt GH, Crandon HN, Yao L, Sadeghirad B, Vandvik PO, Siemieniuk RAC, Lytvyn L, Hunskaar BS, Agoritsas T. Management of chronic pain associated with temporomandibular disorders: a clinical practice guideline. BMJ. 2023;383:e076227. https://doi.org/10.1136/bmj-2023-076227
20. da Silva FW, de Oliveira DV, Zina LG, de Paula JS. Use of hypnosis in the treatment of orofacial pain: a systematic review. J Integr Complementary Med. 2025;31(3):224–32. https://doi.org/10.1089/jicm.2024.0154.